Flashes of Brilliance

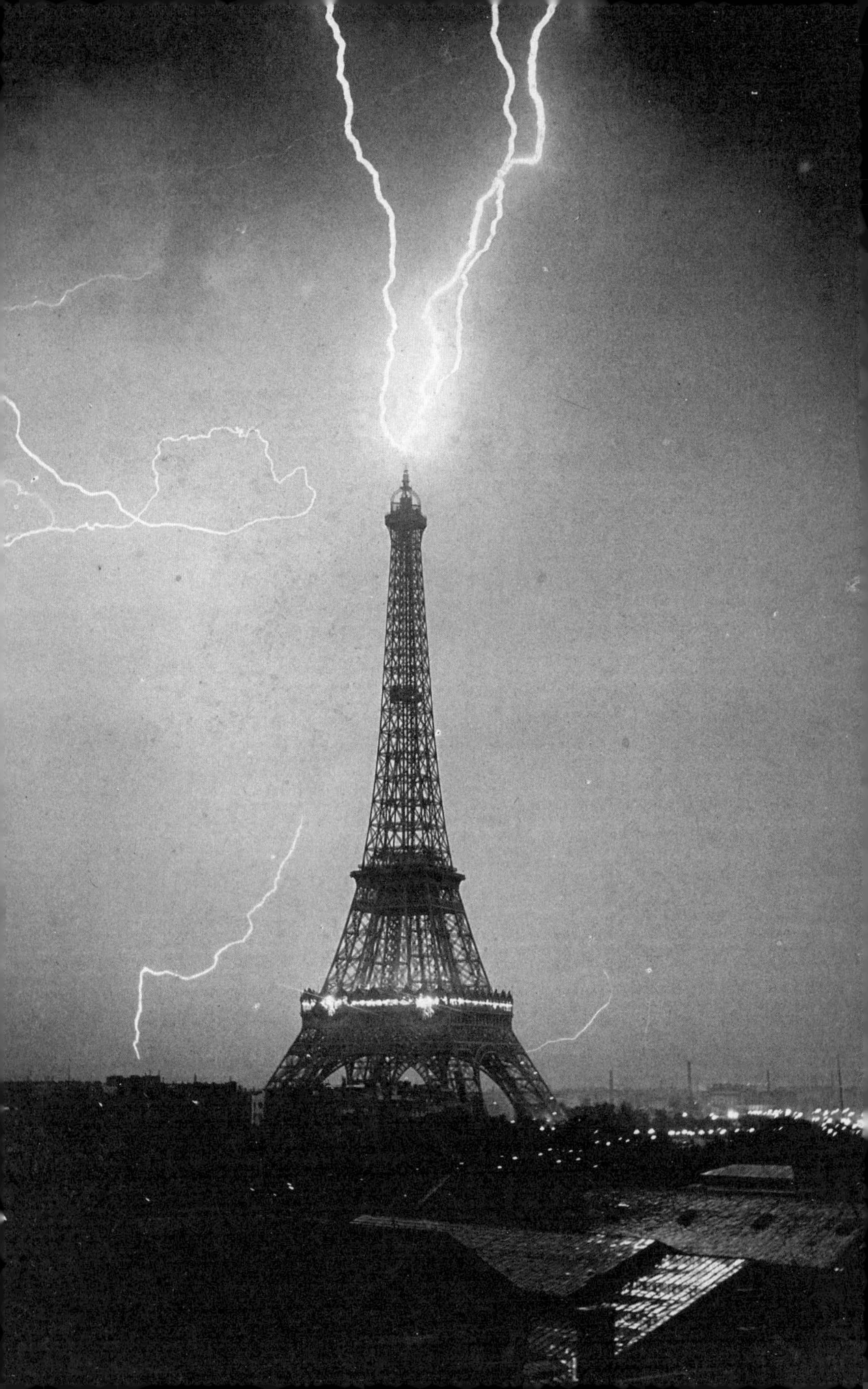

Flashes
OF Brilliance

The Genius of Early Photography
and How It Transformed Art,
Science, and History

ANIKA BURGESS

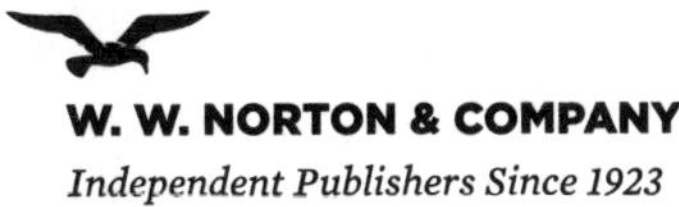

W. W. NORTON & COMPANY
Independent Publishers Since 1923

For information about permission to reproduce selections from this book, write to
Permissions, W. W. Norton & Company, Inc., 500 Fifth Avenue, New York, NY 10110

For information about special discounts for bulk purchases, please contact
W. W. Norton Special Sales at specialsales@wwnorton.com or 800-233-4830

Manufacturing by Lakeside Book Company
Book design by Chris Welch
Production manager: Lauren Abbate

ISBN 978-1-324-05110-7

W. W. Norton & Company, Inc., 500 Fifth Avenue, New York, NY 10110
www.wwnorton.com

W. W. Norton & Company Ltd., 15 Carlisle Street, London W1D 3BS

10 9 8 7 6 5 4 3 2 1

We know that knowledge and the discoveries that make knowledge,
do not proceed from single brains, but are subject as everything
else, to the law of evolution, and are the result of the cumulative
working of many minds.

—A. A. Campbell Swinton, "The New Shadow Photography," 1896

Any man who believes that photography is entirely concerned with
art, science, or, indeed, any *one* branch, to the exclusion of every
other, may safely be written down an ass.

—Walter D. Welford, "The Influence of the Hand Camera," 1893

CONTENTS

Flashes of Brilliance

Sights Unseen

Imagine, for a moment, that you're in London in the mid-1840s, and you've decided to have your portrait taken by a photographer for the very first time. The studio is open only during daylight hours, so in an English winter it closes around 4 pm. It's located in the upper floors of a building, for better access to sunlight; after walking up the stairs and entering the studio, you notice blue-tinted skylights across the ceiling. By some miracle, it's not raining, and there's little fog, which is fortunate: you've read that in duller weather the exposure time would be longer. If the photographer is diligent, the glass will be clean, and not grimy with soot and coal from the city's industries.

If you're a woman, you'll be glad that you had the privacy of a women-only waiting room in which you could check your appearance. You're wearing satin, which you read was best for photographic portraits. You also read that, while posing, to avoid your mouth looking too severe, it helps to say the word *prunes.*

You spot the camera, a wooden, boxlike object with an intimidatingly large brass tube at the front, stranger in person than in the drawings you've seen in the press. Opposite it, on a raised dais, you see a chair and a vertical iron rod with a clamp. The photographer fiddles with a series of blinds to adjust the light. After you are seated in the chair, the photographer, or his assistant if he has one, positions the clamp to firmly hold the back of your head. It keeps you in a still but stiff upright position. He tells you to make a fist or clasp your skirt or the chair, anything to stop the merest tremble in your hands, which are now slightly sweaty because it's warm under the skylights and you're getting nervous. The photographer disappears into an anteroom, returning with something in his hands that he puts into the camera. He says it's a plateholder, and then he fusses with the camera some more and tells you not to move under any circumstances, and now the moment has come. He uncaps the lens and counts the exposure. You've watched all of this without (hopefully) a flicker of movement: you're fastened in position, hot under the skylights, and trying to mouth the word *prunes*. It feels an interminable amount of time, although the photographer reveals that the exposure, which he estimated by the light and his own experience, was a mere twenty seconds. When at last it's over, and the photographer has reappeared from his mysterious darkroom, he presents you with your portrait.

It's a daguerreotype: a portrait on a silver-coated metal plate, protected by a glass cover, mounted in a brass frame, in a case lined with velvet, wrapped in leather. It's small, but for the first time in your life, you have a permanent impression of how you look, or at least, how you looked at that moment, as the light filtered down through the blue glass, into the lens, and onto the plate, capturing you, or a version of you. Maybe you're charmed by it. Maybe you're horrified. (There's a decent chance you're horrified. "The daguerreotype does not flatter," wrote one French commentator, with a degree of understatement. Another said, more brutally, that some daguerreotype portraits "resemble fried fish stuck fast to a silver plate.") Maybe it is a keepsake for your children. You don't have any photographs of your own parents; no one does. Their faces only exist in your memory unless you could afford a painted portrait. Even then, a painting is subjective, more flattering than accurate, whereas this new daguerreotype process is said to be an imprint of nature, drawn by the sun itself. What you're now holding in your hands is the first photographic record of you, conjured up by a camera, chemicals, and light.

Although nineteenth-century photography might be commonly associated with a somber, monochrome world of stiff portraits and slightly dull landscapes, the decades after photography's introduction in 1839 were also a time of bold innovation. Here was a new medium, bursting with potential, ready to be explored with new technology, new approaches, and, occasionally, some highly risky endeavors.

The stories behind these innovations are the subject of this book. Pick any seemingly contemporary photographic genre—street photography, say, or underwater photography—and you'll find its antecedents in nineteenth-century photography. These innovations were sometimes misguided, occasionally obsessive, periodically dangerous, and perpetually fascinating. They also shaped photography as we know it, guided developments in art, and revealed new understandings in science.

But every successful photograph is preceded by failures. This was especially true for photography's beginnings, after years of experimentation, with the daguerreotype in 1839.

The experience of seeing a daguerreotype for the first time was utterly revelatory. "It is hardly saying too much to call them miraculous," wrote the scientist Sir John Herschel, from Paris, in 1839. "It baffles belief," wrote Scottish physicist James David Forbes. An American diplomat, Robert Walsh, wrote in March 1839 that "it would be impossible for me to express the admiration which they produced. I can convey to you no idea of the exquisite perfection of the copies of objects and scenes, effected in ten minutes by the action of simple solar light."

Walsh wasn't alone in his inability to describe the daguerreotype. Even some journalists were lost for words. "Briefly to explain it: it enables him to combine with the *camera obscura* an *engraving power*—that is, by an apparatus, at once to receive a reflection of the scene without, and to fix its forms and tints indelibly in metal in *chiaroscuro*," wrote the British literary journal *The Athenaeum*, not briefly at all. Daguerre's "photogenic drawings" were "drawings from nature through the medium of the rays of the sun," asserted *The Morning Chronicle*. *The Sun* newspaper, after witnessing a demonstration of the process, tried to keep it simple: "This machine, or

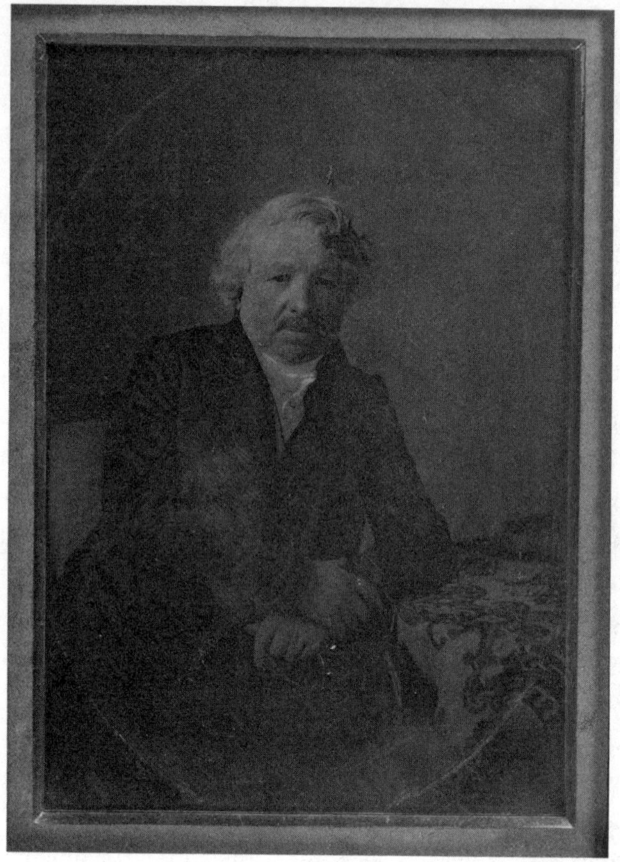

A daguerreotype of Daguerre, taken in 1848, by American photographer Charles Richard Meade during a visit to France.

rather apparatus, for taking, in some fifteen minutes, an exact likeness of any external object." The reality was that no one had seen anything like the daguerreotype before.

The daguerreotype was announced to the Académie des Sciences in Paris on January 7, 1839, by the renowned French scientist François Arago, on behalf of the daguerreotype's inventor, Louis-Jacques-Mandé Daguerre. "In Daguerre's camera, light itself produces the shapes and proportions of the objects outside it, with an almost mathematical precision," Arago said. But he didn't reveal how it worked. The process needed to remain a mystery so that Daguerre and the French government could negotiate a purchase of the rights. Months of speculation followed.

Daguerre was a talented artist who had moved to Paris as a teenager to apprentice as a scene painter at the Paris Opera. In July 1822, at age thirty-four, Daguerre launched the Diorama, an enormously successful visual spectacle that combined vast translucent paintings and clever illumination to give startlingly realistic effects. At some point after this, Daguerre had the idea to capture an image with light rather than paintbrushes. After several years of frustrating experiments, he learned that a man named Nicéphore Niépce had been working on a similar idea for years. Niépce is the creator of the world's oldest surviving photograph, a view from a window that he captured around 1826. The exposure time was about eight hours.

At the end of 1829, Niépce and Daguerre entered into a partnership to share results of their experiments and work toward their common goal. Because Niepce was in Chalon-sur-Saône, in eastern France, and Daguerre was in Paris, they communicated by letter, in code, both as a shorthand and lest anyone should intercept their correspondence and learn about their experiments. Less than four years into their partnership, in 1833, Niépce died. Daguerre continued with the work until at last it was ready. In 1838, he met with Arago about his new "daguerreotype" process.

Once the French government and Daguerre had agreed on their terms, Arago revealed the technical details at the Palais de l'Institut in Paris on August 19, 1839. The event was so popular that crowds stood in the vestibule of the Palais, as more people waited outside. Daguerre published a manual about the process, which became a best seller. In September, he gave public demonstrations in Paris. By December, the clamor for the daguerreotype was such that a French artist drew a humorous cartoon of "daguerrotypomania." It shows a chaotic scene of vast crowds lining up for daguerreotypes, cameras pointing in every direction, bottles of chemicals, and a group of suicidal engravers hanging from gallows.

But the daguerreotype was not the only photographic process that was revealed in 1839. Arago's announcement had prompted an Englishman named William Henry Fox Talbot to announce his own experiments in photography, and rush to claim priority over Daguerre.

Talbot was an exceedingly well-educated Victorian gentleman. His family

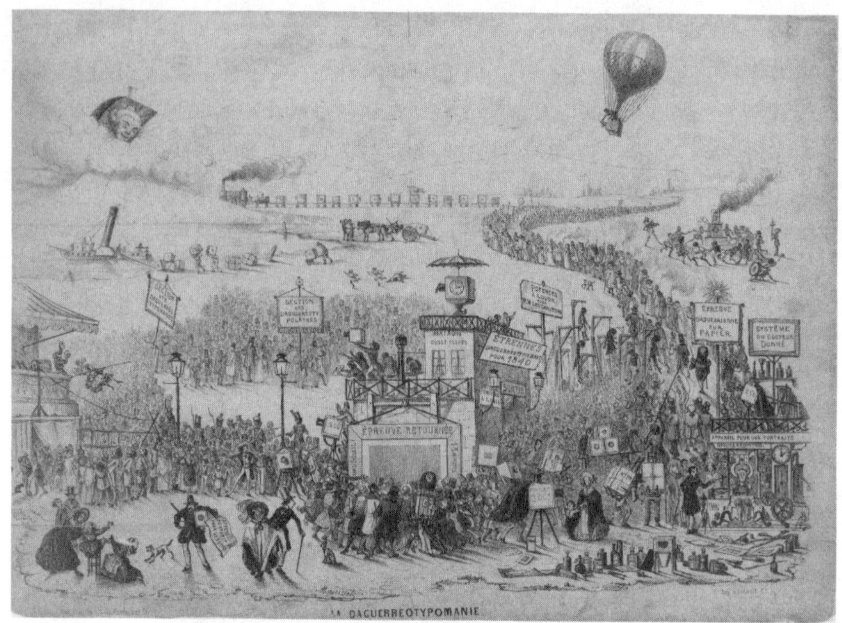

A cartoon of "daguerreotypomania" from December 1839, some four months after Daguerre's process became available for use. Note the figure at bottom right clamped in a torturous-looking posing stand.

home, and the source of his income, was Lacock Abbey, a sixteenth-century manor house in Wiltshire. After attending Harrow School, Talbot studied mathematics at Cambridge University, where he also did things like translate part of *Macbeth* into Greek verse. But what he couldn't do particularly well was draw. While at Lake Como, Italy, in 1833, he attempted a sketch with the help of camera lucida, a prism that artists used to superimpose a scene onto their paper. But it didn't help Talbot: he found his efforts horrendous, and wished that instead he could "imprint" the scene onto paper directly.

Upon his return to England, Talbot experimented with sensitizing paper to light using sodium chloride and silver nitrate. He placed an object—a leaf, for example—on the sensitized paper, and exposed it to the sun. The shape of the object stayed white, and the rest of the paper turned dark. He called this "Photogenic Drawing." He then tried multiple coatings of the solution; when he put this paper in the back of a small wooden box

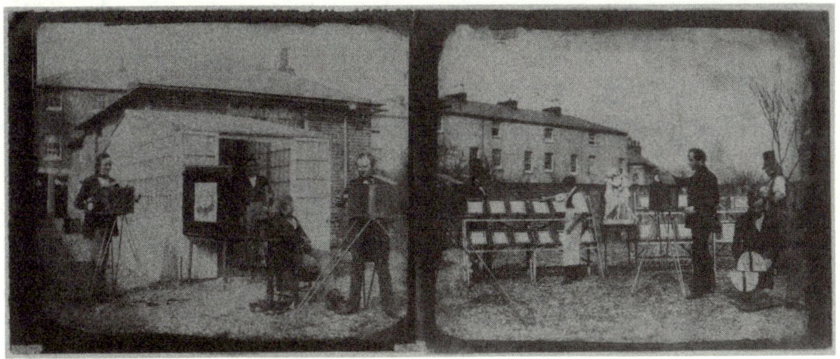

William Henry Fox Talbot, at center with the camera, at a photo printing firm in Reading, 1846. The frames, lined up on racks in the sunlight, show the contact printing process: each one contains a negative and photographic paper.

with a microscope eyepiece, after a long exposure, it produced very small, faint images. But instead of continuing to improve his work, Talbot, in an extraordinary display of shortsightedness, stopped his experiments.

When news of Daguerre's invention reached England, Talbot was alarmed. He later claimed that he had been preparing, at the end of 1838, to share the results of his experiments in 1839. Perhaps this was true, or perhaps he said it to mitigate his embarrassment, for he had not so much been upstaged by Daguerre as he had missed his cue entirely, and was now left floundering in the wings. His own mother even wrote him a letter where she said, "I shall *ever* wish it had been otherwise. This is *at least* the second time the same sort of thing has happened." At the end of January 1839, Talbot's paper on his Photogenic Drawing technique was presented to the Royal Society. But the daguerreotype had already captured the world's imagination.

Of course, the invention of photography also has a prehistory. Before Daguerre, Talbot, and Niépce, there was Thomas Wedgwood, who experimented with coating paper with silver nitrate and exposing it to light in the early 1800s; he couldn't figure out how to make the image permanent. Before Wedgwood, there was Elizabeth Fulhame, whose 1794 treatise included observations of the effect of light on silver and gold salts. But the start of photography as we know it was in 1839, with these two very different processes.

To make a daguerreotype, the operator took a highly polished copper plate coated with silver and sensitized it in a box with iodine fumes. The operator then put the plate in a lightproof holder and inserted it in the camera. To expose the plate, the operator removed the slide that had covered the sensitized plate, and the lens cap. After exposure—which was highly variable and judged by "observation and experiment," according to one manual—the plate was taken into the darkroom and placed over heated mercury, which released vapors to bring out the image. The operator then needed to fix the image, to stop the light from affecting it further; initially this was done with common salt, and then sodium hyposulphate, or hypo, as it became known. (Hypo, now known as sodium thiosulfate, is still used as a fixer in darkrooms.) The daguerreotype was then rinsed in water, dried, covered with glass and sealed to protect the image, and mounted in a case.

A daguerreotype had substance; it was a direct positive image made without a negative, and it was also an object. The only way to reproduce a daguerreotype was to re-daguerreotype it. It was a one-of-a-kind, highly detailed, sharp image with a reflective surface, which is why it was sometimes called a "mirror with a memory." At first, the exposure times for the daguerreotype made portraiture impossible. This quickly changed. Experiments with the photographic chemicals and improvements to the cameras meant that exposure times were soon sufficiently short to allow for the human face to be photographed.

Even so, they were not exactly brief, and also highly weather-dependent: in 1841, a daguerreotypist in London, Antoine Claudet, estimated that exposures varied between ten and twenty seconds in June, and sixty to ninety seconds in September. One of Claudet's subjects recalled the experience of a sixty-second exposure, during which he sat in an open-air rooftop studio on a bright sunny day, while his unblinking eyes streamed with tears.

Early that same year, Talbot patented his own improved process that he called the calotype. To make a calotype, the operator sensitized a sheet of good-quality writing paper with silver nitrate and potassium iodide. The paper was then washed, dried, further sensitized with a solution of gallic acid and silver nitrate, then loaded into a camera and exposed. After it was removed from the camera, the paper still appeared blank: the image only appeared once the paper was washed with a fresh solution of gallo-nitrate of silver. The image was then stabilized with potassium bromide, before

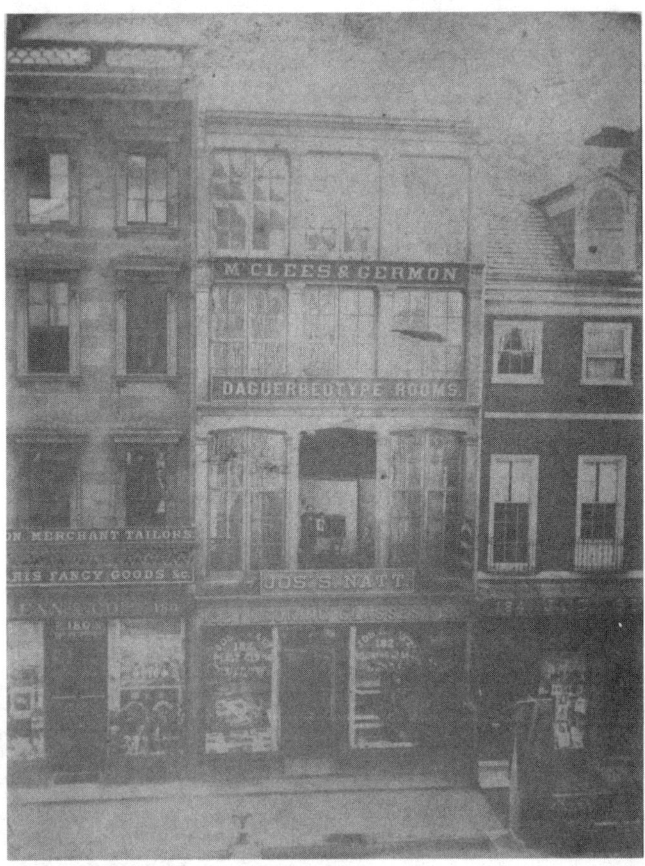

A daguerreotype studio in Philadelphia in the early 1850s. Note the photographer and camera at the second-floor window.

the widespread adoption of hypo as a fixer. The result was a paper negative, from which multiple paper copies could be made through contact printing. The disadvantage was that the fibers and textures of the paper made the image less sharp. Visually, a calotype has a slightly hazy quality, in contrast to the daguerreotype's metallic clarity.

Although the daguerreotype and the calotype were rival processes, in reality the daguerreotype dominated photography's first decade. But the rivalry still stymied English photographers. Daguerre took out a patent in England and Wales, even though he sold the process to the French government, which was, ostensibly at least, making it available worldwide. Talbot, perhaps in reaction to being so publicly bested by Daguerre, rigorously enforced his calotype patent. Together, these patents inhibited photographers in England for about a decade.

Talbot was widely and loudly criticized for his actions. One journal wrote that Talbot seemed to think his patent "secures to himself a complete monopoly of the sunshine"; another described a "practice of intimidation" and a "disgraceful love for patented monopolies." In a portrait from 1864, Talbot poses with a lens in his hands, as though he's trying to claim recognition, even while his face looks out warily; it's a pose of regret. He did not then know that the fundamentals of his process—a negative that produced multiple positives—would be the future of photography.

As always, technology progressed, and newer, better, faster processes were born. One of the most influential was the wet collodion process. Invented in 1851, it combined the best of the daguerreotype and the calo-

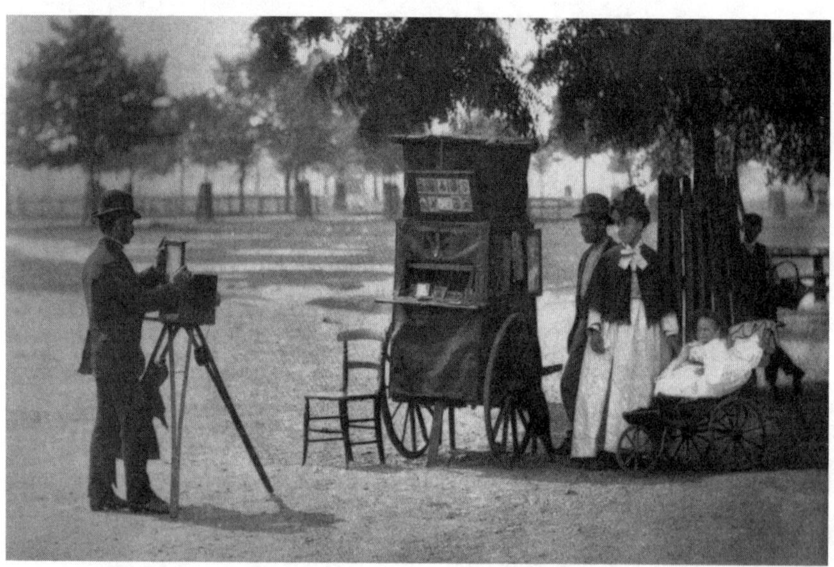

An itinerant photographer on London's Clapham Common, 1877. This photograph, taken by John Thomson, appeared in the book Street Life in London, *which noted, "The work is not always pleasant. The photographer dare not leave his apparatus, for it is impossible to guess when a subject may present himself."*

type: it created superbly sharp images on glass plate negatives that could be reproduced as paper prints. Talbot, in a spectacular example of overreach and poor self-awareness, claimed this new process was in breach of his patent and sued. He suffered some very bad press as a result; he also lost the case. Conversely, the man who invented the wet collodion process, a sculptor-turned-photographer named Frederick Scott Archer, didn't patent it, and never saw a cent.

The wet collodion process dominated photography for nearly three decades. The first step was to create the collodion: gun cotton (cotton soaked in nitric and sulfuric acids and dried) was dissolved in ether and alcohol. The collodion was then mixed with potassium iodide to create a syrupy, sticky liquid, which the photographer poured evenly across a clean glass plate. The next step was to sensitize the plate in a darkroom by dipping it in silver nitrate. Now it was ready to insert into the camera's light-proof plateholder. The photographer had to both expose and develop the plate while it was still wet, so there was no time to be leisurely. If the chemicals dried out, they lost their sensitivity and couldn't create an image. From start to finish, the photographer had perhaps fifteen minutes.

This meant having all the photographic chemicals and equipment to hand. If you wanted to travel or take a landscape view, for example, you needed to carry everything with you to your chosen location—chemicals, water, portable darkroom tent, camera, tripod, developing trays, glass plates, and measuring apparatus. (And then carry them all back again.) One photographer recalled arriving at a train station on a Saturday morning in the early 1860s for a photographic outing with 120 pounds of equipment.

Photography is, of course, based in science, and it evolved alongside scientific knowledge. In the late 1870s, the wet plate process was replaced by the wildly more convenient gelatin dry plate process. It was a technological revolution: it allowed plates to be prepared in advance, and purchased directly from manufacturers, which had far-reaching consequences across the photographic industry.

Once a photographer had a glass plate negative, they produced a positive through contact printing. The negative was placed in direct contact with sensitized paper in a frame and exposed to the sun; as a result, the size of the negative dictated the size of the print. (Although it was possible to enlarge

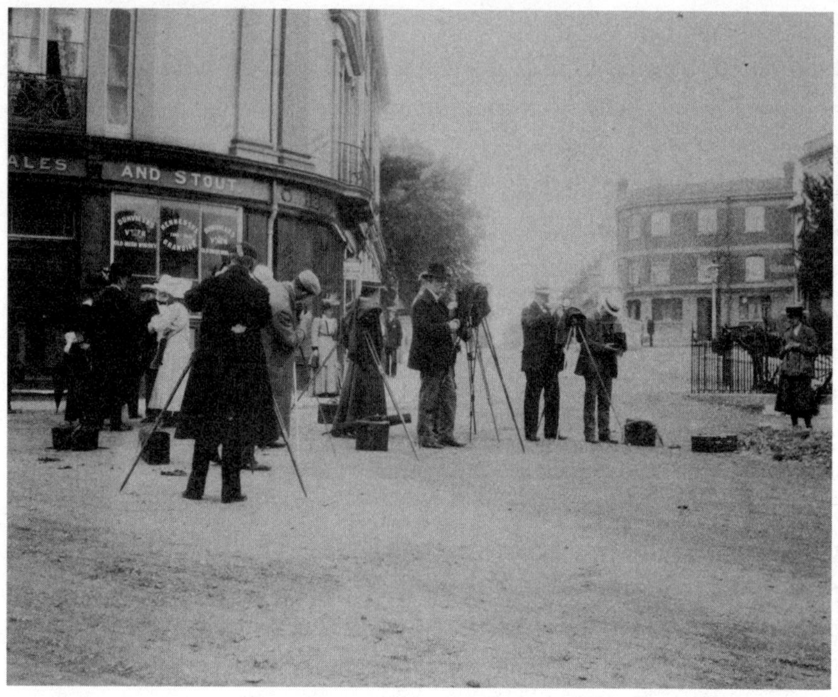

In the 1880s, as photo technology became more user-friendly, vast numbers of amateurs began to shoot anything and anywhere, leading one publication to remark, "The camera is as constant a companion as tobacco to a smoker."

a print, it was difficult, and only in the 1880s, with improvements to paper and artificial light sources, did enlarging become a more workable process.)

One of the most popular and enduring printing materials was albumen paper, which was introduced in the early 1850s by French photographer Louis Désiré Blanquart-Evrard. As the name suggests, albumen paper started with a solution of egg whites, beaten with salt, into which the paper was floated and then dried. When it was time to make the print, the photographer dipped the paper (albumen-side down) into silver nitrate, dried it again, then positioned it in a frame, under a negative, in sunlight. Once the printing was complete, the photographer washed the print, toned it (typically with a gold chloride solution) to enhance its appearance and longevity, and fixed it. The result was a smooth, glossy, sharp print.

One result of albumen paper's popularity—and it dominated until the

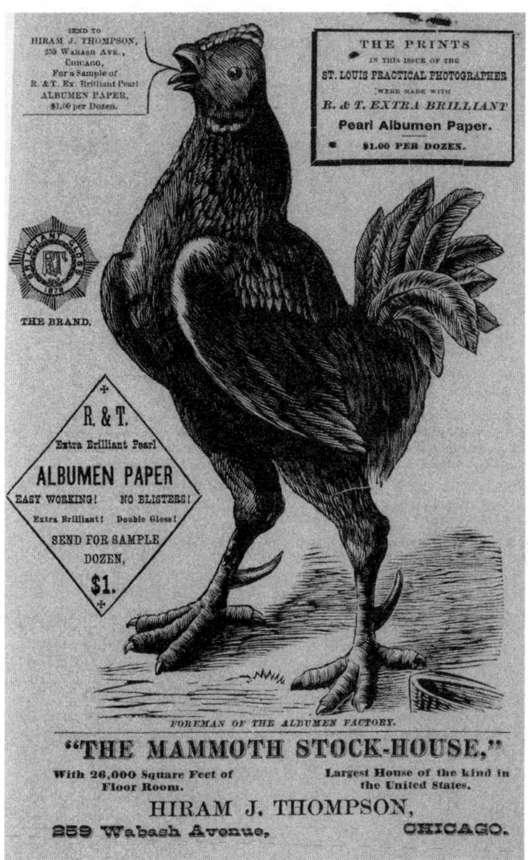

An 1881 advertisement for albumen paper.

early 1890s—was an excess of egg yolks. For photographers who produced their own albumen paper at home and needed to use up egg yolks, one photographic journal printed a recipe for a yolk-filled "photographer's cheesecake."* In the commercial manufacture of albumen paper, the quantities of egg use were staggering. In Dresden, Germany, which was the center of albumen paper production, in the late 1880s one factory alone used over six million eggs in a single year. The separating of those eggs was all done by hand, by women employees, and the yolks were sent to use in leather tanning, or to bakers. The factories in Dresden also initiated a process of

---

*   If you're interested in trying out the recipe, it's in the endnotes.

fermenting the egg white, which made for a glossier paper; it also imparted a distinct smell.

But paper wasn't the only printing surface used, which takes us into one of the weirder sides of nineteenth-century photography. In the 1880s and 1890s, in Paris and Chicago, young women and men had portrait photographs printed onto their skin. They didn't last, which, reported *The Photographic News*, was not a bad thing: "To the fickle the want of permanence will not be regarded as a disadvantage."

Fingernails were also used as a highly unusual (and temporary) surface for printing photographs. One article suggested positioning the portrait so that the head was near the nail bed; otherwise when the wearer cut their nails, the portrait would be decapitated.

With the sheer number of chemicals involved in nineteenth-century photography, practitioners had a host of novel and exciting ways to injure themselves. Daguerreotypists, for example, were exposed to toxic mercury and iodine vapors every time they made an image. An American daguerreotypist named Jeremiah Gurney nearly died in 1852 after continued exposure to mercury. But that was just the beginning.

"I have only to enumerate soluble cotton, nitro-glucose, ether, alcohol, bromine, iodine, ammonia, fulminates of gold and silver, to call up a battery of chemicals the use and combination of which would readily destroy a dwelling-house and all contained therein, either by fire, explosion, or suffocation," wrote R. J. Flower in the *British Journal of Photography*, in 1869. This was not hyperbole.

The components of collodion were gun cotton, ether, and alcohol. Gun cotton is an explosive, and ether is volatile and highly flammable. Fires or explosions in photographic studios and factories occurred with alarming frequency. Just one example, as noted by photo historian Bill Jay in his book *Cyanide & Spirits*, involved an unfortunate photo chemist who experienced two explosions of gun cotton in two years. The first occurred as the gun cotton was drying; it exploded with such force that it blew out both the front and back windows in his studio. The second explosion, two years later, happened as he was packing gun cotton into a cask, and it killed him.

The risks from ether were also considerable, particularly given the abundance of open flames in the nineteenth century. In 1858, *Photographic Notes* recounted "the story of a photographer who enters his laboratory with a lighted candle, a thing which he has foolishly done a hundred times before; he cracks a bottle of ether, and half-an-ounce, not more, is spilled upon the floor; presently the vapour reaches the light, and in two minutes the whole place is a raging furnace."

Some photographers also opted to use cyanide of potassium, a poison, either as a fixing agent for negatives or to remove stains on skin caused by silver nitrate. One 1857 manual, *The Photographer's Pocket Companion*, suggested applying a saturated solution of cyanide on stained fingers and hands. The author then noted, "cyanide of potassium is a very poisonous substance, and if allowed to come in contact with ruptures on the hands, oftentimes produces disastrous results."

The "oftentimes produces disastrous results" is an understatement. Cyanide in an open wound was a ghastly experience. One photographer who cut his thumb on a glass plate and then accidentally rested his hand in cyanide discovered that "the smarting pain was almost intolerable." Multiple photographers suffered from pain, swelling, amputation, and even death from the absorption of cyanide through a skin abrasion. Some photographers experienced "cyanide sores," which they treated with rainwater—both a woefully insufficient and overly optimistic remedy. On an even grimmer note, photographers also drank cyanide, either by accident or as suicide, which meant certain death.

Cyanide—painful, lethal, and surprisingly quotidian—was not the only darkroom risk. One article titled "Photographic Poisons and Their Antidotes" urged photographers to have "not only the antidotes to these poisons always in stock, but also . . . some effective emetic at hand where it can be procured at a minute's notice." After cyanide, the article noted, poisons in use were bichloride of mercury, "the most violent of all those employed in photography"; nitrate of silver; and the various acids used in the darkroom, including nitric acid, sulfuric acid, and hydrochloric acid. The suggested antidotes for bichloride of mercury included drinking an egg white; for nitrate of silver, sipping saltwater. For the darkroom acids, the home remedies included magnesium carbonate or pounded chalk dissolved in water. If that didn't work, to prevent the throat closing the author suggested finding a surgeon to perform a tracheotomy.

For more than a decade, photographic annuals included a handy table of poisons and antidotes for photographers to paste up in their darkrooms. For pyrogallic acid, used as a photo developer, that table of antidotes starkly stated "no certain remedy. Speedy emetic desirable." One misguided photographer applied pyrogallic acid on a patch of eczema, believing it would help his skin; it resulted in an ulcer that took six weeks to heal. Another very unfortunate photographer accidently drank pyrogallic acid, and died three days later.

Even the most safety-conscious photographers had to contend with the generally deleterious effects of poorly ventilated darkrooms. Beyond the terrible smell of collodion, the ether, alcohol, iodine, and bromine produced damaging vapors. Ammonia, if spilled in sufficient quantities, could cause "asphyxia," one journal warned. (In the event of accidental swallowing, another journal proposed the following as an emetic: tickling the throat with a feather.) Photo journals suggested maintaining a well-ventilated darkroom, but one photographer published an honest account of the nine-by-twelve-foot room he had claimed in his house, in which he covered up the window to exclude light (and fresh air). Inside, he wrote, "the air gets loaded with the smell of machine oil, albumen paper and old hypo, and it is sometimes just awful in there, even to me."

This history of photography is about the ways in which the medium was stretched and adapted into the shape that we know today. It has triumphs and failures and a couple of explosions. Part I, "Fields of Vision," looks at how photography was used to reveal unexpected viewpoints, from the darkness of underground and cityscapes at night to the heights of ballooning, from the surface of the moon to the depths of the ocean. Part II, "Perception and Deception," looks at innovations within photography itself, including experiments with the scale of photographs, manipulation of images, and the rise of candid photography. Part III, "Into Focus," looks at how photography made the invisible visible, including the discovery of X-rays and their influence on psychic photography, and what that meant for science.

Nineteenth-century photographic journals provide enormously valuable insights into the day-to-day practices and concerns of photographers, with information ranging from society meetings to new technology to gossip. Given the wide scope of the subjects, I have chosen a geographical focus on Britain and France, where photography was invented. Taking 1839 as the starting point for photography, our time frame extends through to about 1910.

The history of photography has many frames. The ones I've selected here are not only about photography, but about art, science, and the very human desire to improve something that was already revolutionary, and that continues to evolve and inspire.

PART I

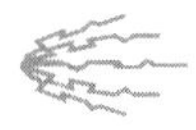

FIELDS OF VISION

Chapter 1

Under Darkness

Toward the end of 1861, a photographer named Nadar descended into the catacombs underneath Paris, armed with his photographic equipment. Nadar, whose real name was Gaspard-Félix Tournachon, had taken up photography in 1854, the year he turned thirty-four; he was also a writer, journalist, caricaturist, and aviation enthusiast. Like most people who operate under a mononym, he was also a talented self-promoter. His name was emblazoned in vast, red, gaslit letters on the outside of his top-floor photo studio on Boulevard de Capucines in Paris, which had opened in September of that year. The studio walls were also red—"a veritable carnival of red," as described by one journalist—and more red graced the borders and signature on his prints. Nadar himself was frequently reported as having a satanic appearance, like "a stage devil," on account of his height and his hair. Portraits show a bright-eyed man with a bushy moustache, slightly twirled at the ends, with hair that flared out at the sides as though it was perpetually energized, like Nadar himself. He was a creative tour de force with enough energy to power an entire city, and he was heading sixty-five feet underground into the catacombs.

He had been brought to this moment by the constraints that daylight forced on photographers and his frustration over how it dictated the rhythms and profits of his studio. In the late 1850s, Nadar had turned his attention

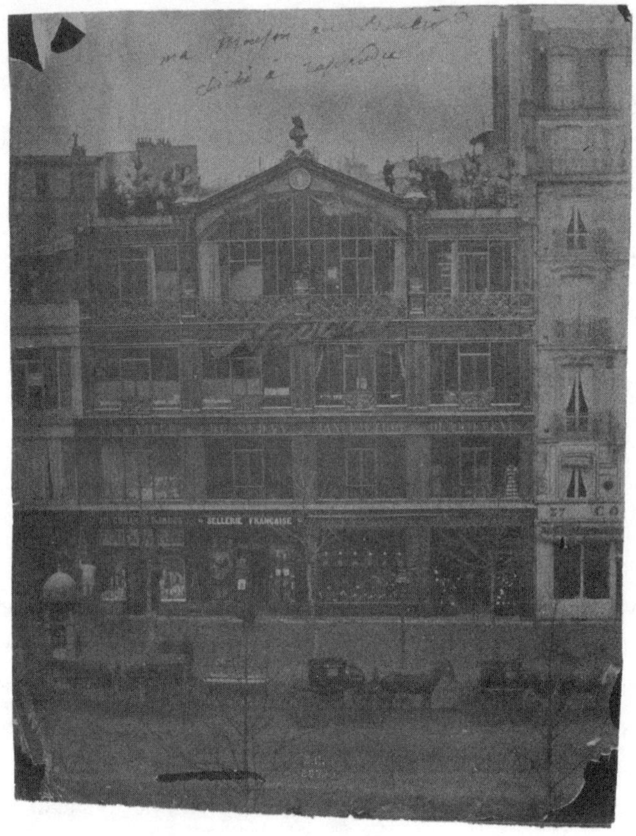

The exterior of Nadar's studio, circa 1860s. The sign outside blazed red, his signature color.

to artificial light. With an electric arc lamp, powered by a stack of carbon and zinc Bunsen batteries and controlled by the recently invented Serrin regulator, he achieved very bright artificial light—too bright, in fact. Early portraits were harsh and unflattering. Nadar experimented with softening the light, eventually using a white linen reflector. Then, ever the showman, Nadar organized demonstrations in his studio. The blaze of electric light that emanated from the glass front of his studio drew in visitors to witness Nadar creating portraits using electricity, a scene that photographer Ernest Lacan described as "the alliance of the two great modern discoveries."

Nadar's electric light photographs brought increasing fame, and, in 1861, a request from the director of the catacombs, Ernest Lamé-Fleury. The catacombs were located in the city's former quarries and stacked with the city's formerly living residents, who had been transferred there, over decades, from overcrowded cemeteries. (In Montmartre, one cemetery was so full with the dead that nearby residents reported, in 1786, that it was "vomiting forth an unbearable stench.") By 1860, the catacombs were the final resting place for some six million skeletons. The bones on display were neatly arranged: skulls were placed in rows, for example, or, in one case, the shape of a heart. The bones that were yet to be displayed were stored in piles.

The catacombs had also become a popular tourist attraction, open four times a year to anyone who wanted to stroll among the skeletons.* In the 1860s, the British press published ghoulish reports of visitors getting lost and never returning above ground, but Nadar recounted seeing 400 or 500 people waiting to enter when it was open. (He also noted, somewhat wearily, "the inevitable band of English tourists.") But, as Lamé-Fleury complained to Nadar, some recently produced lithographs had not done the catacombs justice. He hoped that Nadar could use his electric light to photograph the city of the dead.

Nadar was not one to shy away from a challenge, particularly not one that promised good publicity, but the logistics were immense. Nadar's equipment consisted of the camera, tripod, chemicals, the arc lamp, fifty Bunsen cell batteries, and over 3000 feet of wire to connect the batteries to the lamp. This was the era of wet collodion, so Nadar had to coat every plate right before it was exposed, and then develop it immediately afterwards, right there in the depths of the catacombs.

For Nadar, the catacombs themselves were less spooky than they were inconvenient. Some passageways were so narrow he had to leave his batteries on the street above and pass the cables through the manholes back underground, requiring an assistant to run messages to whoever was guarding the batteries—sometimes with a crowd of onlookers—above.

---

* The catacombs are still open to visitors today; the website warns of the uneven, slippery floor, narrow paths and, of course, "dim light."

When the batteries were in the catacombs, they emitted noxious fumes, which on more than one occasion were so overpowering that his assistants reportedly had to surface for fresh air.

Exposure times were long, although on the plus side, because of the damp underground atmosphere, the plates did not dry out as quickly as they otherwise would. Still, an exposure of up to eighteen minutes was far too long for a living human to keep still. Although Nadar was there to photograph human remains, he wanted to include a human figure to provide a sense of scale, so he incorporated a mannequin in his images. This stiff figure appears in some photos pulling a cart of bones, or propped up by a wall decorated with skulls, part rag doll and part Scarecrow from *The Wizard of Oz*. It is an odd juxtaposition, this imitation human amid for-

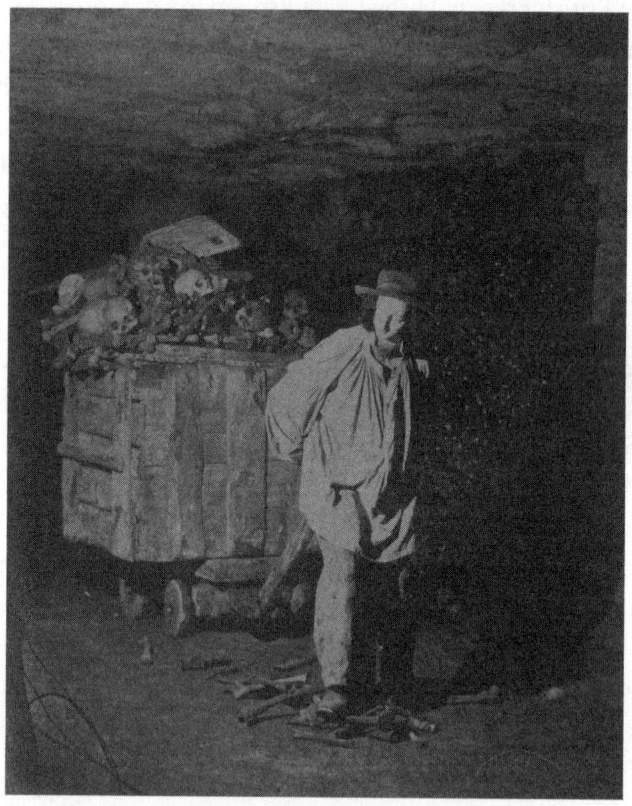

A brightly lit mannequin in the catacombs.

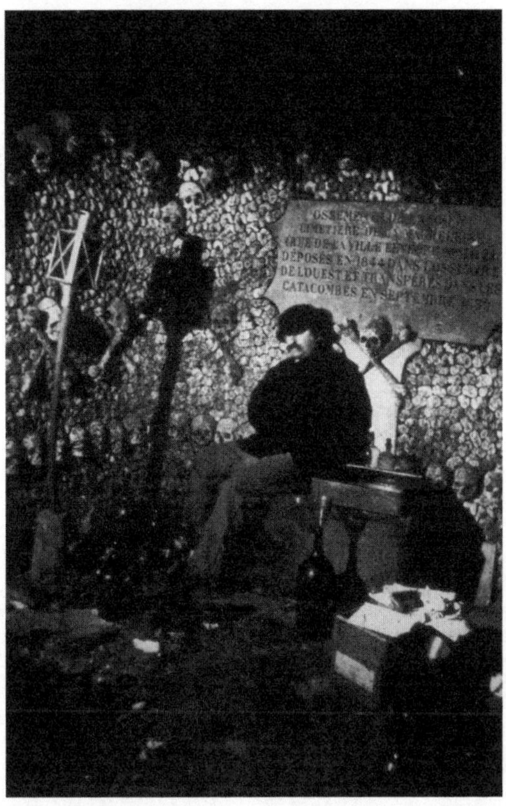

Nadar's self-portrait amid the skeletons. The Photographic News *later said of Nadar, "He is nothing, if not bold, startling, effective, even theatrical."*

merly living humans. The only living human to appear in any of Nadar's catacomb photos is Nadar himself. He sits on a bench, his head next to a skull and crossbones. Understandably, given the long exposure times, his head is slightly blurred from movement. His hat is pulled low over his eyes, possibly to shield them from the arc lamp's fierce blaze. In the bright light, in this place typically shrouded in darkness, the bottles of chemicals cast shadows.

Another photo reveals wires from the arc lamp looping along the stone floor, trailing into darkness. The edges, where the bright electric light can't reach, are dark pools. Almost every photo reveals these opposing forces: light and shade, wires and stone, energy and death.

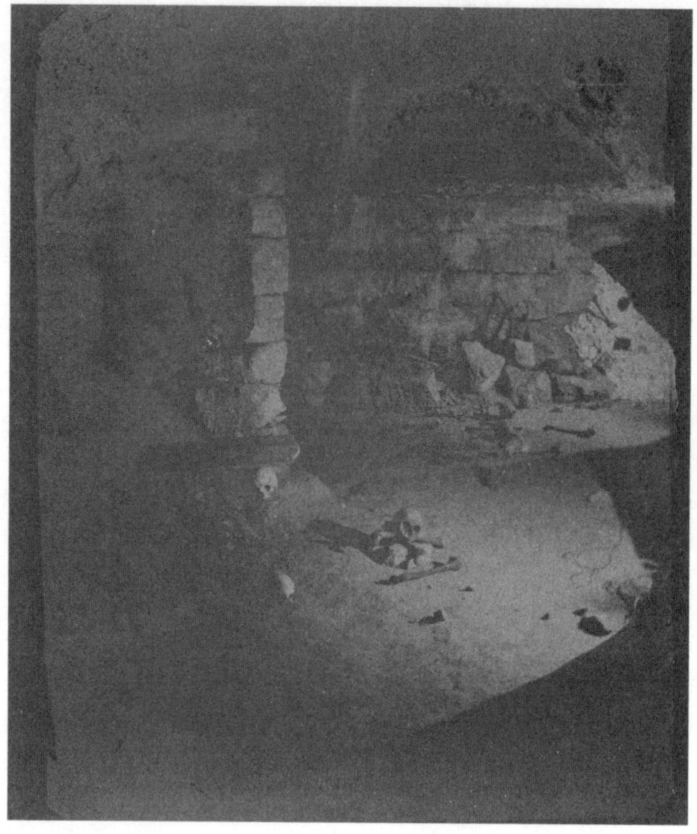

Note the wires to the right, disappearing into the shadows created by Nadar's arc light.

A few years later, in 1864, Nadar went underground again to photograph the recently renovated Parisian sewer system. The logistical issues were as difficult as in the catacombs, although the smell apparently wasn't an issue. (An American visitor noted, a few years later, that the smell of the sewers in Paris was preferable to that of New York City in summer.) But the shoot was so nightmarish that Nadar nearly gave up on the whole operation. It had the same problems as the catacombs—long exposures, endless wires, awkward mannequins—but with the addition of great mists of smoke at unexpected moments, which could (and did) ruin a long exposure in the final minute.

In his memoir, Nadar writes as though the catacombs and sewers happened consecutively, over a three-month period, but historians have shown otherwise. Nadar was, like many showmen, an unreliable narrator. But he is very believable when he wrote that he would not wish those three months on his worst enemy. Nevertheless, Nadar had done it: he had photographed places that were utterly devoid of natural light.

On Saturday January 21, 1871, *Punch* ran a cartoon showing three sad-faced photographers begging for money in a gloomy-looking street, with the caption, "Pity the Poor Fogged-Out Photographers!" More than thirty years had passed since Daguerre's discovery, and despite all the other improvements to processes, plates, and cameras, photographers still lacked a reliable form of artificial light. London fog—once described as "the terror of the photographic studio"—was notorious. But regardless of their location, all photographers were at the mercy of the weather, bound to the sun, and in need of artificial light.

The search for light dictated the design of photo studios. Top floors of buildings were favorable, incorporating glass panels and skylights, like Nadar's top floor at Boulevard des Capucines. In a city like London, the glass had to be kept "scrupulously clean"; if not, the light had to permeate "a villainous compound of concentrated coal smoke, and the victim, impaled on the head rest, is made to suffer double the requisite amount of 'exposure.'" Photographic journals endlessly ran articles about whether north-facing light was better than south-facing light, how to control the light with blinds and reflectors, the best paint color for the walls, and even the exact angle of the skylight, to create the most effective "glass rooms," as they were sometimes known. The advice, although no doubt well-meaning, could be overwhelming and contradictory; one article advised "the prevailing character of the light should be what is called soft or diffused, but at the same time it should be brilliant or forcible." It's no wonder a reader of *The Photographic News* wrote to the editors, in 1862, "I am a little bewildered by the general vagueness of the instructions given."

Photographers needed an affordable, readily available artificial light source. The most promising material to appear, in the late 1850s, was magnesium, which, when ignited, produced a bright white light. (You might

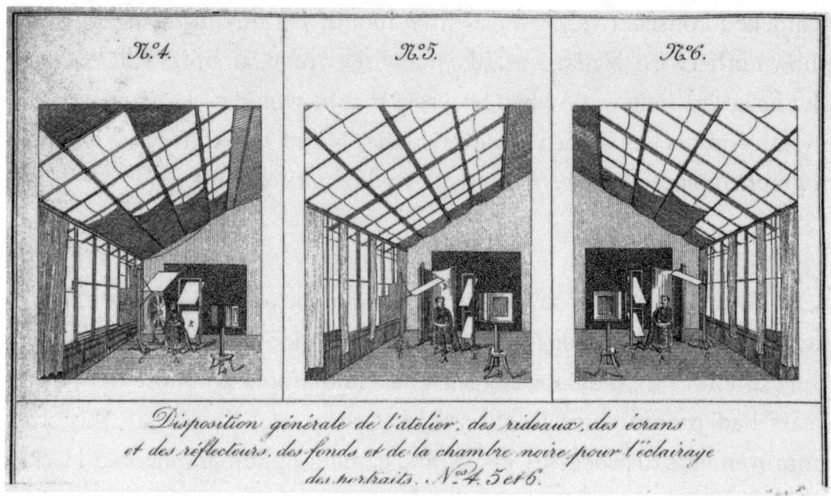

Illustrations from an 1878 French photography manual showing various ways to control and maximize sunlight in the studio. As one English photographer complained during one particularly dull, grey winter, in 1875, "I can draw back my blinds, but I cannot draw back the clouds at the same time."

have seen magnesium in action at birthday parties or July 4 barbecues: it's sometimes used in sparklers.) The downside to magnesium was that it was costly, and produced great plumes of white smoke, which the photographer had to wait to clear until attempting a second exposure. This made it deeply impractical for studio portraits.

Magnesium also needed to be ignited in a way that created a steady bright light. From the mid-1860s, photographers began to use specially designed lamps that spooled the magnesium wire or ribbon into a flame and incorporated a small dish to catch the burned magnesium ash. With later designs, photographers used "puff lamps," where they blew powdered magnesium into a spirit burner. Some photographers got a little too creative with its application: one funneled a mixture of magnesium and sand onto a horizontal soldering lamp to create a flame that was seven feet long.

But even that was relatively tame compared with the invention by a New York dermatologist named Henry Piffard, whose interests included firearms, photography, and skin diseases. In 1888, Piffard patented a "photogenic cartridge" to use with powdered magnesium and an explosive (usually gun cotton), which was ignited when shot from a gun. During a demonstration at the Society of Amateur Photographers of New York, Piffard illuminated a portrait, to great applause, by firing off such a mixture from a "good-sized pistol." In his patent, Piffard stated the device could be used "without the slightest danger or discomfort to the operator," although he failed to note how his patients felt about being photographed by their doctor brandishing a pistol filled with combustible powder.

Although magnesium clearly had potential, it was initially too expensive for widespread use, and too smoky for studios. Photographers also considered some highly experimental sources of artificial light. Bedbug powder supposedly made a "tolerable flash light." So too did firefly phosphorescence, at least according to a Dr. John Vasant, who, in 1887, placed twelve fireflies in a bottle, covered the mouth with a net, and counted the exposure by the number of flashes (fifty), each of about half a second duration. Stranger still, in 1882 the *British Journal of Photography* reported that a photographer had exposed an image by the light emanating from a dead

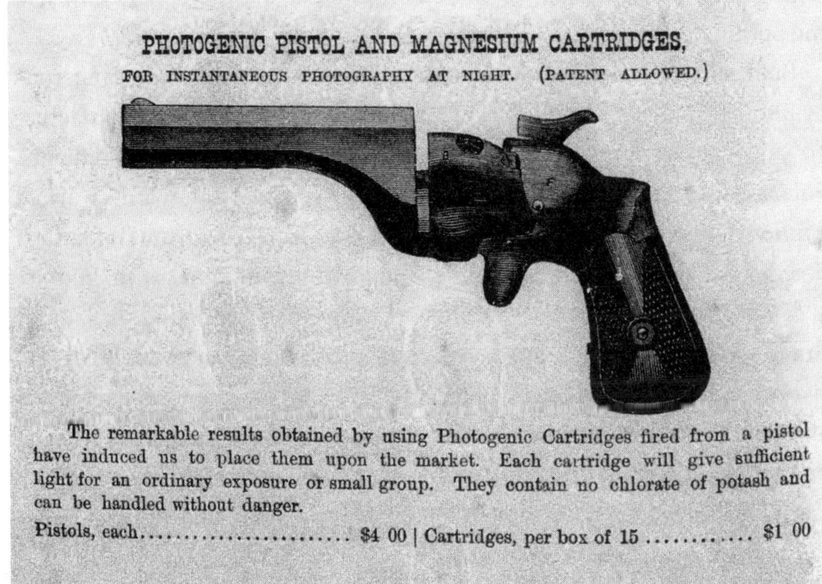

An 1889 advertisement for a pistol that fired magnesium cartridges.

haddock. It was difficult, said one viewer diplomatically, "to discover what it was intended to represent."

Earlier on, photographers had heated quicklime with a hydrogen and oxygen flame to create limelight. (It was used to better effect in the theater, hence the phrase "in the limelight.") They used what was known as "Bengal light," a mixture of various compounds that burned as a bright blue-white light, but also created a lot of smoke. They tried magnesium in the form of lumps, powders, wires, and ribbons. By the mid-1880s, a photographer would be forgiven for despairing over the lack of readily available, reliable artifical light sources. But in 1887, as magnesium became more affordable, photographers had an exciting new form of illumination to test out: *Blitzlichtpulver*—flashlight powder.

As the name suggests, flash powder was invented by two German chemists, Adolf Miethe and Johannes Gaedicke. Flash powder combined magnesium powder with potassium chlorate and antimony sulfide— "pyrotechnic compounds," as one 1890 article phrased it—which ignited the magnesium more quickly and produced a brighter light. Coupled with the faster and more convenient dry gelatin plates that had, in the early 1880s, superseded the old wet collodion process, photographers now had the option of a bright flash for dark days and badly lit interiors.

But flash powder was explosive by nature, rendered even more dangerous by photographers mixing their own at-home formulas, or by manufacturers varying the compounds without any transparency as to the ingredients. Potassium chlorate was "dangerous in every sense of the word," warned *Anthony's Photographic Bulletin*. It regularly appeared in formulas for flash powder as an oxidizer; today, in the United Kingdom, potassium chlorate is deemed too reactive to use in fireworks. Another compound that sometimes appeared in flash powder was picric acid, which is a powerful explosive when dry. It occasionally makes the news today when someone finds an old batch in a warehouse (as discovered in Uckfield, England, in 2017) or a museum archive (in New Zealand's Otago region, in 2021) or a school laboratory (in 2022, at a school in Olney, Maryland); in all those instances the police dispatched a bomb squad.

Flash powder exploded as it was being mixed, when it was being disposed of, and during use. It blew up houses and factories, shattered windows, and destroyed equipment. Photographers lost fingers and thumbs and hands and hair to the explosions, and that's if they were lucky; some lost their lives. In 1890, for example, the president of the Denver Fire-brick and Chemical Supply House was mixing flash powder when it exploded, killing him and destroying the front of the building. In the same year, at the opening of the Pulitzer Building in Manhattan, a photographer brought an "extra quantity" of flash powder to capture the scene, but it exploded and shattered fifty windows. *The New York Times* reported—under the headline "An Unexpected Boom"—"the explosion was felt in all the great buildings along Park Row." (The photographer survived that explosion.) The circumstances of some accidents were so random they seem almost improbable: one explosion, in Germany, was caused by someone accidentally throwing the end of a lit cigar into some flash powder.

In the United States, flash powders had snappy names like Firefly and Violet and Lumino, but manufacturers were opaque about what they contained. An 1897 advertisement for Lumino, for example, claimed that it was free "from the danger attending ordinary flash light mixtures," yet the following year an article reported it contained potassium chlorate. (That same article also included several do-it-yourself flash powder formulas also using potassium chlorate, while also helpfully noting, "these are all dangerous powders.")

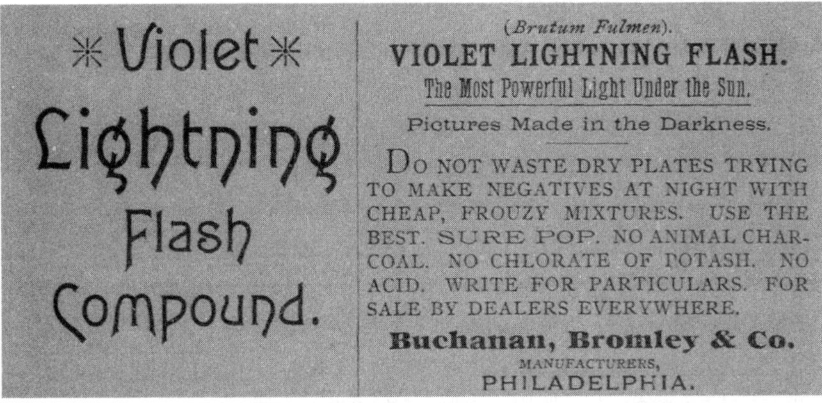

Violet, "the most powerful light under the sun," 1889. That same year, an explosion of Violet flash powder killed three people in Philadelphia.

The flash powder Violet—"the most powerful light under the sun," proclaimed one ad—was the cause of an explosion that killed a sixteen-year-old boy who was handling the powder at the chemical factory in Philadelphia where he was employed. During the trial on the boy's death, a chemical expert testified on the dangers of mixing explosive powders, particularly picric acid and potassium chlorate, with magnesium. "The danger in the materials, he said, is in the handling of them—in trituration, mixing and rubbing. He said that flash powder is more dangerous than gunpowder, being more rapid and violent in action. It may explode without assignable cause." One year after the boy's death, at that same company, one of the company's founders tried to dispose of flash powder down a sink. It exploded and killed him and three other employees.

The accidents in Philadelphia were just some of many covered in the photographic press, accompanied by warnings to photographers. "*All forms of flash powder are dangerous*, the difference between them being only in the danger involved," emphasized one writer for *The St. Louis and Canadian Photographer*. Similarly, *Anthony's Photographic Bulletin* warned, "If this is going to be the record concerning flash-powders, the sooner we see the last of them, the better." The dangers of flash powder reached the non-photographic press, too: one former artillery lieutenant wrote to the New York *Sun* with a list of recommended precautions, noting, "The abuse of flash powders after they fall into the photographers' hands reflects severely upon their intelligence and good sense."

Flash powder continued to be used for about another forty years, until the flash bulb became commercially available in 1929. (One early iteration of the flash bulb also used magnesium.) It remained a risky substance that was liable to explode. In the same year that the flash bulb was introduced, for example, Cunard banned the use of flash powder on its piers and steamships after an explosion at Pier 54 in Manhattan injured nine people.

For all its dangers, flash powder gave photographers a way to light the darkest of places—and there were fewer so devastatingly bleak than inside Manhattan's tenements in the nineteenth century. This is where journalist and social reformer Jacob Riis ventured with a camera and flash powder, to expose, in every sense of the word, the living conditions of the city's poorest residents.

Riis had arrived in New York from Denmark in 1870, at age twenty-one. His first years in the United States were arduous and uncertain: he worked odd jobs for low pay but had no steady income. There were nights when Riis, broke and starving, had nowhere to sleep. Eventually, he got a job at the New York News Association and became a reporter. Riis later recounted his relief, once he had his wages, of sleeping in a boardinghouse. For six of his twenty-three years as a reporter, Riis covered the police beat, spending nights reporting on New York's crowded, unhygienic tenements. But words and drawings could not properly convey the sights, he said, that "gripped my heart." He later recalled sitting at the breakfast table one morning and reading, in the newspaper, a brief report about flash powder. "There it was, the thing I had been looking for all those years."

After a couple of false starts—including with a professional photographer who tried to sell the photos without Riis's permission—Riis purchased a camera and taught himself to take pictures. It's unclear exactly which flash formulae he used in his expeditions, but like other amateurs, he had some dicey moments with it. He set fire to his house twice. He also accidentally blew the flame toward his face, later stating that only his glasses saved his eyesight from permanent damage. Once in the tenements, working in darkness, he ignited the powder on a frying pan. Because of the smoke, and the potential reaction of his subjects, he could only make one exposure within each setting; but one exposure was all he needed.

In Riis's now famous photos, men, women, children, and infants crowd rooms of filthy tenements. As art historian Bonnie Yochelson has pointed out, Riis would have entered each dwelling by whatever dim light existed—if any—and only seen his subjects the moment the light flared. Perhaps this is why some of his subjects look startled; others simply look too tired, too sick, too burdened, to react. Or perhaps they were stunned by this intruder, a stranger with a flaring frying pan and camera who left in a cloud of smoke, another indignity in a life of indignities.

"When the report was submitted to the Health Board the next day," Riis later recalled in his autobiography, "it did not make much of an impression—these things rarely do, put in mere words—until my negatives, still dripping from the dark-room, came to reënforce them. From

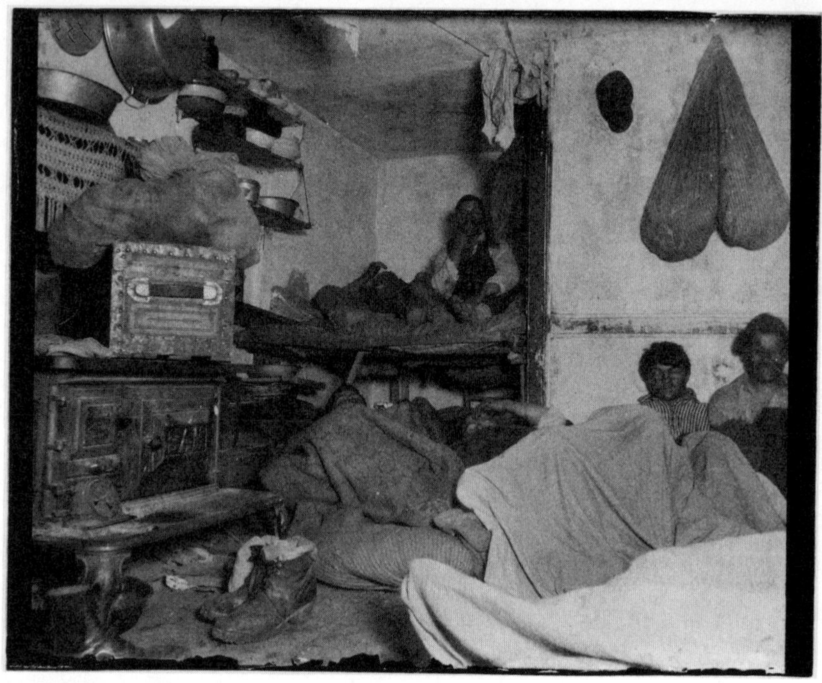

Jacob Riis, Five Cents a Spot. *A flash-lit exposure of a crowded tenement on Bayard Street, Manhattan, 1889.*

them there was no appeal." Riis's lantern slide presentation to the New York Society of Amateur Photographers in early 1888 had a similar effect. "It was hard to realize," wrote a reporter for *The Photographic Times and American Photographer*, "the enormity of the degradation and poverty constantly present in the great city."

Riis continued to publicize what he had seen and photographed. He lectured, with lantern slides, at churches and YMCAs around New York and the Northeast. In December 1889, Riis published an article in *Scribner's Magazine* that included twenty images. Most of these were reproduced as engravings based on the photographs. For six of Riis's images, *Scribner's* used the new, not-yet-perfected halftone process, which involved exposing an image through a screen to translate it into a series of different-sized dots, which was then printed from a metal plate. The following year, Riis wrote what would become a best-selling book, *How the Other Half Lives:*

Studies among the Tenements of New York, featuring, as noted on the title page, "illustrations chiefly from photographs taken by the author." Riis's text focused on living conditions, but also included generalizations, based on stereotype and prejudice, of the different ethnic and racial groups who inhabited the tenements. The majority of the images were reproduced using the halftone process, which had improved since the *Scribner's* article. In its review, under the headline "Matters We Ought to Know," *The New York Times* wrote: "He goes into the slums with his camera and flash light, and in his illustrations he presents what the photograph has produced on the plate."

Riis's photographic series of the city's darkest corners was made possible with flash light powder and stands as one of the earliest examples of social reform photography—work that was continued, to great effect, by Lewis Hine and his photographic survey of child labor. Hine also used flash powder, when he needed to illuminate his portraits of children working night shifts at factories and mills.

Riis, modest about his skills as a photographer, accurately described the sensation of reliving a scene through a photo: "To watch the picture come out upon the plate that was blank before, and that saw with me for perhaps the merest fraction of a second, maybe months before, the thing it has never forgotten, is a new miracle every time."

In 1877, photographer Henry van der Weyde made headlines when he set up electric lights in his studio on Regent Street in London, powered by a gas dynamo installed in the basement, and began to take portraits from noon to midnight. Van der Weyde designed the system himself so he could achieve the right kind of light—light, he said, that would "embrace" the sitter, rather than "strike" them. He used a large lighthouse lens to refract the light onto the subject. *Anthony's Photographic Bulletin* was effusive about the effect: "Not only is every high light picked out crisply and precisely, but every gradation of half-tone is accurately rendered." (The same article also noted the difficulties of working in England, where for so much of the day "sunlight is absent.")

But it wasn't until the 1880s that electricity began to flow into cities.

Central electricity stations began to operate in New York and London in 1882; from this point onward, this invisible force of energy, and with it, light, would slowly but inevitably become omnipresent. In this decade, scientists and photographers also began to experiment with electricity not as a source of illumination for photography but as the subject of photographs.

In France, scientific instrument manufacturer Eugène Ducretet and artist and astronomer Étienne Trouvelot created images of electric sparks without a camera, by positioning an electric current directly onto the surface of a photographic plate. Trouvelot noted how different the positive pole appeared as compared with the negative: "sinuous lines full of branches" for the former while the latter more closely resembled "the lightning bolt in the hands of Jupiter." The visual connection between electricity and a lightning bolt has endured, as historian Kate Flint has pointed out: the charging symbol on your phone or laptop is often an icon of a lightning bolt. The images that Trouvelot and others produced of electric sparks are beautiful: they appear as lush streaks of energy as inviting as they are mysterious.

But it was the older system of street lighting, gas lamps, that inspired the earliest series of that now universally popular subject, a city at night.

The person who most successfully captured London's misty evening glow was an amateur photographer and engraver by the name of Paul Martin. On a dark, rainy evening in February 1896, Martin was walking home from work along the Embankment, a wide promenade that hugs the Thames from the Palace of Westminster to Blackfriars Bridge. As he walked, Martin noticed how lovely the city's gas lamps looked reflected on the wet ground. The following rainy night, he returned with his camera.

Although an amateur, Martin had already received some acclaim, earlier in the 1890s, for his series of street photographs. He had benefited from the new photographic technology that became available in the 1880s—smaller, more unobtrusive cameras, and dry plates—but night photography still presented huge challenges. The low light meant long exposures, which, in turn, required stillness. His early attempts revealed more problems: halos of light around the street lamps, which Martin likened to Saturn's rings, and long streamers from the lights of passing carriages. He began to use plates with an opaque backing that absorbed the light and stopped it from bouncing back to the surface and creating the halo. To fix the streamers, he covered the lens with his bowler hat when he saw a cab approaching and resumed the exposure once it had passed.

It was not a comfortable experience. Martin's advice was that, in night photography, "a good stock of patience and perseverance is necessary." He wasn't exaggerating. Even when he adjusted his start time to when there was just, in his words, "a ghost of daylight left in the twilight," exposure times averaged between ten and thirty minutes. On a frosty night by the Thames, or a damp evening in a London park, Martin must have felt every second of those exposures down to his bones.

He also encountered gawkers, lurkers, curious passersby, and nosy policemen. It would have been perturbing, for almost any Londoner, to see a man standing around with a camera on a cold, wet night. Martin had frequent interactions with the police, who were either skeptical or suspicious—as when, in Trafalgar Square, Martin patiently explained to one officer that his camera was not an explosive. Only one of Martin's photos shows a person, and it's a policeman; he had presumably fallen asleep, given how still he had to have been to register in the long exposure. One of the reasons Martin preferred to photograph on wet nights, in addition to the lovely reflections he captured on pavements, was because he was less likely to be harassed.

Martin photographed some of London's most recognizable, loudest sights—Trafalgar Square, Piccadilly Circus, Leicester Square—but in his images, they're hushed and elegantly lit. No one had yet produced night images so resoundingly, moodily beautiful. For maximum effect, he produced his photos as lantern slides, to project at lectures, and tinted them blue and yellow. The atmosphere of the photos, even some one hundred and thirty years later, is enchanting. He gave the series the poetic and evocative title "London by Gaslight."

Martin won a medal for the series at the Royal Photographic Society Exhibition in 1896. His nighttime photographs, and the articles he wrote about his work, inspired many other photographers to try their own hand at nocturnal photography. News of Martin's work made its way across the Atlantic, to a photographer in New York by the name of Alfred Stieglitz, who became one of the first to immortalize the radiance of New York City at night.

In photos like Stieglitz's and Martin's—and the long string of night photographs that have come since—every twinkling city light is a promise, every shadow an unexplored mystery. How many people have been drawn to a city after seeing it festooned with light in a night photograph? With gas lamps and electricity, cities had an entirely new landscape after dark; photography made those glittering landscapes iconic.

Chapter 2

To the Moon

How the moon appears each night varies with its phases, which hemisphere you're in, the time of night, and the weather. It can appear as a ripe orb, a Bauhaus half-circle, an elegant sliver, or not at all. It can be luminously bright or shrouded, starkly detailed or fuzzy. Even over the course of a single night, the moon's appearance is reliable only in its inconsistency. For the nineteenth-century lunar photographer, who required stillness, this was an extremely inconvenient fact. But the moon's allure still tempted those who relished photographic challenges and tolerated sleepless nights; of those photographers, two in particular captured the moon in ways that were innovative, imaginative, and extremely effective.

On May 1, 1851, Queen Victoria opened the Great Exhibition of the Works of Industry of All Nations. This vast spectacle in London's Hyde Park, housed in a sprawling glass-and-iron building so huge it encased entire elm trees, showcased over 100,000 objects from Britain and various countries around the globe, including the United States. Over six million people had visited the Great Exhibition by the time it closed in October, and they gazed on everything from a 31-ton broad-gauge railway engine to a fruit dish decorated with figurines playing cricket.

Britain was, at this point, the richest country in the world, and the Great Exhibition was an extravaganza of its industry and might. Visitors flocked most of all to see the "Machines for Direct Use," home to the railway engine as well as carriages and hydraulic machines. (This exhibit, the *Official Catalogue* preened, "indicates in what direction the current of industrial activity has been most successfully conducted in Great Britain.") In the "Manufacturing Machines and Tools" section, a printing machine churned out 5000 copies of *The Illustrated London News* every hour. In the "Stationery" section, a machine folded envelopes at a rate of 45 a minute, a job that had previously been done by hand. This particular machine had been co-invented by a British businessman and amateur astronomer named Warren De La Rue.

De La Rue's family business, and the source of his wealth, was printing and stationery, but his passion was astronomy. At his house in London he'd set up a telescope with a ten-foot focal length, through which, before he turned to lunar photography, he observed and sketched Saturn. In the same year as the Great Exhibition, he was elected a Fellow at the Royal Astronomical Society, where gentlemen scientists and astronomers met regularly to share their research and observations.* At some point during the Great Exhibition, in those streams of people *oooh-ing* and *aahh-ing* over hydraulic presses and machine-folded envelopes, De La Rue saw a daguerreotype of the moon.

It was part of a series created by Boston daguerreotypist John Adams Whipple, working alongside William Cranch Bond, the director of the Harvard College Observatory. One of Whipple's early-1850s lunar daguerreotypes still exists today at the Harvard and Smithsonian Center for Astrophysics. It's small, around three-by-four inches in size, and on its silvered surface, emerging from a shadow, is the moon: rotund, cratered, ancient.

To nineteenth-century visitors at the Great Exhibition, the majority of whom would have never seen a photograph of the moon before, it must have been an astonishing sight. Lunar photographs were not only diffi-

---

* Except for a few women who were accepted as "Honorary Members" of the Royal Astronomical Society, women were only permitted to be elected as Fellows in 1916.

cult to produce but limited in how they could be shared. Newspapers and magazines did not yet have the technology to reproduce photographs; *The Illustrated London News* that rolled off the presses at the Great Exhibition featured engravings, not photographs. For most people, the moon only existed as it appeared in the sky each night, not duplicated with photographic chemicals at a fixed point in its ebb and flow. But at the Great Exhibition, visitors could see a moon photograph up close, with their own eyes. Whipple's moon daguerreotype made a strong impression on De La Rue, and prompted his own attempts at lunar photography, which began the following year.

It made an impression on the Great Exhibition jury too, who awarded Whipple a medal. The jurors praised the moon daguerreotype as perhaps "one of the most satisfactory attempts that has yet been made to realise, by a photographic process, the telescopic appearance of a heavenly body, and must be regarded as indicating the commencement of a new era in astronomical representation."

When François Arago described the daguerreotype's many applications, in July 1839, he specifically mentioned its potential in mapping the moon. With photography, Arago said, "it will be possible to accomplish within a few minutes, one of the most protracted, difficult, and delicate tasks in astronomy." He had a point. Prior to photography, the only option for astronomers was to sketch what they saw through a telescope and hope for the best. Photography had the potential to create a permanent visual representation of the moon, but it was a potential that took some time to be realized.

Not that photographers didn't try. John William Draper, for example, a chemistry professor at New York University, experimented with lunar daguerreotypes in the early months of 1840. (Daguerre also had created a moon daguerreotype in 1839 that was subsequently lost, but it seems that he too had trouble with tracking the moon: it was described as "like the tail of a comet.") That spring, at New York's Lyceum of Natural History, Draper reported that he had been successful in creating a lunar daguerreotype. But a daguerreotype could not be easily shared or duplicated; at best, it could be translated into an engraving.*

By the time De La Rue started his moon photography experiments, in 1852, he had the advantage of using the wet collodion process. But he still had to figure out how to track the moon. During the exposure, he had to move the telescope with the moon's trajectory, which required two people: one to uncap and recap the end of the ten-foot-long telescope, and one to guide the movement. He thus had to rely on someone who was "willing to lose the greater portion of a night's rest, often to be disappointed by failures resulting from the state of the weather," he later recalled, which would have tested the strongest of friendships, along with other "numberless impediments." De La Rue resolved to only try again once he could move the telescope with a mechanism.

It wasn't until 1857, five years later, that De La Rue restarted his lunar photography experiments. In the intervening years, spurred on by Whipple's moon daguerreotypes and the new wet collodion process, more photographers attempted to capture the moon, and with greater success. In June 1854, members of the Liverpool Photographic Society reported that they had captured about 150 moon "portraits," one of which they projected, during a later meeting, to about twenty feet in diameter, "as large as the room would allow." This was not even the largest moon of the year: during a presentation to the British Association for the Advancement of Science, members of the Liverpool Photographic Society projected moon photographs at nearly fifty feet in diameter, which was achieved with a magic lantern (a type of projector), and limelight. This astonishing size was made possible by the size of the host venue, the impressively large St. George's Hall in Liverpool.

The Illustrated London News covered the British Association event, and in keeping with the theme of enormous moons, published a huge engraving based on one of the Liverpool photographs that ran almost across an entire double-page spread. It showed a beautifully detailed, resplendent

---

* Draper's daguerreotype was considered lost until 1960, when some daguerreotypes were found in an attic of an NYU library and were subsequently attributed to Draper. That small lunar daguerreotype, showing the moon as it appeared over Manhattan from a rooftop, in the earliest days of photography, now lives at NYU's Special Collections, half a block from where it was first taken.

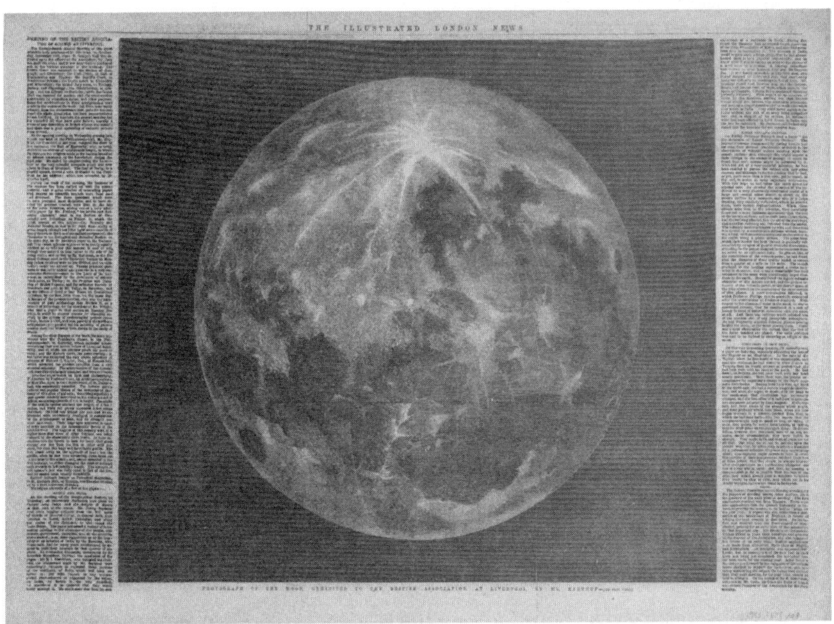

An enormous moon, engraved from a photograph, in The Illustrated London News, *September 30, 1854.*

full moon, barely contained within the pages of the paper. To anyone who opened the paper on Saturday, September 30, 1854, who only knew the moon in the night sky, unphotographed and unfixed, it would have been like putting on new glasses to see an old friend.

But an engraving was not a photograph. An engraving was sometimes based on a photograph, but it started life as a wood block, onto which a draftsman drew an image, reversed from left to right. Engravers—sometimes known as "woodpeckers" because of the sound of their sharp tools against the wood—carved the image into the block, to be inked and then reproduced in print. Later technology allowed photographs to be transferred onto wood blocks, cutting out the need for the draftsman; eventually, once photographs could be reproduced in print, there wasn't a need for engravers either. (Years later, for engraver Paul Martin, the slowing of work ultimately had a silver lining: it allowed him to leave work early one February night in 1896, where the damp twilight inspired him to photograph London by gaslight.)

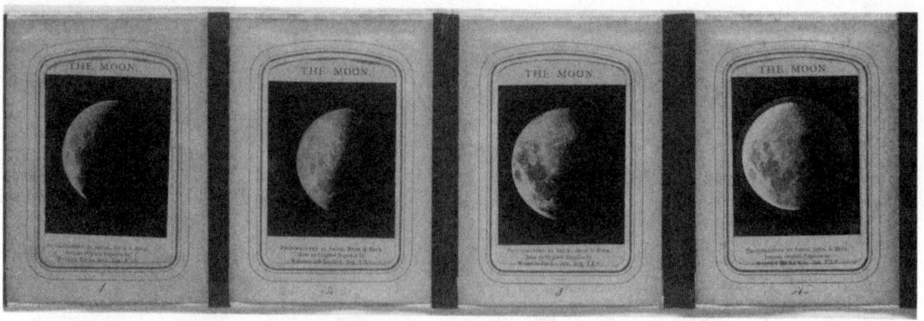

The moon's phases, photographed by Warren De La Rue, circa 1863.

The enormous *Illustrated London News* moon image was made up of several blocks bolted together; if you look closely, you can see faint lines where they were joined. Nevertheless, *The Illustrated London News* captioned the image as a "photograph of the moon."

Once De La Rue resumed his experiments in 1857, he was working from his new observatory, situated above his photo darkroom, at his new estate in Middlesex, removed from London's smoggy skies. Crucially, he had also created a clockwork mechanism for his telescope to accurately follow the moon. But the impediments were still, in De La Rue's words, number-less. A slight change in conditions—an "almost imperceptible mist," for example—could double the exposure time. De La Rue was working with wet collodion, which he described as "very capricious in its properties," and which required the plate to be rushed from the darkroom to the telescope, exposed, and then rushed back again. He estimated that his success rate was four or five plates out of seven. De La Rue was not exaggerating when he said, in 1859, that success with astronomical photography could only come through "perseverance, long practice, and a strong determination to overcome obstacle after obstacle as it arises."

De La Rue's moon negatives were small, just over one inch in diameter, but he was able to make bigger positive copies on glass, using an enlarg-ing camera. The risk with enlargements was a loss of detail, so De La Rue also examined his moon negatives under a microscope. A meeting of the Royal Astronomical Society in 1857 reported that the beauty of De La Rue's moon photographs "gave great promise that at a future period photogra-phy will be considered as the only correct means of mapping down the lunar surface." But even so, De La Rue wanted to show the moon as a solid object with mass, believing it would provide more topographical information. The answer to his conundrum came in the form of an optical device known as a stereoscope.

"Seems, Madam! Nay, it IS!" advertised the London Stereoscopic Company in the mid-1850s, turning a line from *Hamlet* into memorable, slightly shouty sales copy, and accompanying it with an illustration of a family crowded around a stereoscope, eager to look inside. A stereoscope was a wooden viewing device that displayed two similar images side by side, approximately eye-width apart. When the viewer peered into the stereoscope, they saw a single three-dimensional image. It was a neat optical trick and, to nineteenth-century eyes, utterly transporting. "You are not sensible of a picture—do not

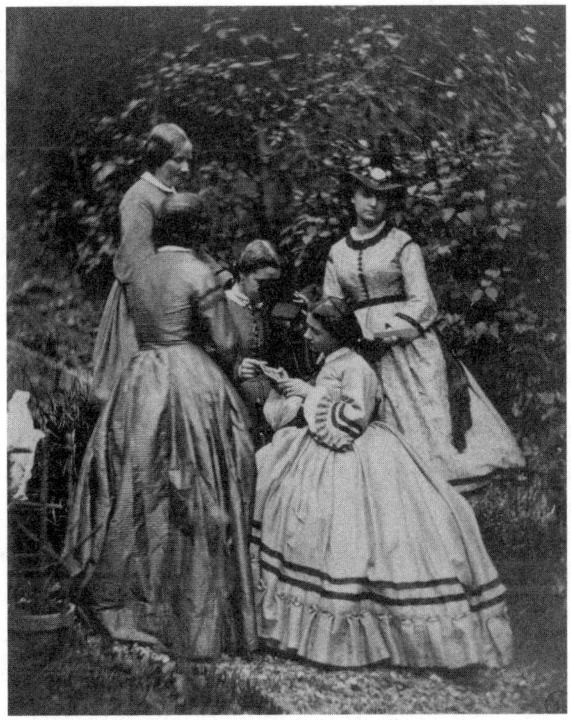

A group of women using a stereoscope, circa 1865. These devices were so enormously popular that the journal The Leisure Hour *stated, in 1858, they were "in use and demand in almost every town and hamlet in the country."*

think of a picture: the illusion is complete, and the mind, as well as the eye, dwells on the actual scene," reported the journal *The Leisure Hour* in 1858. Stereoscopes' popularity had grown after the Great Exhibition in 1851, when Queen Victoria had examined a portable stereoscope and, according to *The Leisure Hour*, "expressed her gratification on beholding it."

By 1856, the London Stereoscopic Company was selling upwards of a thousand different scenes in stereoscope format. Those slides made their way into people's homes, dispersing photography in an engaging and accessible format, in contrast to magazines and newspapers, which could only print engravings. In 1859, Charles Baudelaire observed that "a thousand hungry eyes were bending over the peepholes of the stereoscope, as though they were the attic windows of the infinite." De La Rue had been looking for a "globe-like representation of the moon," as he later put it; with the stereoscope, he found it.

Late in 1857, De La Rue first publicly exhibited his stereoscopes, and at the end of 1858, he presented enlarged stereoscopes to the Royal Astronomical Society. The response was effusive. "The appearance of rotundity over the whole surface of the moon is perfect; and parts which are as plain surfaces in the single photograph in the stereoscope present the most remarkable undulations and irregularities," reported the Royal Astronomical Society's journal. Scientist Sir John Herschel had only the highest praise. "Nothing can surpass the impression of real corporeal form thus conveyed by some of these pictures," adding that their production "is one of the most remarkable and unexpected triumphs of scientific art."

In September 1858, *Photographic Notes* published a letter from an enthusiastic fan of De La Rue's stereoscopes, who wrote that the moon was "presented to us with a convex disc as protuberant and spherical as that of a terrestrial globe or a cricket ball." Sales were reportedly enormous. Moon stereoscopes, wrote *The Art-Journal*, "bring this luminary so near our sight that we can examine it as a real globe placed on our table, showing all its mountains, craters, and valleys distributed and expanded on its round and solid form." The moon no longer just traversed the sky each night; with the stereoscope, it had been caught, fixed in glass, and mounted on cardboard, available to gaze upon at any time of night or day.

To a modern viewer, De La Rue's moon stereoscope images still retain a

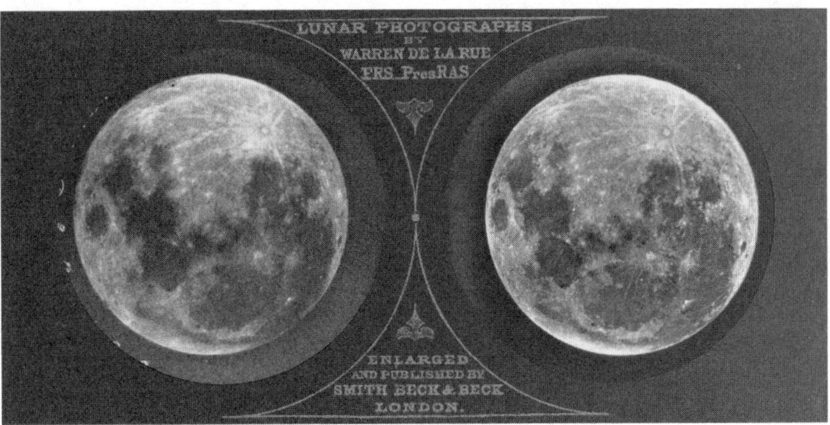

One of De La Rue's lunar stereoscope slides, showing a full moon, circa 1858–59.

delicate luminosity, almost appearing to glow from within; it's easy to see how viewers were so knocked out by the three-dimensional sight of them in a stereoscope viewer. Even when displayed as prints, De La Rue's photographs are mesmerizing. One sublime set of his photographs shows all the phases of the moon, emerging from the shadows of space with its own weighty beauty.

De La Rue published the moon phases along with a booklet and chart, which awed a reviewer in the *British Journal of Photography.* "No drawing or engraving, however perfectly executed, can for a moment vie with these photographs in conveying to the mind the actual condition of the lunar surface." The *Journal* also linked the value of moon photographs to the study of astronomy: "No books of instruction connected with the science of astronomy can be deemed complete without being accompanied by some at least of these or similar photographs."

In 1862, the president of the Royal Astronomical Society presented De La Rue with a gold medal for his work in astronomy and photography. He singled out De La Rue's stereoscopes, which "brought to light details of dykes, and terraces, and furrows, and undulations on the lunar surface, of which no certain knowledge had previously existed."

The following year the editor of *Photographic Notes*, Thomas Sutton, took the opposing view. This was not uncommon for him. When Sutton died in 1875, some obituary writers awkwardly tiptoed around the more confrontational aspects of Sutton's personality: one wrote that Sutton's "zeal was most intense and uncompromising," another that he was a "keen and warm controversialist," a third, that his writing had a "degree of trenchant vigour." The delicacy with which these writers memorialized Sutton was at odds with how Sutton had served his opinions in his journal: forcefully, sometimes with hostility, and an occasional side of acid. He was once sued for libel after publishing allegations of financial impropriety by a photo exhibition jury member. In addition to antagonizing other photographers and editing *Photographic Notes*, Sutton was a photographer, a writer, and inventor of photographic equipment. It was Sutton who had recalled the tears streaming down his face as he sat for a rooftop daguerreotype in 1841. (True to form, he described the resulting portrait as a "caricature.") He also later suffered health problems that he believed were caused by inhaling ether from collodion.

An engraving shows him with dark wavy hair and a bushy bib-like beard that concealed not only his entire mouth and chin, but also his neck, shirt, and (if he was wearing one) tie. It's easy to imagine that beard twitching as Sutton's mouth convulsed with disapproval over some photographic process or lens. It was perhaps doing exactly that when Sutton vented, at a meeting of the Photographic Society of Scotland, that De La Rue's stereoscope views "merely illustrate the fact of the rotundity of the visible part of the moon, which was known before, and they do *not* prove that the mountains stand out in actual relief, as some people suppose they do." Similarly, German astronomer Johann Heinrich von Mädler, who had co-published a complete map of the moon in the 1830s, was quoted as saying, in 1868, that the details achieved in lunar photographs were still "inferior to those which an able observer can determine." Just like its subject, lunar photography was in a state of flux.

More than one hundred years later, on Sunday, July 20, 1969, astronaut Buzz Aldrin took one of the most famous photographs of the Apollo 11 moon landing mission. It shows an imprint from one of his boots, carved into the fine grey dust of the lunar surface. Before he took the photo, using a Hasselblad camera, Aldrin had been kicking up moon dust to watch it land in its peculiar, powdery way. The purpose of the photo, according to a later NASA report, was to study the moon's "surface bearing strength." So unknown was the lunar surface that some scientists had feared it might swallow up the landing pads, or the astronauts themselves.

Understandably, this was also on Neil Armstrong's mind as he descended from the lunar module. As he climbed down, Armstrong released the Modularized Equipment Stowage Assembly (MESA) in which a video camera was stored, and footage of his first steps was beamed around the world. (The MESA also carried a stereoscopic camera for surface close-ups.) In the video, before taking a full step, Armstrong gingerly tests the surface, the same way a swimmer might dip their foot into water from the safety of a poolside ladder.

The aura of the moon landing is such that, in our minds, Aldrin's boot photo and Armstrong's first step are sometimes conflated. Conspiracy theorists, on the hunt for that long-awaited piece of incontrovertible evidence, have also seized on this photo, erroneously proclaiming that the tread shown in the dust didn't match that of the astronauts' boots. But long before Aldrin's photo took on new and unintended meanings, long before paranoid doubts were cast over the Apollo 11 mission, an amateur astronomer created, and photographed, his own elaborate fake lunar landscapes.

Like his friend De La Rue, James Nasmyth was a wealthy amateur astronomer. He was also an engineer who had invented, and became rich from, the steam hammer, an enormous machine used to forge metal on an industrial scale. (At the Great Exhibition, Nasmyth exhibited a fifteen-foot-high steam hammer whose anvil weighed eight tons and hammer weighed a ton and a half. In his autobiography, written some thirty years later, he listed at least thirty-three more of his inventions.) He built his first telescope in 1827, when he was in his late teens, crediting his father, a successful Scottish landscape painter, for his love of astronomy. When he was working at his foundry in Manchester, on some clear nights he would get out of bed and go into his garden, in just his nightshirt, to look at the stars through his telescope. But with his steam hammer wealth, he retired in 1856, at age forty-eight, moved to Kent to a large house he named Hammerfield, and devoted his days and nights to astronomy.

Like his father, Nasmyth had a talent for art. In 1842, he began to sketch the surface of the moon, translating the barren distant shapes as seen through his telescope onto grey paper with black and white chalk. "I endeavoured to illustrate the landscape scenery of the Moon in like manner as we illustrate the landscape scenery of the earth," Nasmyth later wrote. "The telescope revealed to me distinctly the volcanoes, the craters, the cracks, the projections, the hollows; in short, the light and shade of the moon's surface."

Nasmyth had seen his father create plaster models of landscape scenes for his paintings, and he began to do the same. They measured around 50 by 50 centimeters in size, and showed, in hand-sculpted detail, craters

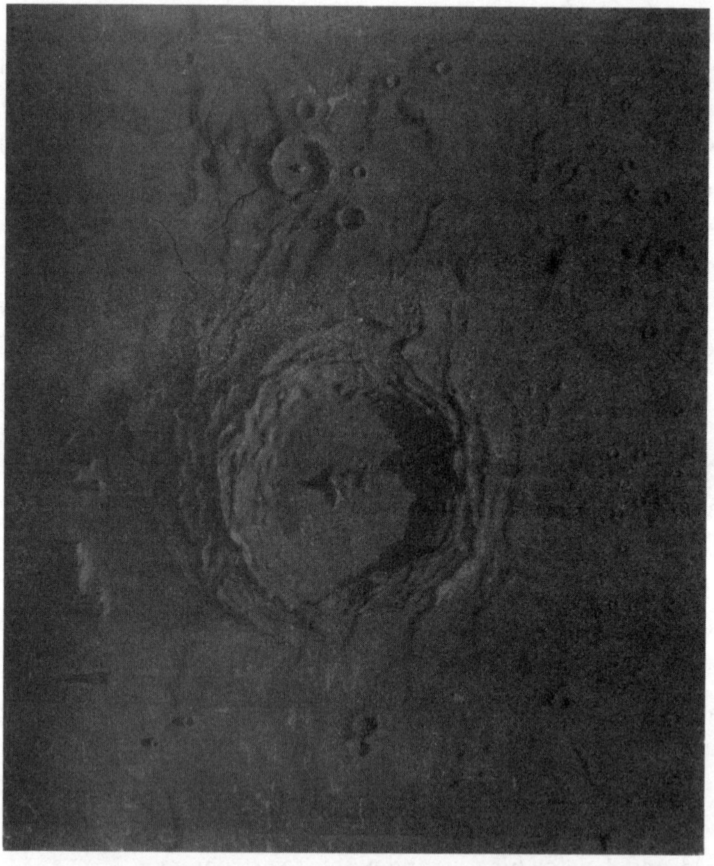

A sketch of the lunar surface by Nasmyth, circa 1860, using chalk and grey paper.

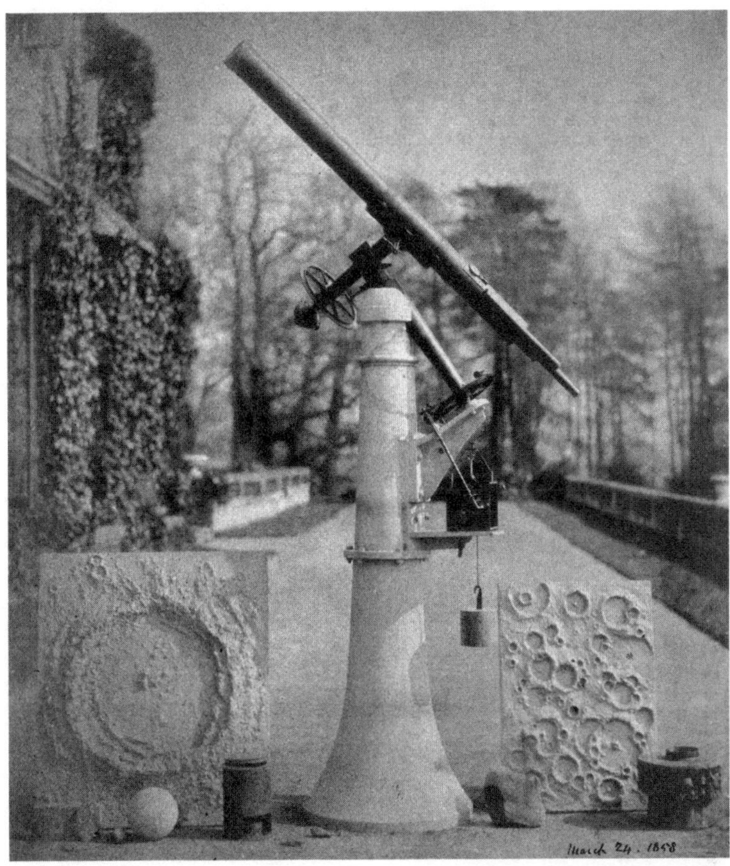

Nasmyth's plaster moon models, set up next to a telescope at his house in Kent, 1858.

and ridges, valleys and peaks—the close-up views that a camera was not yet powerful enough to capture. Nasmyth didn't have to spend countless nights trying to photograph and track the moon through a telescope, only to be thwarted by a cloud, or an ill-timed exposure, or a poorly prepared plate. Instead, he created his own miniature three-dimensional lunar landscapes.

But photography still had a role to play. It occurred to Nasmyth that if he photographed the moon models in sunlight, the results would "faithfully reproduce the lunar effects of light and shadows" and, in turn, "produce most faithful representations of the original." Photographed in the light of the sun, his models took a new shape: their surfaces became littered with shad-ows, adding dimension and form. He had created what De La Rue had also been looking for: the appearance of an object. His photographs, Nasmyth later wrote with an almost admirable lack of modesty, "were hailed by the highest authorities in this special department of astronomical research as the best examples of the moon's surface which had yet been produced."

Nasmyth's goal was to publish photographs of his models in a book, but he was reluctant to do so until the photographs could be reproduced in print. He didn't wait entirely in vain: photographs of Nasmyth's moon models did appear in a journal in 1855, but not in a way that would reach a mass audience. *The Strines Journal: A Monthly Magazine of Literature, Science and Art* was produced between 1852 and 1860 by two employees of the Strines Printing Company, located near Manchester. It contained articles, drawings, and photographs, which they achieved by producing

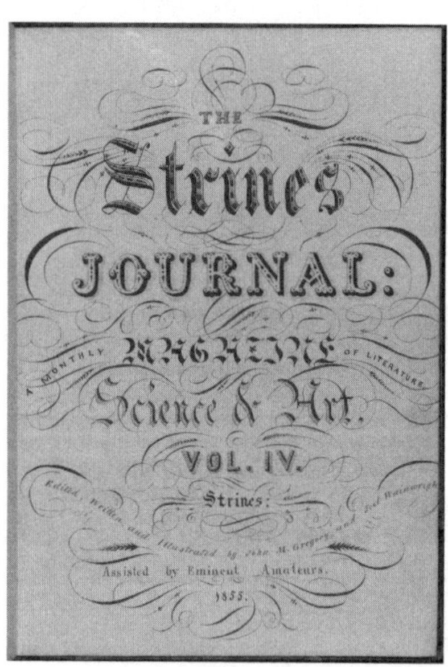
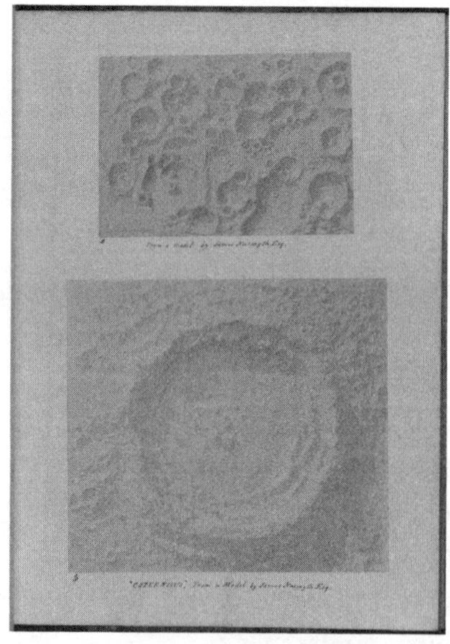

The gorgeous hand-lettered title page to The Strines Journal *and, from an interior article, photographs of Nasmyth's moon models, taken by a partner of the Strines Printing Company, 1855.*

just one copy of each issue. It was handwritten, with photographs either pasted onto pages or bound as a page of the journal. Because each copy was unique, it was read by employees in turn, who would have to wait impatiently to read about topics such the Duke of Wellington's obituary, an overview of an American telegram office, and, in the case of Nasmyth, an article titled "The Moon and Its Surface Inequalities," which included two photographs of his models taken by a Strines employee.

By the early 1870s—almost thirty years after Nasmyth had started his lunar observations—innovations in photomechanical printing had made it possible to accurately reproduce images in books. These innovations included the heliotype process, which was used in Charles Darwin's 1872 book *The Expression of the Emotions in Man and Animals*. Seeing an opportunity for his own project, Nasmyth wrote to Darwin's publisher, John Murray, the following year.

In 1874, Murray published *The Moon: Considered as a Planet, a World, and a Satellite*. By this time, Nasmyth had acquired a coauthor, James Carpenter, an astronomer at the Royal Observatory in Greenwich. The book's twenty-four images were printed with several different photomechanical processes as well as engravings. It also included a reproduction of a full moon photograph, taken by Warren De La Rue.

In the book's preface, Nasmyth wrote of his "thirty years of assiduous observation" and then explained his plaster model process. He wasn't deceiving his readers; he emphasized that it was "as near an approach as possible to the absolute integrity of the original objects." To Nasmyth, the models were the only possible way to truly depict the wonders of the lunar surface. After his book was published, Nasmyth's moon photographs—created with crayon, paper, plaster, and then a camera—were highly praised for their authenticity.

"No more truthful or striking representations of natural objects," wrote the astronomer J. Norman Lockyer in his review for the journal *Nature*, adding that "rarely if ever, have equal pains been taken to ensure such truthfulness." The views of lunar scenery are "wondrous," wrote *Popular Science Review*. Nasmyth's peers also heartily complimented his work, which Nasmyth shamelessly included in his autobiography. "The illustrations are magnificent, and I am persuaded that no book has ever been published before which gives so faithful, accurate, and comprehensive a picture of the surface of the moon," wrote the amateur astronomer William Lassell.*

It's a rather wonderful experience, looking at Nasmyth's photographs today. One model fills the entire frame and shows a rough, cracked surface, peppered with craters and peaks casting spiky shadows. It's possible, just for a moment, to feel the wonder this utterly unique view would have inspired.

Other views feel distinctly homemade. An image of Vesuvius mid-eruption includes some smoke created with fingerprints. A view of lunar peaks has an oddly distorted perspective reminiscent of a 1950s sci-fi B movie, an aesthetic that was a long way off when Nasmyth was molding his plaster sets.

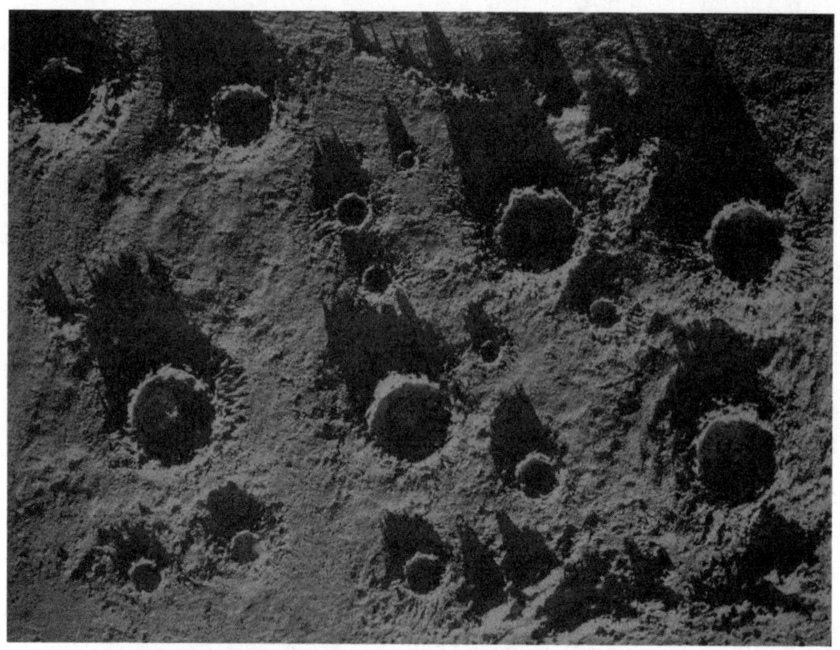

The cratered (plaster) surface of a (fictitious) moon, molded and photographed by James Nasmyth.

* Lassell was no slouch as an amateur astronomer himself: he discovered some of the moons of Saturn, Neptune, and Uranus, and one of Neptune's rings is named after him.

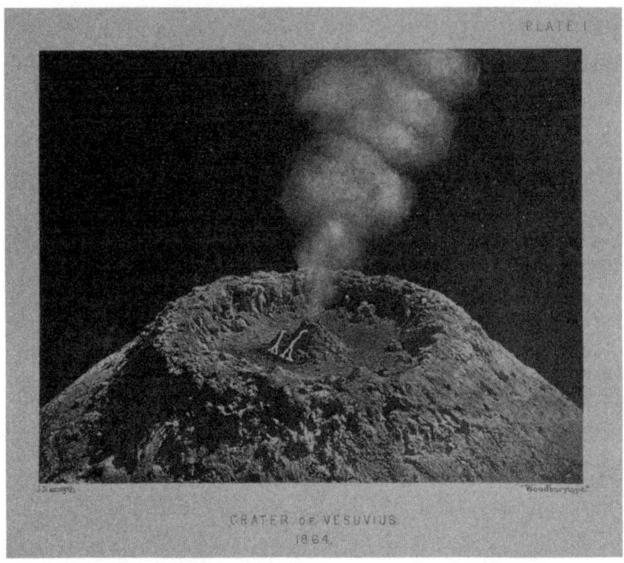

On the photograph of his model of Vesuvius, Nasmyth added smoke using fingerprints, 1864.

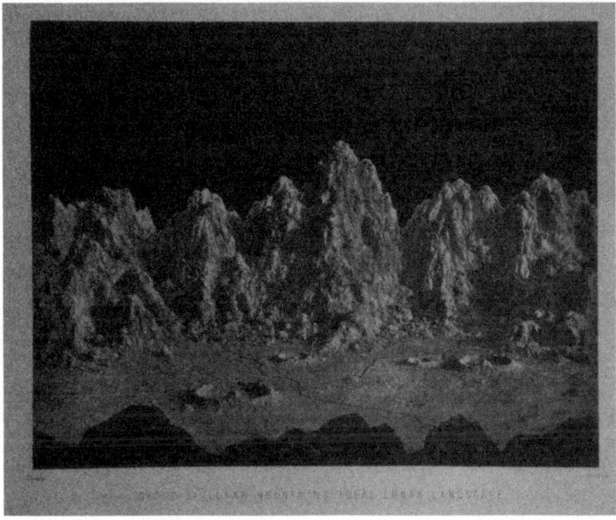

Nasmyth wrote in his autobiography that his lunar photographs "were hailed by the highest authorities in this special department of astronomical research as the best examples of the moon's surface which had yet been produced."

Both Nasmyth and De La Rue sought to show the moon's heft and form, so it's fitting that they're forever linked to the moon's geography. Some of the moon's features have evocative names, like the Sea of Tranquility, which De La Rue captured in his full moon stereoscope views. Others are named for scientists, like the Einstein crater, or the Copernicus crater, which Nasmyth molded in plaster. Near Endymion, a crater named from Greek mythology, is the De La Rue crater; another, farther south, is called Nasmyth.

Chapter 3

Up in the Air

On the afternoon of Monday, May 29, 1882, at Alexandra Palace in North London, Cecil V. Shadbolt climbed into the basket of a gas balloon named "Reliance." He was an adventurous twenty-three-year-old photographer who was eager to capture balloon views for himself, despite the decidedly inconsistent results from earlier aerial photographers. Shadbolt barely had time to fix his camera to a tilting base at the edge of the basket before the aeronaut untethered the balloon. As they rose high above London, Shadbolt felt a "curious cracking sensation," which any twenty-first-century plane traveler will recognize as the *pop* of ear pressure.

The tilting base was to capture different angles; he'd also left the seats behind, to provide more space for his equipment. (As he became more experienced—he took balloon photographs for about ten years—his standard equipment would come to include not only plates, lenses, and shutters but also a barometer, a compass, a map, and, for once he'd landed, a train timetable.) On that first trip, as with others, he started to photograph as soon as the balloon was airborne. This was because, he later explained, "there is no telling what emergency may arise."

On that first trip, as the balloon rose to a height of 5000 feet into the air, Shadbolt was struck by the beauty of the panorama beneath him. He felt like the balloon was "hanging motionless in space" as the earth glided past, revealing below the sprawling city as he had never seen it before.

The descent was slightly less serene. As Shadbolt packed his equipment in newspaper to prevent damage during the landing, he glanced over the side of the balloon and was "horrified to see the various objects upon the earth growing larger and larger with unpleasant rapidity." The balloon's aeronaut instructed him to hold on tight and threw out the grapnel hook. The balloon bounced once before the hook latched to the ground, bounced again, and came to a stop.

In his aerial photography, Shadbolt's goal was to obtain "map pictures" of earth; he even called one of his more successful vertical images an "instantaneous map-photograph." (A vertical aerial photograph is a straight overhead shot, akin to the perspective seen on a map. An oblique aerial photograph shows the ground at an angle—the way you usually see the earth from a plane window.) Aerial photography had enormous potential across a spectrum of endeavors, including meteorology, reconnaissance, agriculture, and archaeology. But the daguerreotype did not lend itself to the constant motion of a balloon; only with the wet collodion process and later, dry gelatin plates, did aerial photography become feasible. Even then, it had many obstacles.

The first obstacle was getting the camera into the air. A ladder or a rooftop provided high-angle views, but a true aerial view required vertical height, which, in the mid-nineteenth century, meant a balloon. And balloons were unpredictable.

The "lighter-than-air" gas that filled a balloon's envelope and kept it aloft was hydrogen or coal gas, so balloon departures were typically from a gas works. Some balloon photographers recounted feelings of stillness once airborne, a peaceful sensation of being suspended above the revolving earth. But in reality, the basket was in constant motion: it vibrated and trembled and rotated, jeopardizing every photograph. One aeronaut instructed his passengers not only to keep still while a photographer made an exposure but also to hold their breath.

Another complication was the weather. A strong wind made ascents and descents more hazardous; Shadbolt recounted one attempted departure where the wind blew with such force the balloon's aeronaut was tipped out of the basket. (Fortunately, he managed to clamber back in before they were airborne.) Mist, fog, rain, cloud cover, or too little sun all jeopardized the photographer's work.

But the biggest problem with balloons was navigation, or rather, the total lack of it. The aeronaut could make the balloon go up, by dropping ballast, or down, by pulling the valve line and releasing gas, and that was it. Without navigation, the balloon photographer had to take whatever views were presented to him. Far more concerning, if a balloon drifted out to sea, there wasn't anything anyone could do about it. English meteorologist James Glaisher preferred to use the gas works at Wolverhampton, a city in the West Midlands, because it was so far from the sea.

In 1862, Glaisher and aeronaut Henry T. Coxwell survived one of the most hazardous ascents in the history of ballooning. According to Glaisher's barometric readings (he had brought with him a host of meteorological instruments), less than an hour after they departed Wolverhampton gas works they had shot up to an altitude of five miles, or over 26,000 feet. Glaisher also tried to take photos as they climbed, but the balloon's rotation made it impossible. The purpose of this flight was to study the atmosphere at high altitudes: how high they ultimately reached is impossible to know, given that Glaisher passed out. (The last reading he remembered before losing consciousness was 29,000 feet.) As for Coxwell, he had to climb into the frozen balloon ring to disentangle the valve line, after which he realized that his hands were frozen. He swung back down into the basket using his arms and managed to release the valve line with his teeth.*

Landings could also be appalling. The balloon's grapnel hook was its anchor, but an anchor in search of an object. If it failed to catch the earth, the balloon continued to bounce and roll along, not very merrily. Shadbolt was once dragged in an upside-down basket toward a river before the grapnel hook finally gripped the ground and stopped the balloon, and its passengers, from ending up in the water.

One of the most dangerous landings in the history of ballooning occurred in the monstrously large balloon built by Nadar, the Parisian

---

* Such were the logistics of ballooning that once they landed, Glaisher had to walk to the nearest train station to get home, a distance, he said, of some seven or eight miles.

photographer whom we previously met venturing into the catacombs. He experienced balloon flight for the first time in the 1850s, and fell in love with it, describing a feeling of "levitating into the silent immensity of welcoming and beneficent space." Nadar predicted he could survey the whole of France from a balloon, with a camera, in thirty days. This proved to be wildly overconfident, particularly because his earliest attempts at aerial photography ended in utter failure, with blank, sooty plates.

According to Nadar, the solution presented itself by accident. One morning in a village outside of Paris, after the previous day's fruitless attempts at aerial photography, Nadar tried to ascend in the tethered balloon. But it had partly deflated overnight, and barely moved even after Nadar removed the ballast, his darkroom tent, and his equipment. In his memoir, Nadar claimed he then unhooked the basket, removed his clothes, and, holding the hoop, ascended in the nude with his camera. He took an aerial photo of the village below—"feeble," he later wrote, and no version seems to have survived. But he did realize his previous error: in earlier expeditions, as the balloon rose, excess gas had vented out the neck and onto the plates, ruining them. But on that morning, the morning of his supposed naked ascent, the valve was closed. Vindicated and victorious, Nadar proclaimed himself to be the first successful balloon photographer.

Several years later, after his adventures in the catacombs, Nadar hatched a plan to build a balloon so gargantuan that crowds would flock to see it ascend. The money raised from such a spectacle would be used to construct a heavier-than-air flying machine, which Nadar believed was the future of flight. (He was still alive in 1903 when the Wright brothers flew their aircraft over a beach in North Carolina.) He called this enormous, publicity-generating balloon Le Géant (the Giant).

When inflated, Le Géant stood 196 feet high—taller than the Statue of Liberty from her toes to the tip of her torch. The two-story basket, painted Nadar's signature color red, was absurdly well appointed. The lower space could hold a printing press, darkroom, galley kitchen, triple-decker bunk beds for twelve, a toilet and, perhaps most crucially, a wine store. A ladder linked it to the open upper deck, where passengers could take in the view. In illustrations from the era, it looks like a small flying house appended to a bulging silk ball of unfathomably large dimensions.

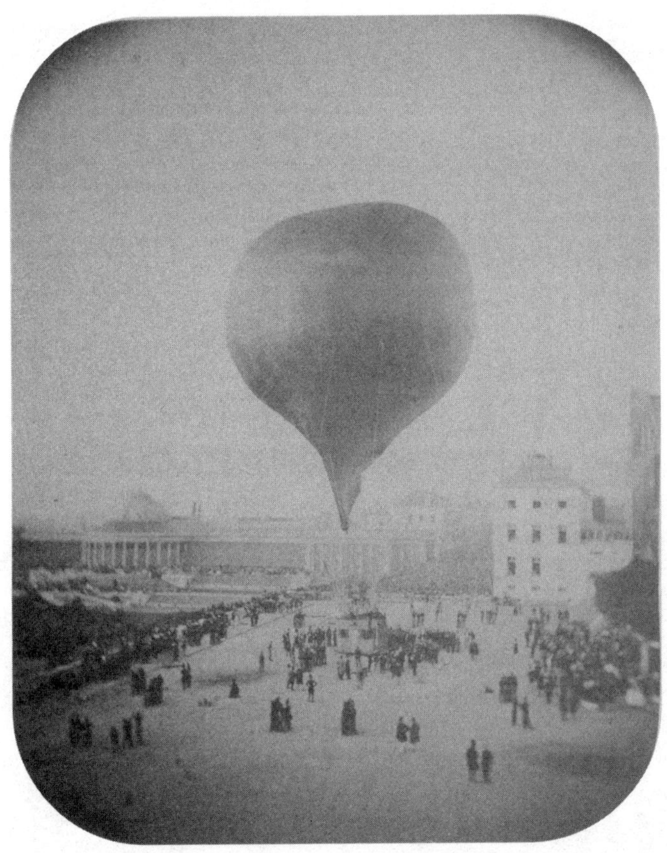

Nadar's balloon behemoth, Le Géant, ascends in Brussels, September 1864.

Always with an eye toward promotion, Nadar took photographs of himself in the studio, in a prop basket. His wife Ernestine appears in one photo, looking slightly resigned, and no wonder: Nadar was obsessed with something that was dangerous, expensive, and complicated.

Le Géant's infamous landing occurred at the end of its second journey, in October 1863. After flying eastward through the night, the nine passengers (including Ernestine) enjoyed coffee and croissants on the deck. The serenity did not last. The balloon's aeronauts, Louis and Jules Godard, began the descent, for reasons that were later disputed. One account stated that Nadar thought a cloud bank was the open sea and pulled the valve line; another was that the Godards took the balloon down too quickly.

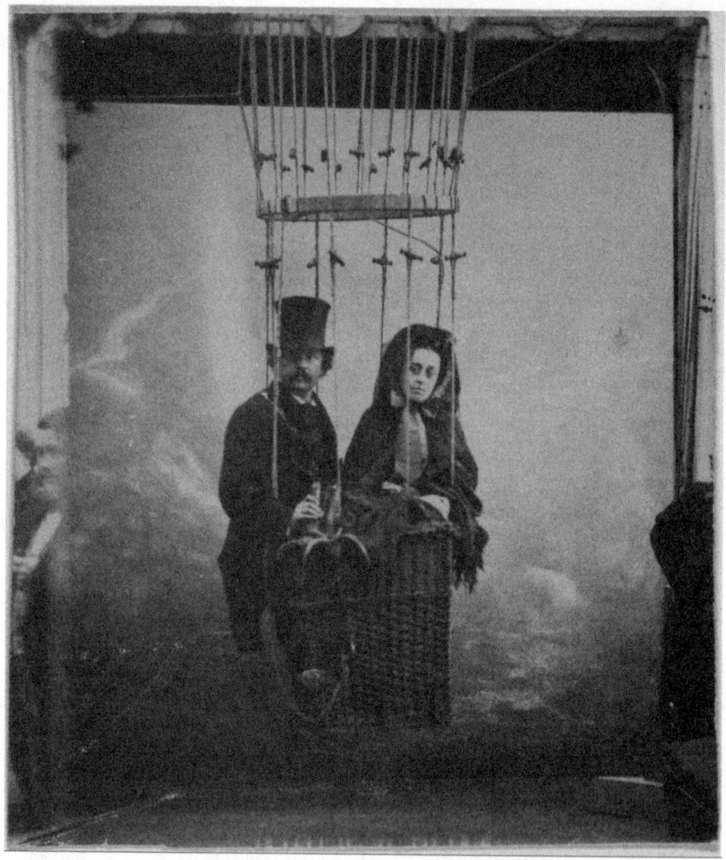

Nadar and Ernestine feigning flight in the photo studio, circa 1865.

Either way, they descended into a strong wind with a broken valve line and without enough ballast to re-ascend.

The Godards released the balloon's two iron grapnel hooks, but the basket grazed the ground and bounced with such force that the grapnels tore off. The balloon ricocheted high into the air several more times—somewhere between 25 and 40 meters (around 80 to 130 feet), according to one passenger's estimation—before slamming back to the earth. For the next 25 miles, the basket was dragged by the partially inflated balloon, crashing through trees and hedges. At one point in its frenzied journey, the balloon approached a train track, where an express train was hurtling toward them. They avoided a collision only because the engineer saw them and slowed the train down. When the balloon finally came to rest in a woodland, half the deck had been ripped off, and all the passengers were injured, sprawled on the ground, with Ernestine landing in a river. Amazingly, no one died.

Any photographs Nadar took from Le Géant have been lost. Although Nadar repaired Le Géant and took more voyages, it failed financially and he sold the balloon in 1867. Nadar's most successful aerial photo was from 1868, in a tethered gas balloon some 1700 feet in the air above Paris. He used a multi-lens camera, in the hope that at least one of the frames would be clear and sharp. Nadar was using the wet collodion process, which pre-

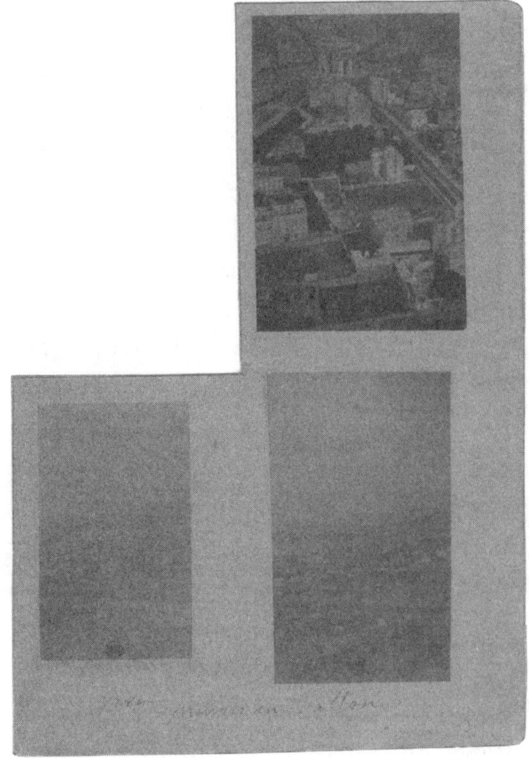

Nadar's aerial photographs taken from a balloon tethered some 1700 feet above Paris, 1868.

sented its own set of problems. He later recounted how, during his aerial photography attempts, he hung his darkroom tent from the balloon's hoop, but underneath it got so hot he had to keep his chemicals in an ice bucket.

Aerial photographs were a strange sight. The French newspaper *Le Petit Figaro* published Nadar's aerial view as an illustration with annotations so its earthbound readers could decipher this alien view. After James Wallace Black captured an aerial view of Boston in 1860 (also using a multi-lens camera), Oliver Wendell Holmes wrote about it in *The Atlantic Monthly*: "Boston, as the eagle and the wild goose see it, is a very different object from the same place as the solid citizen looks up at its eaves and chimneys."

But it's exactly because aerial photography showed a perspective that was unnatural to humans that it held such appeal. It offered an elevated, expanded, and mobile viewpoint. After the start of the Civil War, Black wrote to the *New-York Tribune* to suggest aerial photography. Although there's no evidence that photography was used by the balloon corps, its military potential was obvious. "If lawn tennis grounds, clean clothes hanging out to dry, and individual rows of root crops in a field can be photographed from an altitude of something like 3000 feet in dull weather, the value of balloon photography, especially for recording observations in time of war, is now clearly established," wrote Shadbolt, in 1884. "That there are obstacles in the way of success, however, and pretty big ones, too, is a well-known fact; but like most other difficulties, they are not altogether insurmountable."

Shadbolt's photographs are the earliest surviving aerial photographs in England, but they were only rediscovered after seventy-six of his lantern slides turned up at a secondhand market in 2015. The collection is now part of Historic England, a government agency that protects and promotes English heritage. One slide is a photo of Shadbolt and aeronaut William Dale. They're posing with what looks like an oversized picnic basket; Shadbolt's camera is on the edge, and ropes trail out of the frame. Shadbolt perches inside the basket; Dale leans on the outside, ankles crossed, exuding confidence. The overall impression is one of promise, adventure, and risk.

If Shadbolt seems to have a glimmer of a smile on his face, it is perhaps because his balloon photographs were so well received. At the end of 1883, his images were highly praised as "the best pictures taken from a balloon in

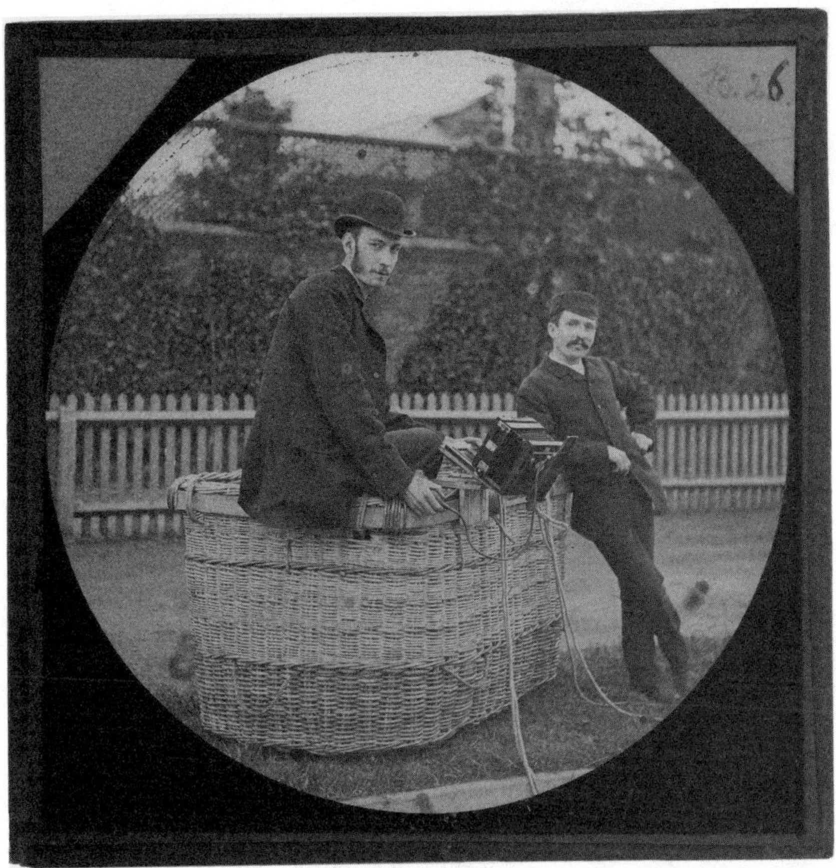

Shadbolt (left) and Dale with their balloon, circa 1880s. Note the camera attached to the edge of the basket.

this country." In 1884, the French publication *La Nature* wrote that Shadbolt's photographs were "beyond doubt, the most remarkable obtained up to the present time from the car of a balloon." At a photo convention in Minneapolis in 1888, his balloon photographs "excited a lively interest in all who beheld them."

On June 29, 1892, Shadbolt and Dale ascended from the grounds of the Crystal Palace with two other passengers. Shortly after takeoff, when they had reached about 600 feet, the envelope burst, and the balloon plunged to the ground. The two other passengers survived; Dale and Shadbolt did not. The dangers of ballooning were very real; those less inclined to take such a risk in the name of photography had to try other ways to get their cameras into the air.

Kite photography was cheaper, more convenient, and less dangerous than balloons, provided that a person holding a camera was not also attached to a kite. (Naturally this was also attempted.)

The biggest logistical challenge of kite photography was exposing a photograph from the air while the operator was on the ground. In France, in the late 1880s, a photographer named Arthur Batut attached a fixed-focus camera to his homemade kite. Rather than glass plate negatives, he used the newly available strip film, which exposed more quickly and, more importantly, wouldn't shatter in a crash landing. Batut also devised a fuse to trigger the shutter: after calculating the desired burn time, he ignited the fuse, launched the kite, and began the countdown. Once the fuse had burned, it released the shutter mechanism and exposed the film. It also prompted the descent of a small pleated piece of white paper, a signal to Batut to bring down the camera. It was still complicated, but—unlike balloon photography—at least it wasn't going to kill him.

Kite photography was a delicate balance. The line needed to have enough pull to get the kite and camera airborne, but not so much that it would snap. Losing the kite altogether was also a risk. In 1901, *The Photographic News* in London obliged a request they had received from a kite photographer based near Calais, in France. "He let the string slip from his fingers, and the kite sailed off with the camera towards the English coast," the journal reported. "If anyone on this side of the Channel has picked up the flying camera, M. Delcourt will be obliged if he will return it."

In Batut's 1890 book, *La photographie aérienne par cerf-volant*, he set out the advantages of kite photography for the military, exploration, and agriculture. He didn't mention news photography, but the famous aerial panorama of San Francisco in ruins after the 1906 earthquake was shot using kite photography by George R. Lawrence. And even before that, in New York City in 1896, journalist and avid kite photographer William A. Eddy photographed a parade on Broadway using a line of kites from the top

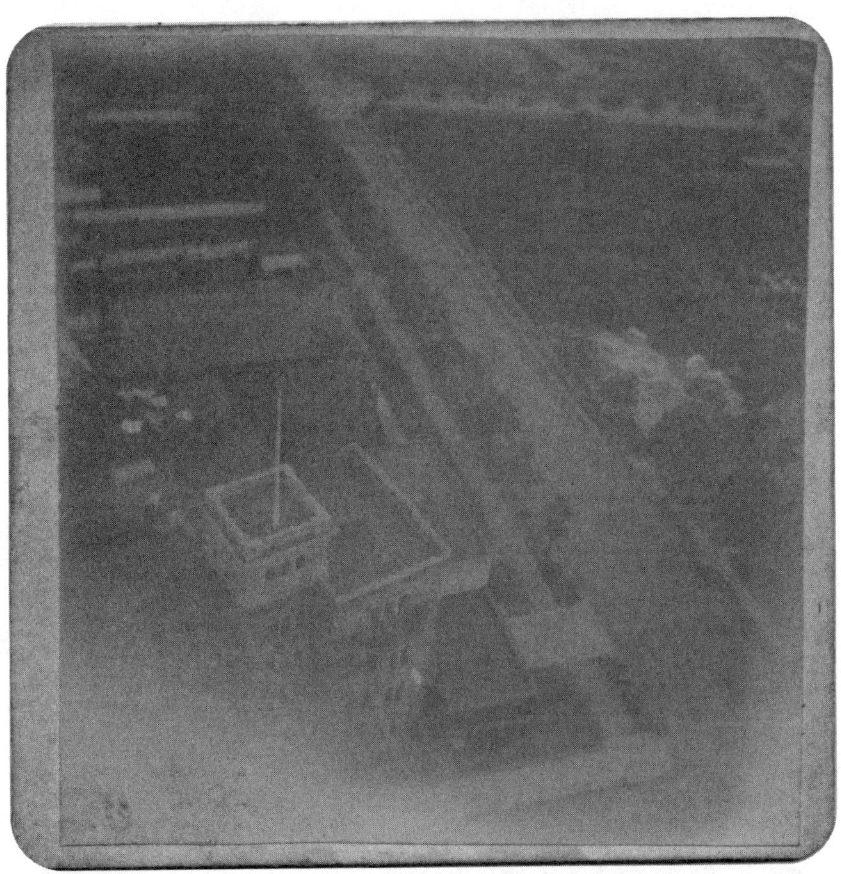

William Eddy experimented with kite photography in the 1890s. In this photo, taken in 1897 in Washington, DC, he stands on the roof at top left, amid a scattering of kites, and no doubt holding tight to the kite line.

of a building. Eddy had regularly used six to nine kites in a train, but on this occasion, the main kite line broke and he had to retrieve his camera from a neighboring rooftop.

Eddy promoted the use of kite photography for many purposes: he offered to teach aerial photography to the navy in Cuba; worked at the Blue Hill Observatory, near Boston, helping with its meteorological kite observations; and is possibly the first and last person to use kite photography to try to solve a crime. In 1908, according to *The Jersey Journal*, an ice cream delivery vanished from the back porch of Eddy's house. He used his kite and camera to survey the area, and the developed plates reportedly showed a group of men under a tree. When Eddy later visited the site, he found an empty tub of ice cream.

It was perhaps inevitable that someone would try to devise a kite that was strong enough to lift both a camera and a person. In fact, several people did, including Captain Baden Baden-Powell of the Scots Guards. (This Baden-Powell was the brother of Robert Baden-Powell, founder of the Boy Scouts.) In 2017, historian Christopher de Hamel related a story, told by his grandfather Alex, of a visit by Baden-Powell to see Alex's father in 1895. Baden-Powell brought with him a new man-lifting kite he'd been developing called the "Levitor," but the only person light enough to ascend in the basket was six-year-old Alex. As de Hamel wrote, "My grandfather, as he recounted later, was more frightened of dropping his father's camera than of falling himself." The photograph, an aerial view of the house from an altitude of 400 feet, later appeared in an 1897 book, credited to Baden-Powell.

The Levitor kite was intended for wartime reconnaissance and observation. So, too, was another man-lifting kite, invented by an American expatriate in the UK, Samuel Cody, although it is more accurate to say human-lifting, as Cody's wife, Lela Cody, also ascended while strapped to a train of kites. Cody had come to aviation by way of Wild West vaudeville shows; he had a goatee and a moustache with the ends waxed into fine upturned points, like a small pair of hirsute wings attached to his upper lip.

His invention, which he named the "Cody Aeroplane," consisted of a train of bat-shaped kites. (To contemporary eyes, and when rendered in a black-and-white photograph, they resemble the bat signal from *Batman*.) From the kite that was closest to the ground, Cody attached what he called a "basket chair," which looked much like a bucket made of straw. From here, the kite flyer could control the height and had access to a camera, a telescope, and, as Cody once noted, a "quick-shooting" firearm. During trials with the British Admiralty in Portsmouth, both of Cody's sons took photographs from the heights of around 200 feet and 800 feet; one of the photos shows a vertiginous perspective of the deck of a battleship.

Ultimately, both Baden-Powell's and Cody's heavy-lifting kites were superseded by airplanes. Cody himself devised several heavier-than-air

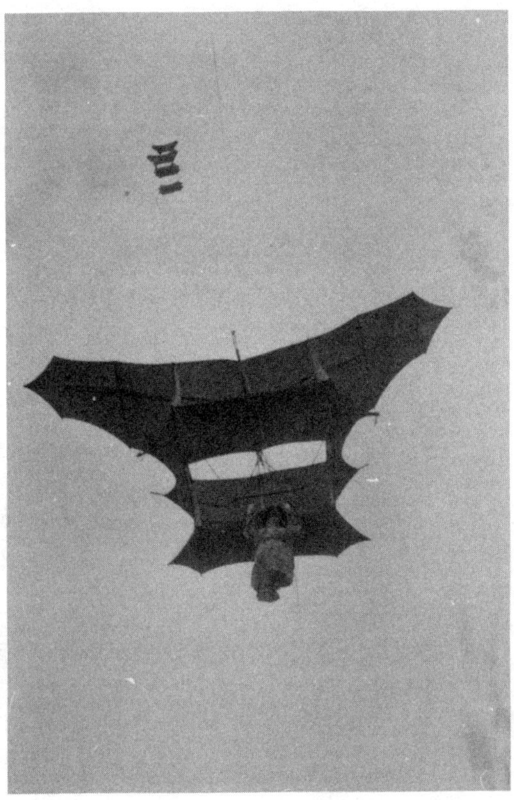

A human-lifting kite also had potential for aerial photography. Here, the human in question is likely Lela Cody, 1909.

flying machines, but he never saw the full realization of aerial photography that was to come in a few short years, during World War One. Cody died during a test flight of his "Cody Floatplane" in 1913.

Although he was not the first to attempt photography from a rocket, Alfred Maul, a German engineer, had the most success with this explosive form of aerial imagery. The potential benefits of rocket photography for the military were clear: it was more affordable and faster to execute than balloon photography, and the rocket itself was almost impossible for the enemy to shoot down. But Maul's early experiments, in the spring of 1901, were plagued

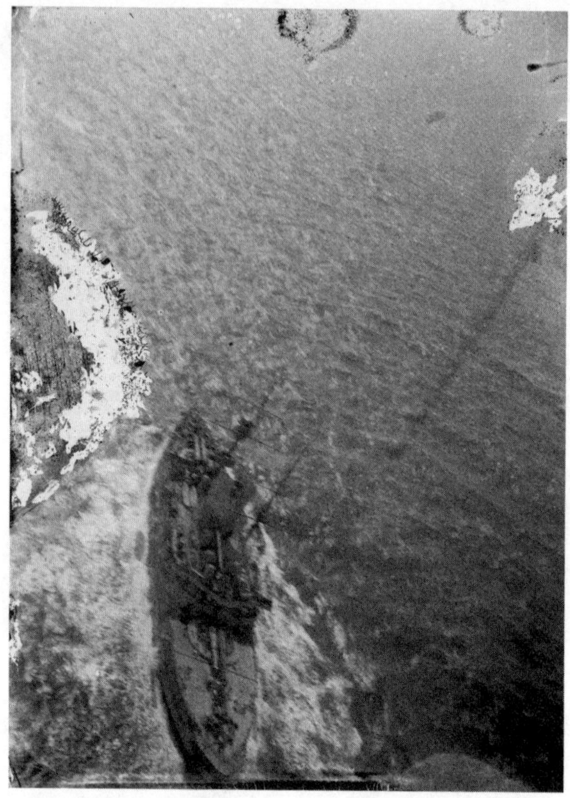

In 1908, Samuel Cody demonstrated his human-lifting kites to the British Navy; this photograph shows a vertiginous view of the HMS Revenge.

with problems specific to launching a camera inside a rocket, including how to release the camera to make the exposure and how to prevent crashes and explosions. Maul later wrote, with what sounds like a hint of pride, that there were "an impressive number of expensive apparatuses destroyed."

Maul built several versions of his rocket, but the essential components were a camera inside the rocket's nose cone, and a parachute for a soft landing. After the rocket was fired, just before it crested, a pneumatic-electrical contact opened the shutter and the parachute. This triggered the nose cone to break from the body of the rocket. They remained connected by a cord, so that the rocket hit the ground first, leaving the camera to descend gently with the parachute.

"The exposure, the opening of the parachute and the detachment of the camera follow in rapid succession, and in about a minute the apparatus comes safely to earth not far from the starting point," reported one observer. "The whole apparatus is carried on a small hand-cart, which can be easily taken to the front, and the rocket is in little danger of being shot down in its brief flight." Maul was so confident in his rocket camera that during a test, he reportedly invited 120 soldiers to open fire as it landed.

As he fine-tuned his apparatus, Maul tried different-sized rockets, some of which were purchased from a fireworks factory, and different methods of stabilization, eventually using an internal gyroscope. The largest rocket he built, in 1912, reached a staggering altitude of 2600 feet, and produced photographs around eight-by-ten inches in size.

Even as Maul was developing his highly mechanized rocket photographs, another German photographer was experimenting with aerial photographs that were far more fluid in nature. "The pigeon is the first bird to become a photographer," reported the *New-York Tribune* in 1908,

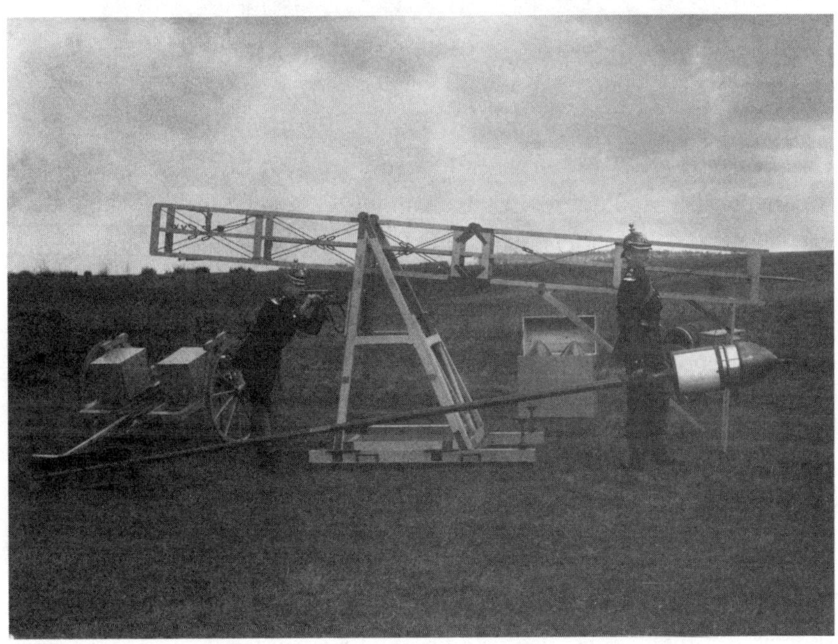

Alfred Maul's rocket camera getting ready for action, circa 1906.

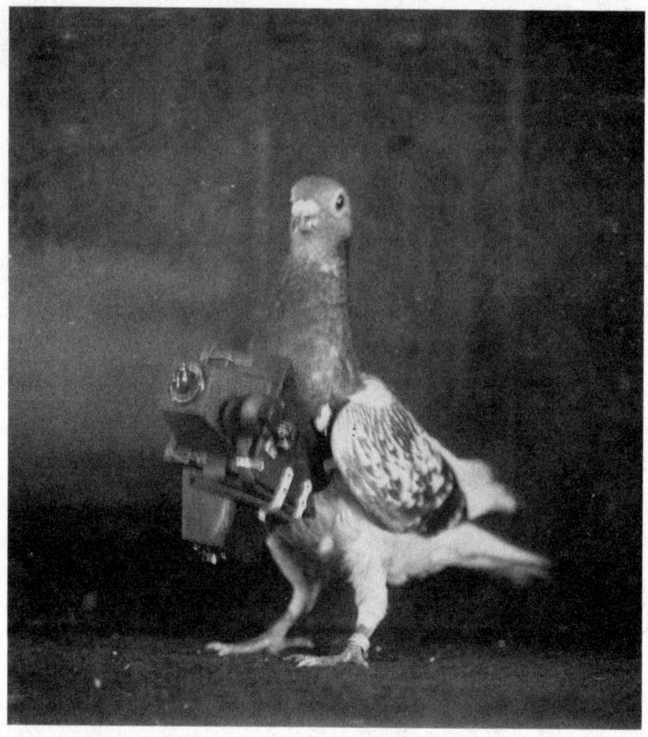

One of Neubronner's pigeon photographers.

the year after Julius Neubronner lodged his patent titled "Method of and Means for Taking Photographs of Landscapes from Above." Neubronner was a German pharmacist who used pigeons to receive prescriptions and send out small quantities of medicine. After one of his pigeons, released in a fog, didn't return to its dovecote for four weeks, Neubronner sought what he later described as a "photographic record of its journey."

Neubronner designed several lightweight cameras that he attached to the pigeons with harnesses. One camera had two lenses, facing forward and backward, and produced images about 1.5 inches square. To fire the shutter, he devised a mechanism using a ball filled with compressed air and a lever. As the pigeon flew, a small perforation in the ball slowly released air; once the ball was deflated, the lever moved and tripped the shutter. Another camera had a single lens, and could make eight exposures at pre-timed intervals, also using a ball of air and a clockwork mechanism.

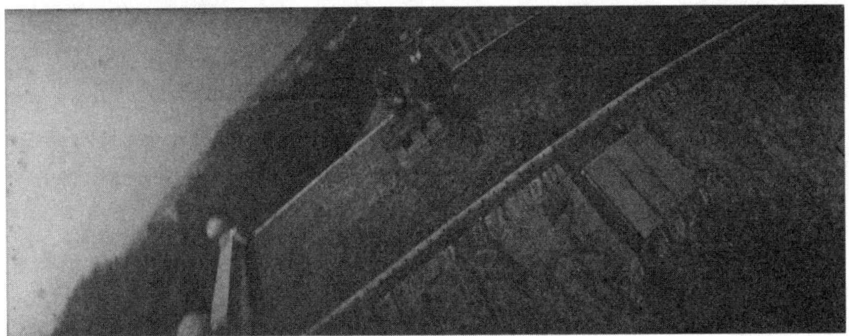

Neubronner, seen on a rooftop as he releases his roving pigeon cameraman.

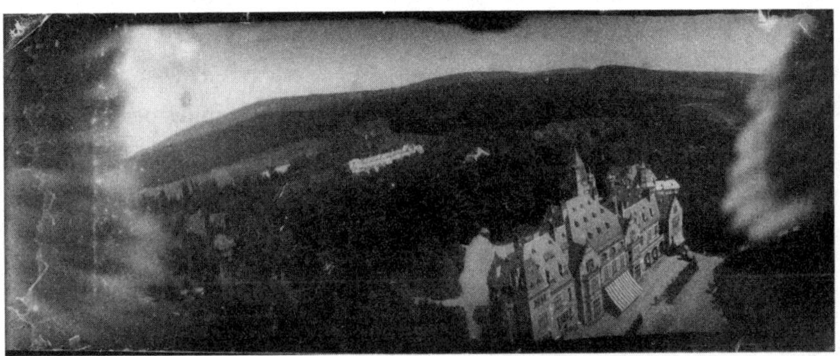

The thrill of flight, framed by the photographer's wingtips.

In the history of aerial photography, Neubronner's pigeon photos offer the truest and most charming representation of a bird's-eye view. Horizons tilt and curve, and rooftops fall away at oblique angles; some photos are framed by the feathery wingtips of the flapping photographer. It is impossible to know what it is truly like to fly like a bird, but Neubronner's pigeons, gliding home with their photographic cargo, provide a permanent, soaring glimpse of flight.

Neubronner hypothesized about the usefulness of a pigeon photographer in war. He proposed releasing pigeons from airships above enemy lines, and the use of mobile, portable dovecotes. (He made one of his own, which also had a darkroom.) The German war office took an interest in Neubronner's work, but both his and Maul's aerial photography methods were short-lived.

After Europe exploded into war in 1914, all sides in the conflict began to photograph enemy positions from planes. Wartime aerial photography proliferated so rapidly that between January and September of 1918, the British Royal Flying Corps produced more than five million aerial photographs. That same year, German planes were photographing the entire Western Front—about 450 miles—every two weeks.

For all its potential uses—cartography, agriculture, archaeology—it was military reconnaissance that propelled aerial photography into an established practice, one that was nevertheless still fraught with danger. Major Edward J. Steichen, Photographic Section Chief of the US Air Service, American Expeditionary Forces, described World War One aerial photography in stark terms: "The military aerial photograph is made with a heavy, cumbersome camera more or less successfully suspended in the fuselage of an airplane traveling through the air at a good hundred miles per hour. The fuselage receives all the vibrations of a powerful motor, whirring off more than a thousand revolutions in a minute, as well as the thumps and bumps of air-pockets and exploding enemy anti-aircraft shells."

Just a few years before the war began, in 1909, an Italian aeronaut named Eduard Spelterini journeyed across the Alps in a balloon. He took photographs of the landscape stretched out below him, including the Mer de Glace, a vast glacier in the French Alps. In 2017, Dr. Kieran Baxter from the University of Dundee used Spelterini's photographs to assess changes in the glacier. After mapping Spelterini's route, Baxter and his team rephotographed the glacier from a helicopter to create a time lapse video. The project, wrote Baxter, "allows for a visual comparison of the landscape then and now to reveal the staggering reduction in the ice surface that has taken place over the last century."* The purposes of aerial photography have changed with the very terrain it surveys.

---

* To find the video on YouTube, search "100 Year Time Lapse Kieran Baxter."

Chapter 4

Into the Water

On a bright September morning in 1898, in a small coastal village called Banyuls-sur-Mer in southern France, a marine biologist named Louis Boutan climbed on board a small boat. He was with his colleague Joseph David, a mechanic at the Arago Laboratory, the marine station where they both worked. But the experiment they were about to conduct had nothing to do with either man's day job. David wore a helmeted diving suit, and Boutan had with him a new underwater camera that he was trying out, after several years of unsuccessful attempts with other cameras. Boutan wasn't alone in his difficulties with underwater photography. Over the previous decades, photography had expanded upwards, into the air, and outward, toward the moon; photographers had even started to conquer artificial light. But descending with a camera into the world beneath the waves had been a decidedly more elusive proposition.

Previous attempts by various photographers in the 1880s and 1890s didn't inspire confidence. One photographer in Scotland tried unsuccessfully and subsequently rationalized that submarine photography, as it was sometimes known, didn't have any practical value. Another defeated photographer summarized his attempt simply as "not great." Others were skeptical of the entire enterprise. Even after Boutan had begun his experiments, a writer at *The Photographic News* dismissed underwater photography as an "idle dream."

The challenges of underwater photography were significant. At the most obvious level, photographers were working in an entirely different medium that would not only flood a camera but, with sufficient water pressure, crush it. (Even today, underwater cameras have a depth rating to indicate how far they can descend before they cease to function.) The other major complication faced by early underwater photographers was light, or lack thereof. One 1882 article hypothesized that, although underwater photography was seen "almost as an impossibility," it could work—provided there was an underwater electric light.

Boutan was not a photographer, but he was a scientist, familiar with considering new hypotheses and testing out the unknown. After graduating from the University of Paris, in 1880 he journeyed halfway around the world to Australia, to attend the Melbourne International Exhibition with representatives from the French government. He spent much of his time in the countryside, studying local species and helping winegrowers identify and treat phylloxera, a nasty parasite that destroyed grapevines. An entertaining but unsubstantiated later report said he traveled around with two outlaws.

Once back in France, Boutan completed his doctoral thesis on *Fissurella alternata*, a type of limpet, at the University of Paris. He also took a lecturer role working under the biologist and zoologist Henri de Lacaze-Duthiers, which turned out to be pivotal to Boutan's underwater photography experiments. He now had access to the Arago Laboratory, which Lacaze-Duthiers founded in the early 1880s, in Banyuls-sur-Mer. Situated on the Mediterranean close to the Spanish border, Banyuls-sur-Mer offered diversity of marine life and seasonable weather, an ideal combination for underwater photography.

Arago was a place for students to live and work, with access to a library, offices, and a mechanics workshop, all situated in a large building on the water's edge. (In a lovely loop of photographic serendipity, the laboratory was named after François Arago, the scientist who introduced the daguerreotype in 1839.) Arago is still a marine station today, renovated and expanded, and known as the Observatoire Océanologique de Banyuls-sur-Mer.

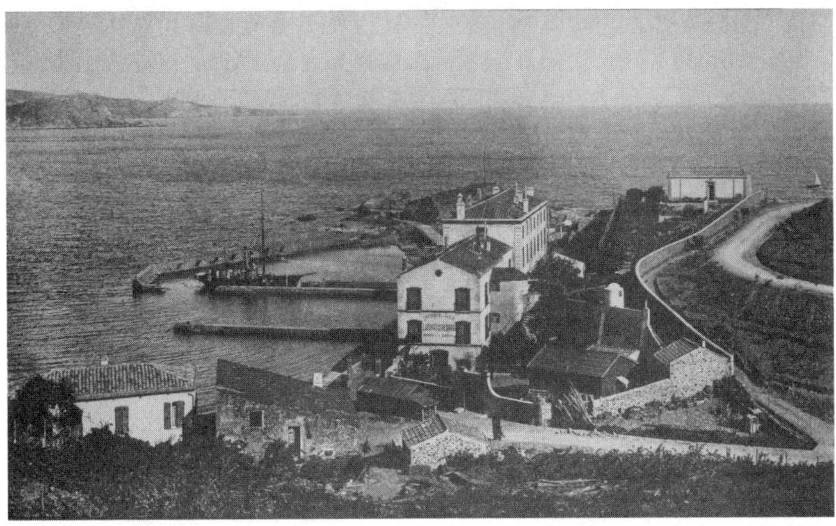

A view of the Arago Laboratory in Banyuls-sur-Mer, circa 1909.

At Arago, Boutan was studying the development of the *Haliotis* mollusk in the laboratory aquarium when he was faced with an intractable problem: in the lab, the mollusks died once they reached adulthood. The only way to study them alive was in their natural environment, which Boutan was able to do, having learned to dive after he started working at Arago in the mid-1880s. As he looked for mollusks in the water surrounding Banyuls-sur-Mer, he was struck by its resplendent loveliness—"the beauty, the extreme beauty, of the submarine landscapes." He wished he could properly capture what he saw—and the solution, he realized, was underwater photography.

Astonishingly, the earliest known attempt at underwater photography had taken place almost forty years earlier in the coastal town of Weymouth, southern England, in 1856. The photographer was a solicitor named William Thompson, and the idea struck him as being enormously helpful to assess water damage to underwater infrastructure. The idea came to him as he was sheltering in bad weather at an inn, watching a nearby river churn under a bridge. Unlike Boutan, his impulse was not to share a glorious underwater scene; instead, like a true lawyer, he was thinking about accurate documentation.

There wasn't a blueprint for Thompson's experiment: as far as we know, the underwater camera that he created was the world's first. He built a wooden box for his camera, with a glass front covered by a weighted shutter that was attached to a length of string. The box was made watertight—at least in theory—with a series of iron screws. Thompson's plan was to secure the box to an iron tripod and lower it by rope into the water from a boat, pull open the shutter with the string, and wait for the exposure.

If it sounds simple, it wasn't. Thompson was using the wet collodion process, which, as we have seen, created beautifully sharp negatives but had one particularly significant downside: everything had to be prepared and developed on the spot, which was hardly ideal when the location was a damp beach in southern England in February. Aided by a friend, Thompson prepared the plate under the darkness of a tent by Weymouth Bay. Once that was done, the clock was ticking. He put the plate in the camera, put the camera in the box, screwed it closed, climbed into the boat, rowed out to deeper water, lowered the tripod and camera to the seabed about eighteen feet below, and pulled the string to open the shutter. All of this would have been done in about the same time as it might take to boil the water for a cup of tea.

Thompson later admitted that he'd had to do this experiment twice, because the first attempt, with a five-minute exposure, yielded no image whatsoever. But the second attempt, with a ten-minute exposure, did, even though water had leaked into the camera. To contemporary eyes, the photo resembles more an abstract wash of charcoal than an undersea landscape. But *The Photographic and Fine Art Journal* described it as "a very good photograph of the rocks and weeds." It also caused quite a stir when it was shared at a meeting of the Liverpool Photographic Society: "This remarkable picture excited considerable attention as it was handed round to the members."

Thompson wrote to the *Journal of the Society of Arts* about his experiment, arguing that underwater photography could be an "incalculable benefit to science." But he was an amateur photographer and a lawyer, and this was the end of his underwater experiments. (Still, his hometown hasn't forgotten his contribution to photography. In Weymouth, Thompson's underwater achievement has been commemorated with a plaque outside his former residence.)

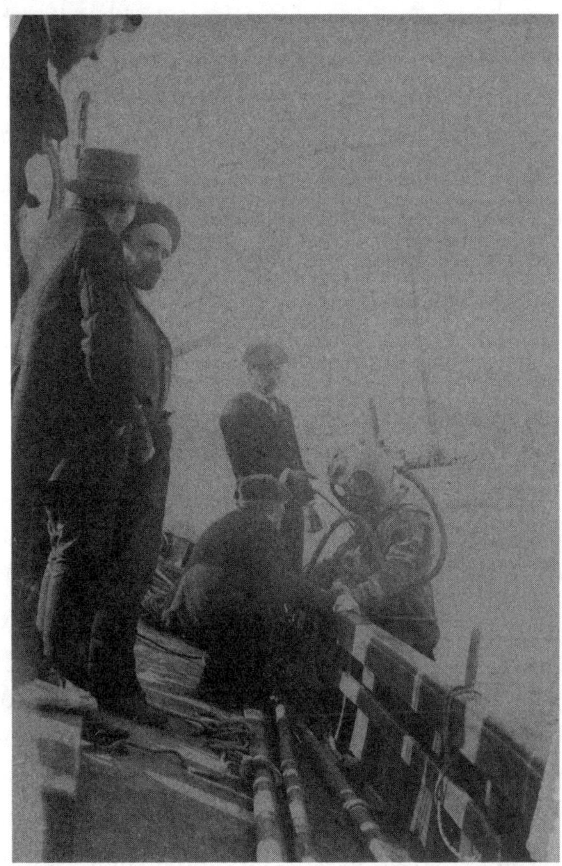

A diver descends. Boutan stands on deck, with a cigarette, looking toward the camera.

Some four decades later, Boutan had a few advantages over Thompson. First, he was in the Mediterranean in summer rather than the Atlantic in winter, unquestionably a more pleasant setting for any activity whatsoever. Second, he could operate the camera from a boat, or experiment with underwater photography as a diver, opening the shutter himself. Lastly, by the time of Boutan's experiments, the wet collodion process had been superseded by the far more convenient gelatin dry plates. Photographers no longer needed to carry an entire darkroom with them any time they wanted to take a picture, which was especially helpful when the shoot location was the sea.

Boutan's first camera, from 1893, was designed with his brother, Auguste. It consisted of a waterproof copper box for the camera, with a rubber-lined lid that was clamped shut with eight copper screws. Outside, they added levers to operate the shutter and plates, and a glass porthole for the camera's lens.

Inside the copper box was a small camera known as a detective camera, which had become popular in the 1880s. The name derived from its unobtrusive boxlike shape, which made it ideal for candid, and sometimes unwelcome, photographs. It had a fixed focus, and its six plates, measuring 3.5 inches by 5 inches, could be rotated between exposures, which meant Boutan wouldn't have to resurface after every snap of the shutter.

But waterproofing the camera wasn't enough; Boutan also had to think about water pressure. His solution was beautifully simple. He inflated a rubber balloon about three-quarters full of air and connected it by a tube to the interior of the box. As the copper box descended into the water, the water pressure squeezed the ball and sent the air into the box, equalizing the pressure.

The camera was now ready, but it needed an operator. Boutan would have known the risks of diving. Professional divers—those who worked underwater to repair bridges, tunnels, or wells, or salvaged shipwrecks or searched oysters for pearls—knew that every descent held the possibility of death, and often a gruesome one at that. Suffocation, drowning, the bends, or, potentially less lethally, encounters with sharks and other "enemies of the fish tribe," as described by an 1897 article, were all part of the experiences of a nineteenth-century diver—not to mention that the average weight of a late nineteenth-century diving suit was about 170 pounds.

Boutan likened his rubberized canvas diving suit to an enormous, full-body pair of underpants. The sleeves were clamped shut with wristbands; lead-soled boots encased his feet; and a large leather collar encircled his neck. If this doesn't sound claustrophobic enough, next came the metal helmet, which screwed onto the breastplate on the outside. It had four glass portholes, three bolts, two lead weights at the front and back, and one breathing tube. It also had a communication tube, whereby the diver could hear and respond to instructions from the boat above. Once the helmet was locked in place, the crew above water who were operating the piston pump got to work. Their job was to push the air into the breathing tube as Boutan climbed down the ladder on the side of the boat and descended to the sea floor. For the duration of his time underwater, they pumped air into his helmet to keep him alive.

In these early tests, Boutan developed a methodology for getting his shots. Once he was in the water, one of the crew in the boat lowered the tripod and camera box with a rope. On the seabed, Boutan positioned the camera and waited for the water to settle. Before opening the shutter, he tugged on a length of rope connected to the boat. This was the signal to

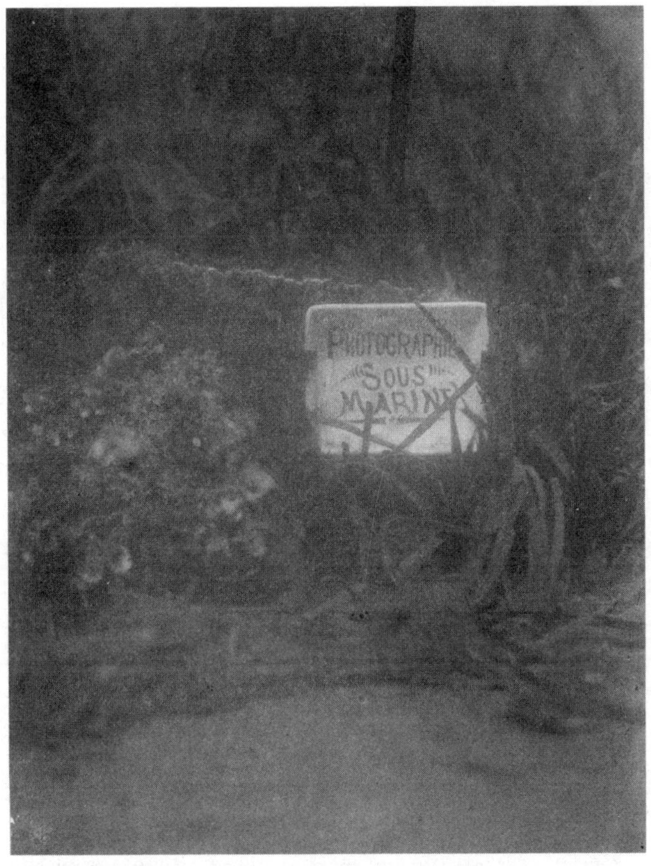

A sign propped up in the seagrass, in the water around Banyuls-sur-Mer, 1898.

start timing the exposure. (Boutan later wrote about the impossibility of underwater timing without some kind of special device. He'd have to wait several decades: diving watches were not available to buy until the 1930s.) When the exposure time was up, one of the crew tugged the rope to indicate to Boutan that it was time to close the shutter.

The experiments with this first camera were entirely trial and error. In Boutan's own frustrated words, the photos showed "an extremely vague outline that was unsatisfactory in every way." They were cloudy, indistinct, and lacked depth. Boutan varied the exposure times and the sensitivity of the plates. He relocated his experiments to a bay with better visibility. He tried various filters in front of the lens, eventually discovering that a blue filter produced sharper images. (Underwater photographers still use camera filters today.)

But the real problem was light. Exposure times varied from ten minutes to more than half an hour, depending on the depth. When you consider Boutan might want to expose all six plates in his camera, that's a lot of time waiting around in a diving suit. He needed some kind of artificial light for underwater. And so, with the help of an electrical engineer, a Monsieur Chaufour, he built one.

It was the first and smallest of three different underwater lighting devices Boutan would try over the years. In a glass bottle filled with oxygen, Chaufour placed a spiral of magnesium wire, and a platinum wire that was connected to a battery. When he ran a current from the battery, it lit up the platinum wire, which in turn ignited the magnesium, creating a bright flash of light. But Chaufour and Boutan discovered the same problems with magnesium as land-based photographers: the magnesium created a vapor that could cloud the bottle, dimming the light. More alarmingly, on more than one occasion the heat caused the glass bottle to explode.

Boutan, with the help of Joseph David, tried another type of artificial light, similar in principle but not in scale. They put an alcohol lamp on the top of a large upright barrel and covered it with a bell jar, which was then screwed in place. Inside the perimeter of the bell jar, on the barrel top, they notched holes, so that when they filled the barrel with oxygen, it seeped through the holes to feed the flame. The major advantage of this design was that it could remain lit during the descent from the surface to the

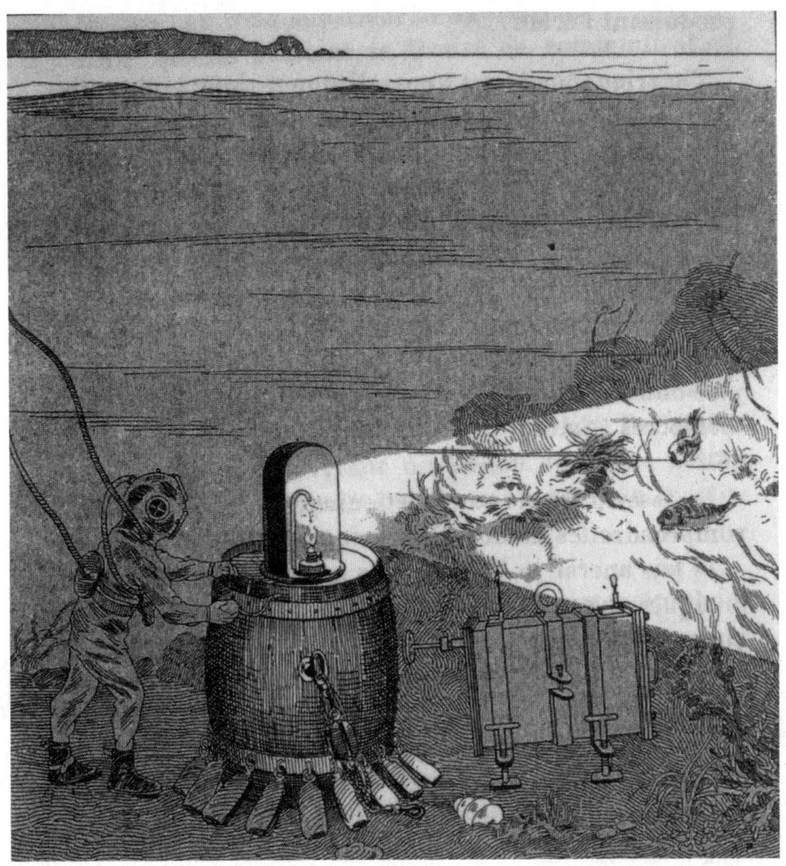

An illustration of Boutan's barrel lamp, designed to remain lit as it descended to the sea floor. In reality it was unwieldy, and liable to malfunction or cause bodily harm (or both).

seabed. Boutan and David then turbocharged the lamp: they took a tank filled with magnesium powder and connected it to the lamp with a tube and a switch. When spritzed with magnesium, the lamp would flare with an intense light.

In practice, this lamp was unreliable, cumbersome, and dangerous. It tipped over. The flame died out. The switch was faulty. It was immensely arduous to operate. Boutan weighted the barrel with both iron and lead to help it descend vertically and more easily, but to get it into the water required the boat's hoist. On one occasion, it came dangerously close to smashing into Boutan underwater. He had set up ropes with a cork to indicate where he wanted the boat crew to drop the barrel. As he tried to guide its descent, the ropes tangled with his breathing tube. The barrel continued to sink as Boutan scrambled to free himself to avoid getting smacked in his helmeted head. Once it landed on the sea floor, and not on his head, he heaved the wayward barrel into position. As he was about to press the magnesium switch, the flame went out. He had to resurface, relight the lamp, and start again.*

It wasn't until 1898 that Boutan began to make what he considered real progress with his underwater photos. It's worth pausing here to note that for Boutan, whose day job was zoology lecturer, underwater photography was essentially a hobby. During the 1890s, he published around twenty different scientific papers, one of which was about pearl production in the *Haliotis* mollusk and one on the use of diving equipment in the study of both marine biology and underwater photography. He also knocked out a 510-page zoology coursebook for the University of Paris.

After the first unsuccessful device, and before he started to make progress later in the 1890s, Boutan had tried a completely different tack: a camera that was submerged directly into the water without a waterproof case. The device was built at the Arago workshop at the end of 1896, but it was short-lived. The results were, Boutan said, very mediocre, which is not surprising given that the water would have sloshed about every time the shutter opened. In 1898 Boutan assessed what he knew, what had—and hadn't—worked, and reconsidered the concept of the waterproof box.

At about 11 am on that bright September morning in 1898, Boutan sat in the boat while, some ten feet below him, his colleague Joseph David waited underwater in his diving suit. Across from David, resting on the sea floor, was a vastly enhanced waterproof camera.

Boutan's third and final device was similar in concept to the first ver-

---

* Boutan did fix some of the problems with the barrel lamp, but he still found the results to be mediocre. On a more poetic note, the boat's captain, who watched the magnesium flame flare from above the water, described it as "a thunderstorm in the water."

sion but better in every conceivable way. The box was significantly larger, and built from iron, with a rubber-hinged door that screwed shut. It held six plates, measuring about seven by nine inches. On the outside, it had levers to flip the plates and operate the shutter, and a visor to reduce reflections from the surface. It also came with an accessory: a large iron stand more closely resembling a table, with a tilting shelf to capture different angles. Instead of a fixed-focus lens, Boutan asked an optical company in Paris for a symmetrical anastigmatic lens, which helped to correct the distortion underwater and created a clearer image. This new lens was a game-changer: Boutan could now choose the focus point of his underwater images.

But focusing the new lens was an elaborate process. At a dry dock, Boutan and a small team of men used pulleys to lower the heavy iron box partially into the water. In front, at the required distance, Boutan had already positioned a screen with writing. With the back of the camera open, but facing the screen, he focused the lens on a piece of glass and fixed the plateholder into position. Once that was set, they hauled the box out of the dry dock and later replaced the glass with the sensitized plate, ready for the shoot.

This was the camera that now sat about thirteen feet in front of Joseph David on that bright morning. From the boat above, Boutan held a cord that trailed down into the water and connected to the shutter lever. Through the clear water, he could see David some ten feet below. He pulled the cord and the shutter opened.

Boutan later wrote that this photo was one of his more interesting results; looking at the photo now, you find yourself yearning to know if he felt, after all the years of trying, some delight. Did he do a little dance in the darkroom or merely jot down the details of the experiment? More than a century later, it holds an otherworldly allure.

In the photo, David, anonymous in his diving helmet, appears framed in a window of filtered sunlight. He is crouched low in a bed of seagrass, whose gentle motion is, for this eternal moment, stilled. In the foreground, the sunlight that was dancing on the seabed appears as a static pattern of light. The bubbles escaping in a rush from David's helmet are frozen in their frantic ascent. This is more than a photo; it's another world. Whether he acknowledged it or not, Boutan had captured a scene of underwater beauty, just as he had intended all those years before.

At first glance, it's easy to miss a charming detail in this image. Maybe because Boutan pulled the shutter too soon, or maybe because David was more focused on diving than posing, but if you look closely, you'll see that the small sign he is holding, inscribed with the words *Photographie Sous Marine*, is upside down. If every experiment has the potential for an unforeseen variable, this is surely one of the more delightful.

With this new camera Boutan was finally hitting his stride. One photo features Boutan, unencumbered by a diving suit, his cheeks puffed as he tries to keep still while David snaps the exposure. Another, taken in waist-deep water, shows a lineup of his colleagues' swimsuit-clad legs. This was Boutan's response to a newspaper cartoon published after he had first presented his underwater photography experiments. It had showed

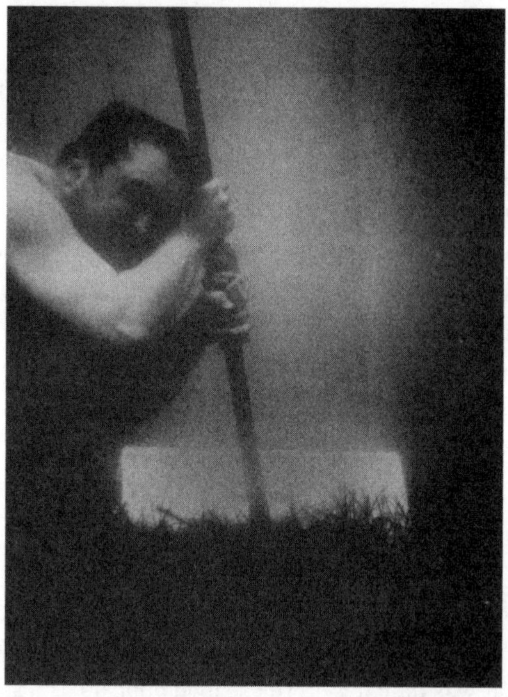

A photo of Boutan taken by Joseph David. The iron bar was intended to help with the camera's focus.

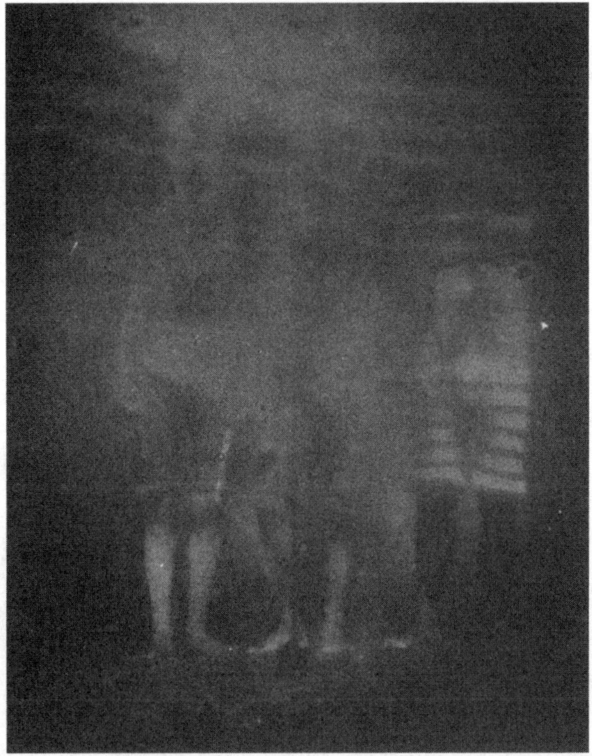

Swimsuit-clad legs, photographed in jest by Boutan.

a bespectacled scientist hiding in the water, using his camera to furtively photograph the legs of bathers. In Boutan's version, the legs are a collection of knobby knees peeking out from long swimming trunks.

Although Boutan, with the help of the mechanic David, had built the most successful underwater camera the world had yet seen, it was also one of the least maneuverable devices in photographic history. It was so heavy it had to be moved by three men. To get it into the water from the boat required a hoist. Underwater, it wasn't a lot easier: once the camera had been lowered to the sea floor, Boutan had to drag it into position, in his diving suit. On more than one occasion, Boutan tried to move the box where he wanted it, only to find himself sweating so much that he fogged up the portholes in his helmet. Unable to wipe them clear with his hands, he rubbed his tongue and nose over the glass, like a human windshield

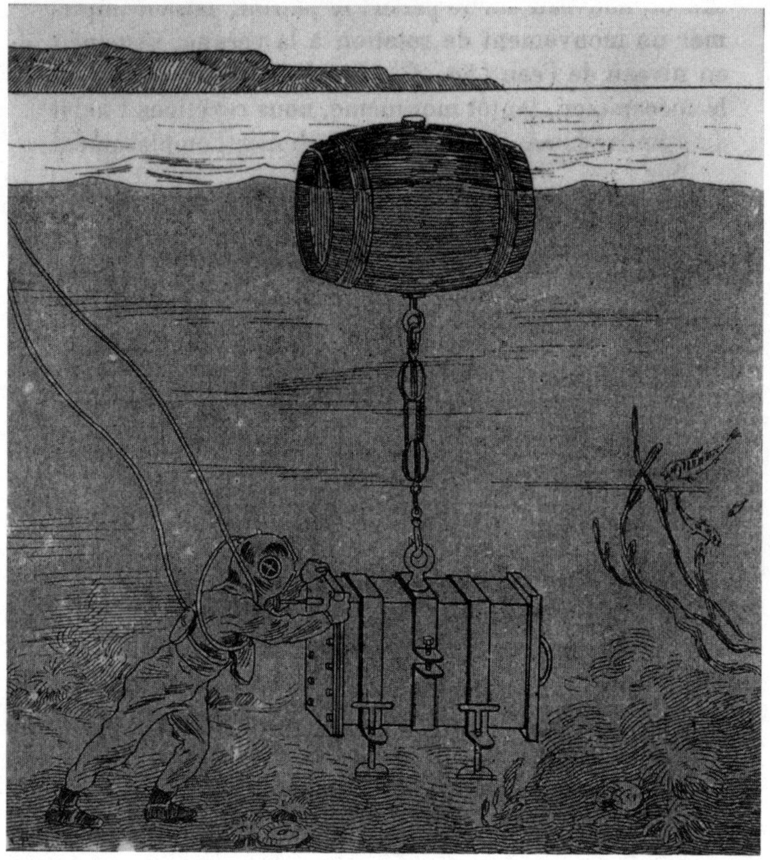

An illustration of the flotation device that helped to maneuver Boutan's third, final, and heaviest camera into position on the sea floor.

wiper. Eventually, Boutan created a flotation device using a barrel and a chain, which he attached to the box. Boutan could then push the box into position, rather than wrestle with it on the sea floor.

Buoyed by the success of this third device, Boutan theorized that he could take photographs at far greater depths, provided the camera had a powerful light source. But to test this theory, he needed to build the lights, and that required money. The laboratory's finances covered its operating costs; they would not stretch to building a very expensive, entirely hypothetical lighting system for an underwater photography experiment that was only tangentially related to marine biology.

Once word got out about his proposed experiment, it didn't take long for a benefactor to appear. François Deloncle, a former politician now working for an optical company, and a man with an appetite for grand scientific experimentation, saw an opportunity in Boutan's experiments. Since 1892, Deloncle had been building the world's largest telescope, with a focal length of 57 meters, and which he planned to be the highlight of the upcoming Paris Exposition of 1900. During a meeting with Boutan in Paris, Deloncle offered Boutan everything he needed to build his underwater lights, with one proviso: that he could display some of Boutan's photographs at the exposition, next to those captured with his enormous telescope. The sea and the stars, together. Who could say no to that?

By the end of August 1899, the lights were ready. Boutan wanted to photograph at depths of 50 and 100 meters,* and he'd once again turned to Chaufour for his expertise. Chaufour built two spherical arc lamps of twenty amps each, cast in iron. They each had a bronze porthole containing glass about half an inch thick. Each sphere also had a hermetically sealed outlet for the operating cords, which combined into one single steel-framed cable that would, once underwater, trigger the lights and shutter.

Powering the lights was a staggering undertaking. Boutan had a specially designed sixty-element battery built for the experiment. The batteries were stored in sets of five, divided across twelve wooden cases to transport into the boat's hold, although this required more than one person: each case of batteries weighed 65 kilograms (over 140 pounds). To fully charge the batteries using a steam engine took nearly three days.

Boutan first tested the lights in a tank of seawater, and then again, on a calm night anchored in the bay at Banyuls, with the camera. During this experiment, to a depth of 6 meters (about 20 feet), Boutan saw the dark water light up below the boat as he counted a five-second exposure. Once the camera was back on board, he rushed to develop the negative. It

---

* About 164 feet and 328 feet.

showed a gorgonia, a type of soft coral, brightly lit against a sea of black, satisfying Boutan enough to try greater depths.

But at this last, critical moment, with the stage set, the lights tested and ready, at the end of Boutan's time in Banyuls before he had to return to Paris for the academic year, the experiment stalled. The mistral, a fierce wind that periodically rampages through southwest France, had begun to blow. It whipped through Banyuls-sur-Mer, churned up the water, and stopped any attempts to get a boat full of photographic equipment out of the bay. For nearly eight days, Boutan waited impatiently. Finally it died down, and Boutan and his crew stepped on board the boat to conduct the two deep-water photography experiments at 50 and 100 meters.

At least, that was the plan.

The crew rowed out of the bay in calm water. Once they found their spot, they didn't anchor. Boutan didn't want to be tied to the seabed in the event of a sudden swell, given the weight on board of the camera and the cases of batteries. The steel cable for lowering the equipment was also heavy: 200 meters of cable weighed 225 kilograms. That meant the approximate weight of just the cable and all sixty batteries was just over 1000 kilograms—more than a ton.

Boutan had another piece of equipment, a metal frame, roughly rectangular, to hold the camera and lights. At one side, on a small ledge, he positioned the camera between the two lights; directly opposite the camera's lens, about eight feet away, he stretched a white canvas with the words *Photographie Sous Marine*. The wires to operate the camera shutter and the lights were encased in the single steel cable that also held the frame. Once submerged, this cable would be the only lifeline between every piece of equipment and the boat.

The crew operating the hoist moved the frame with the camera and lights from the deck to the water's surface, where they were slowly engulfed. It's easy to imagine Boutan standing on the boat, watching as every element was swallowed up by the water until it disappeared from sight. All that was left was the cable, stretching from the air into the water, reaching down into the abyss. Boutan triggered the lights and shutter. The ten-second exposure began.

During those ten seconds, Boutan's brain must have been ticking through the possible outcomes of the experiment that was happening below him, unseen. Perhaps the water pressure had crushed the camera, or maybe the lights had failed. Perhaps the canvas had ripped, or the frame had buckled. The only thing Boutan knew for sure was that he had just dropped more than a ton of photographic equipment into the sea, where it was now dangling at the end of a single cable, fifty meters below the surface of the water.

When the exposure ended and it was time to wind up the cable, Boutan faced a new, unforeseen problem. The hoist was controlled by two men who were crowded next to the battery elements in the boat's hold. When they opened the ratchet to bring the cable back up, it unraveled, overpowering them and sending the equipment deeper into the water. They closed the ratchet, but Boutan and his team were now in a decidedly precarious situation: adrift, alone, and, trailing far below them, an underwater camera and lights.

They could sever the cable and jettison the camera, lights, and frame, leaving the equipment to sink to the sea floor—abandoning them, and the culmination of years of experiments, to the currents, to end up a photographic wreck in the sand. Or they could try to haul it all up by hand.

Every person on board took hold of the steel cable and began to heave. The weight of the equipment was immense. Slowly, slowly they pulled, and slowly the equipment was dragged up through the mass of water. It was nearly an hour before it broke the surface.

Boutan, relieved and sweaty, cast aside any thoughts of attempting the experiment at 100 meters; he was done. The boat had drifted south to the coast of Spain. A light breeze had picked up, so they set sail for Banyuls-sur-Mer with the photographic cargo safely back on board.

As for the experiment, despite one of the lights filling with water, it was successful. In the photograph, the sign is clear and sharp against a featureless void of darkness. It's not exciting, but the image is so crisp you can see the stitching that holds the canvas to the frame and the serifs on each letter. With huge effort and an unforeseen amount of risk, Boutan had proved his hypothesis: it was possible to take a photograph remotely, at great depths, using electric light.

Boutan's experiments in underwater photography went unmatched for decades. The next major milestone, underwater color photography,

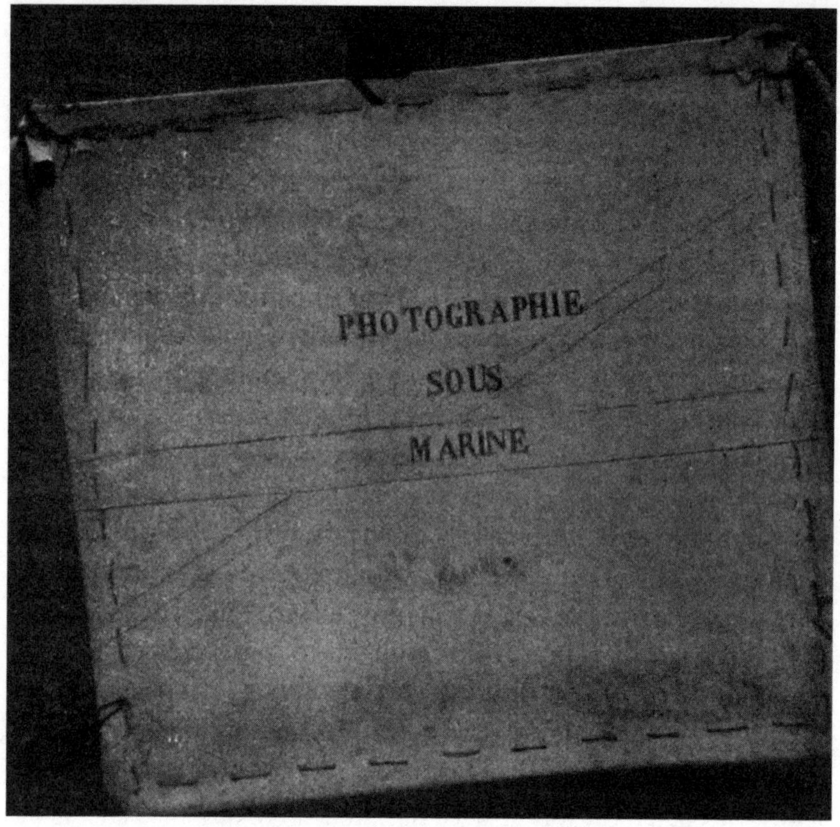

Likely the last underwater photograph Boutan ever took, and certainly the riskiest.

didn't occur until 1926, nearly thirty years later. (Those underwater color photographs were also lit by magnesium powder—a pound of it, ignited on a pontoon.)

In 1900, Boutan published a book about his underwater photography experiments, but after that, he moved on. He lived in what was then Indochina, and then relocated to Bordeaux. He researched the cognitive abilities of gibbon monkeys; his subject was a gibbon that he had adopted and named Pépée. During the First World War, Boutan and his brother Auguste developed a self-contained diving apparatus that subsequently became regulation equipment on some submarines. In the 1920s, Boutan's lifelong study of mollusks once again led him down an unexpected path and into the world of fine jewelry, where he testified in a case involving Kokichi Mikimoto's cultured pearls. (Today, Mikimoto is one of the world's leading pearl producers.) Boutan published a book on pearls and moved to Algiers to work as a professor of general zoology, which is where he died in 1934, at the age of seventy-five.

If there was a guiding principle behind Boutan's experiments, it was the belief that photography could help reveal the secrets of the seas. Using a camera properly equipped with a powerful artificial light, he mused in 1898, "Will not such photographs contain a most precious fund of information?"

Boutan was right: in the deepest oceans, discoveries never end. In 2021, extraordinary footage of the rarely seen glass octopus flooded the internet, revealing a translucent, otherworldly creature floating elegantly through the darkness. It was filmed in a way that Boutan could have only dreamed: by an underwater robot, equipped with a suite of lights, that can reach depths of more than 14,000 feet.

PART II

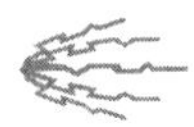

PERCEPTION AND DECEPTION

Intrigue and Impact

Adaguerreotype was available in a range of standard sizes. The largest was known as the "whole plate" daguerreotype, and it measured 6½ by 8½ inches. From there, the sizes decreased—half-plate, quarter-plate, and so on—until you reached the smallest, the diminutive sixteenth-plate size. This measured about 1⅜ by 1⅝ inches, roughly the size of a passport photo. So you can imagine the astonishment when one photographer, based in Liverpool, reduced a daguerreotype using a microscope lens to the tiny size of an eighth of an inch, or three millimeters.

That photographer was an enthusiastic amateur who had traveled all the way to London in late 1839 to see Daguerre's extraordinary new process for himself. He was in his mid-twenties, and working as an optical instrument manufacturer, selling items such as spectacles and microscopes; once he returned from London, he also began to sell daguerreotype cameras. His name was John Benjamin Dancer.

Microphotography hit its stride with the wet collodion process, and Dancer was once again at the forefront of these experiments. (Microphotography, the photography of regular-sized objects in minuscule format, is separate from photomicrography, photographs taken through a microscope of the otherwise tiny and invisible—the hairs on a flea, for example.) In April 1853, Dancer, now based in Manchester, took a standard four-by-five-inch photo of a newly completed memorial plaque for the inventor William Sturgeon. The plaque held nineteen lines of text, 680 letters in total. Dancer reduced his photo's size to one-sixteenth of an inch, or approximately one-and-a-half millimeters. That's only slightly larger than the point of a sharp pencil. But under a microscope, those nineteen lines of text, those 680 letters, were crystal clear.

To the naked eye, a microphotograph was a mere speck, an image hidden in plain sight, whose secrets were only revealed with magnification. The subjects of Dancer's microphotographs, which he began to produce and sell, included portraits of famous leaders and scientists, religious texts, paintings, and landmarks. Microscopes were an increasingly popular item in mid-Victorian households, and it is easy to imagine the scene in which an eager viewer took a slide with what looked like a small dot in the center, placed it on the microscope stage, peered down the tube, brought it into focus, and presumably made some excited exclamation like "A-ha! It *is* the Tintern Abbey Cloisters! Hurrah!"

Dancer's microphotographs had some high-profile admirers. Queen Victoria owned a ring with microphotographic portraits of her family,

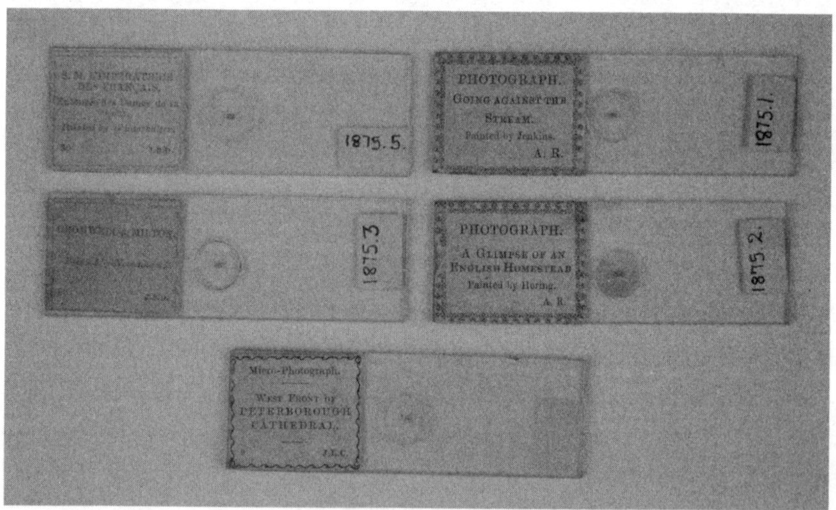

Some of John Benjamin Dancer's microscopically small photographs mounted on slides.

produced by Dancer, with a jewel lens for magnification. The scientist Sir David Brewster was also a fan. "A family group of seven complete portraits occupies a space the size of the head of a pin; so that ten thousand single portraits could be included in a square inch," he explained, for the "Microscope" section of the 1857 *Encyclopædia Britannica*. "They are executed upon films of collodion as transparent as glass; so that a family group could be placed in the center of a brooch, a locket, or a ring, and magnified by the central jewel, cut into a lens sufficient to exhibit the group distinctly when looked into or held up to the light."

The following year, Thomas Sutton—the combative editor of *Photographic Notes* who would later criticize De La Rue's moon photographs—struck a sour note in his new book, *A Dictionary of Photography*. Microphotographs, he wrote, were of "little or no practical utility," before adding, somewhat contemptuously, that they "must strike any reasonable person as somewhat trifling and childish, when he considers how many valuable applications of photography remain yet to be worked out."

Sutton was wrong, of course. Microphotography formed the basis for microfilm and, in Sutton's own lifetime, would prove immensely valuable for wartime communication. But the most trifling uses of photomicrography were perhaps the most interesting and widespread: as a form of hidden, sometimes erotic, entertainment.

During a visit to Paris in the summer of 1857, Brewster suggested embedding microscopic photographs in jewelry to various opticians and photographers. For that trip, he brought with him some of Dancer's microphotographs, along with a small handheld lens for viewing. After all, a microscopically small photo is only useful if it can be seen.

Brewster's visit came to the attention of René Prudent Patrice Dagron, a Parisian merchant who had also been learning photography. By 1858, Dagron was experimenting with the production of microphotographs as a novelty item. The challenges were significant: he had to figure out how to reduce photographs in size, what kind of viewing lens to use, where to embed the microphotographs, and how to make them commercially viable.

Photos had been incorporated into items of jewelry since the daguerreotype. These were not microscopic images, but they were small enough to fit in a brooch, say, or a bracelet. But what Dagron produced, after more than

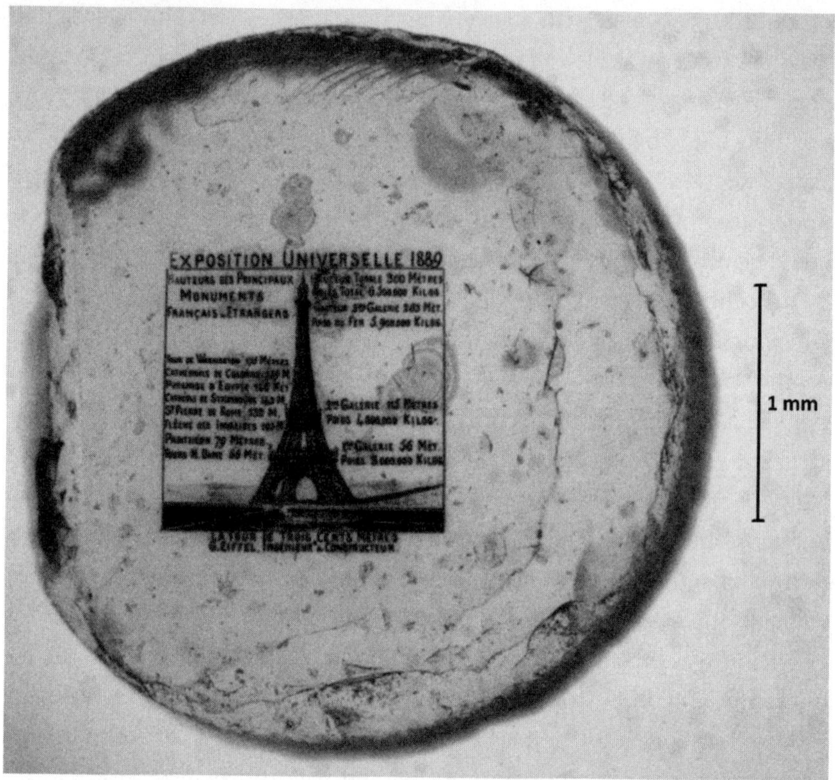

The view through a hidden Stanhope lens, revealing the Eiffel Tower. Not all Stanhopes were as innocent.

a year of experimentation, became known as *bijoux photomicroscopiques*, or Stanhopes, after the type of convex magnifying lens used to view the image. He filed his first patent in 1859. Within a small personal item—a ring, for example, or a pipe—Dagron embedded a microphotograph and covered it with a small lens. The item's owner could peek into the lens and see the magnified photo, but it was otherwise out of sight.

To produce microphotographs en masse, Dagron used a long wooden box that contained, at one end, a glass negative of the image to be reduced. At the other end was the reducing camera with up to twenty-five small lenses and the sensitized plate. When the end with the negative was held to a light source, the image was projected into the lenses and onto the sensi-

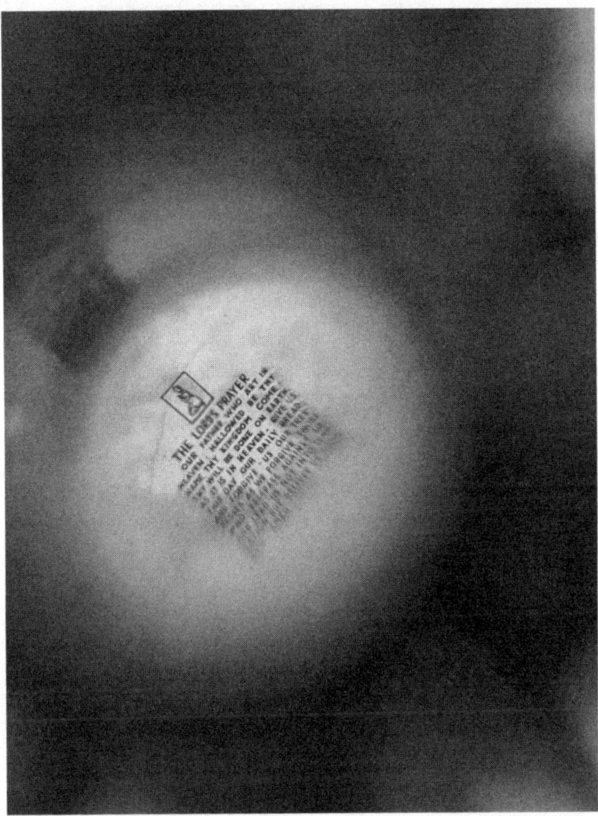

A microscopic photo of the Lord's Prayer. This one was hidden in a crucifix pendant.

tized glass plate to create multiple positive transparencies, each measuring about two millimeters square. Dagron employed the Taupenot dry collodion process, which used albumen to keep the plates sensitive after they had dried. The exposures took up to six times longer than wet collodion, but this mattered less in microphotography than it would for, say, a portrait. Dagron had also cleverly made the plateholder adjustable, so he could reproduce up to 450 microphotographs onto a single plate.

Each of those minuscule images was inspected with a hand magnifier, and those that were sufficiently clear were cut out and glued to the flat end of a cylinder of optical glass about eight millimeters long—the average length of a human eyelash. The opposite end of the cylinder held the small

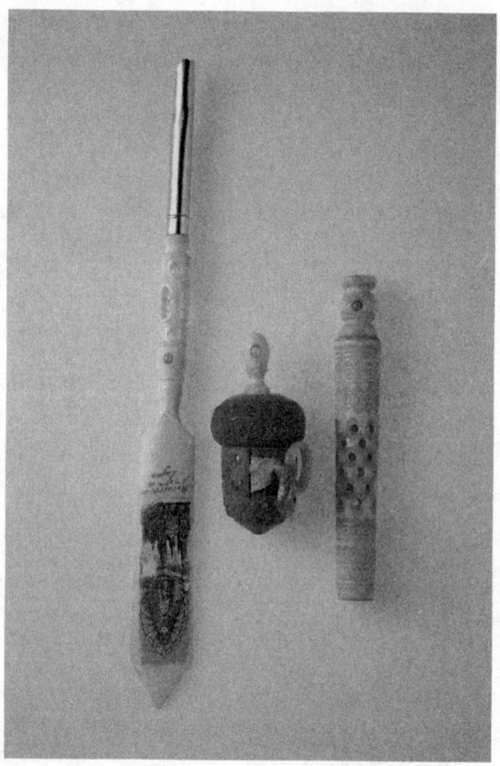

Some examples of Stanhope objects, including, at middle, a tape measure. Stanhopes could be anything from a ring to a baby's rattle.

magnifying lens. The cylinder was mounted into the jewelry or trinket, inconspicuous to everyone except its owner.

"By placing the eye behind the convex side, these photographs, invisible almost to the eye, were seen so distinctly and so highly magnified that they excited general admiration," wrote Brewster, in 1864, of Dagron's Stanhopes. He added, "The members even of a very large family may have their portraits grouped in one of these photomicroscopic *bijoux*, and the mother or the sister, or the father and the brother, may carry about their persons in a locket, a ring, a bracelet, a gold sphere, or a watch-key, this interesting family group, and smile or weep even in the social circle over these cherished representations of the living or the dead."

Although his patent was successfully challenged by his competitors in the early 1860s, Dagron created a thriving business. In 1861, he exhibited at the Salon de Photographie, and by the end of the year, he'd expanded his workforce to 150 employees. Once he lost his patent, he expanded to sell do-it-yourself microphotograph equipment and an instruction booklet. At an exhibition of his photomicrographs in 1865, the *Photographic Journal* called his process "rapid, simple, and ingenious."

With the enormous popularity of Stanhopes, and the increased number of manufacturers, it was possible to buy microscopic photographs in almost any item. These included watch keys, bracelet charms, brooches, needle cases, manicure sets, letter openers, pipes, rings, crucifixes, pocketknives, rattles, and even, in one reported case, a bullet. The images showed the typically popular subjects of the era—important landmark

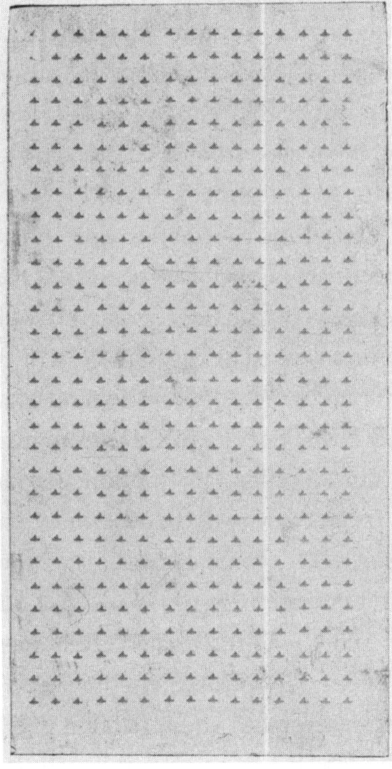

A single glass plate holding 450 microscopically small photographs from one of Dagron's competitors.

buildings, for example, or portraits of family and royalty. But a microscopic image hidden in a personal item, known only to its owner, naturally lent itself to one other subject, a subject that the photo journals of the day described as "indecent," "lewd," or "objectionable." In a perfect match of format and subject, some Stanhopes also contained tiny nude images and depictions of sex.

"Several Customs seizures have taken place at Port Adelaide of late," reported the *South Australian Register* in September 1869. In one, the article continued, "the case was detained for examination, and was found to contain, amongst other articles, a packet of watch-keys each containing a diminutive photograph with magnifying power, which, upon being held close to the eye, represented the most obscene and disgusting pictures." Customs laws prohibited any importation of "obscene and indecent pictures," though a letter written into the *Adelaide Observer* that same month makes clear that it was not uncommon: "I happen to know of some similar goods having been imported some time ago," the author wrote, advocating "a strong effort to prevent such an abominable trade." In the case of the watch keys, they had been sent from the defendant's father in Germany, unrequested by him (or so he claimed).

By their very nature, erotic Stanhopes were meant to be concealed, which means that records on their seizures, like the one from Australia, are few. One notable collection survives entirely by chance. One day in 1959, a cardboard box arrived at the Kinsey Institute at the University of Indiana. It contained around 2000 erotic Stanhope lenses, which had been seized by the US Postal Service in 1924, en route from the supplier to the distributor. The Kinsey collection of Stanhopes has been dated to the late nineteenth century or early twentieth century, and consists of photographs and drawings of naked women, erotic rather than explicit, although the latter were also in circulation.

According to Jean Scott, who researched Stanhopes extensively for her book *Stanhopes: A Closer View*, Dagron's company also produced erotic Stanhopes; one of the earliest was an illustration of a bare-breasted scene of seduction. Like the watch keys in Australia, the items that contained erotic Stanhope views were those typically worn and used by men. That's also the case for another Stanhope collection that was discovered by acci-

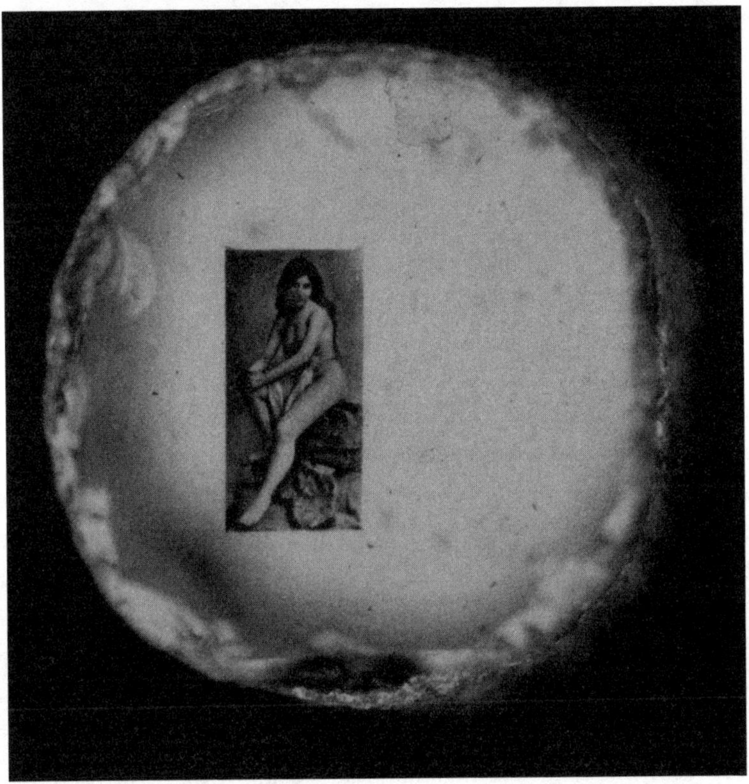

Through a peephole: one of the microphotos that arrived at the Kinsey Institute in 1959.

dent. In 1865, a steamboat, the *Bertrand*, bound for gold mining towns in Montana Territory, sank in the Missouri River. More than one hundred years later, the wreck was excavated; some thirty years after that, in 1996, a researcher decided to take a closer look at some salvaged pocketknives. Concealed in the handles of some of the pocketknives, he saw Stanhope lenses that revealed tiny drawings and a photograph depicting couples having sex, and nudity.

For those who sold erotic or pornographic images in regular, non-microscopic size, the penalties could be severe. In England, in 1867, for example, one Sidney Osborn Fowler (alias Dr. John Galt) was convicted of publishing "twelve obscene and indecent photographic pictures" and for selling "lewd, wicked, scandalous and obscene prints, to the corruption of good morals and good manners." He received two years' hard labor. Sellers were prosecuted under the 1857 Obscene Publications Act, also known as Lord Campbell's Act, which was zealously upheld by the members of the Society for the Suppression of Vice. Their usual mode of operation was a nineteenth-century version of catfishing: they posed as a customer who wanted to purchase erotic materials and, once the offender had produced the goods, often in a back room, they called in the police. (In an irresistible if admittedly juvenile detail from an 1872 case, the officer who carried out the search warrant and arrest was an Inspector Cocks.)

At the SSV's annual meeting in 1870, the secretary crowed that over the previous sixteen years, they had seized 134,670 "obscene" prints, pictures, and photographs. (It's worth considering that there could well have been even more material right under the SSV's nose, which they were unable to truffle out because the photos were microscopic in size.) Yet the material still proliferated, and the SSV must have felt less smug when, in 1874, a police search of a single proprietor unearthed 100,000 "obscene" photographs and 3000 negatives "of the same class."

Beyond the use of terms such as "obscene" or "lewd" or "wicked," it's difficult to know the nature of this material because it was destroyed during the court process, and defied description in the press. However, a letter to *Photographic Notes*, in 1865, lingers perhaps a little too long in its objection to "photographs of young ladies (?) at their toilette—some lacing their corsets, some exposing their legs while lacing their boots and arranging their garters, some stooping just to exhibit their bosoms, others reclining on couches in exciting postures. . . . In brief, woman may be seen in every conceivable attitude that a lewd imagination could suggest, with every conceivable sin pictured on her face."

Photos were also in circulation of artworks featuring naked bodies—the type of classical art that hung on the walls of galleries and museums around London, and which some felt was more acceptable than photographic nudity. Others did not. In 1877, a woman and her fifteen-year-old son were prosecuted for selling "indecent" photographs, the majority of which were well-known works of art. The judge deemed that they were still indecent by law, the images were destroyed, and she was fined. (Her son was acquitted.)

Some three years later, two artists took the stand as witnesses for a defendant who had been accused of selling prints and photos "of a very objectional description." They testified that, as they couldn't afford real models for their work, they used the images as art studies and that they viewed the images as not indecent "but entirely of a classical description." The defendant was sentenced to eighteen months' hard labor.

None of these cases involved microscopic photographs. But consider how much easier it would have been to hide tiny erotic or explicit photographs from the police or, for that matter, the SSV. And while microscopic photos could (and did) contain "obscene" imagery, they could also contain far less "trifling" material—such as secret messages in a time of war.

When the Prussian Army laid siege to Paris in 1870, Dagron left the city in a balloon and set up an operation in Tours, where he produced microscopic copies of correspondence to be flown into the city by carrier pigeon. He reduced the images to such tiny dimensions that a single collodion film contained about 3000 messages and still only measured about 1.1 inches by 2.3 inches. (That's less than half the size of a standard three-by-three Post-it note.) When rolled up, some fifteen films fit into a goose quill roughly the length of a ballpoint pen cap, which was then attached to the pigeon with a silk thread. Dagron sent the same dispatches repeatedly until he received confirmation of receipt via balloon post; he later estimated that he sent about 2.5 million dispatches. (The pigeons carrying the dispatches didn't always meet their intended destination: some were shot, some killed by trained birds of prey sent up by the Prussians, and some were lost in the winter weather.)

Dagron also copied, in microscopic format, entire editions of newspapers. It was a glimpse of what was to come—the storage of entire archives on microfilm. In 1857, Brewster seemed to have predicted as much, showing the foresight that Sutton lacked. "Microscopic copies of despatches and valuable papers and plans might be transmitted by post," Brewster wrote prophetically, adding, "and secrets might be placed in spaces not larger than a full stop or a small blot of ink."

By and large, when it came to printing a photograph, the size of the collodion negative dictated the size of the print. Although it was possible to enlarge an image, it was a complex process and one usually left to the experts. What that boiled down to was that if you wanted a large print, you needed a large negative, and if you had a large negative, you needed a large camera. And if you wanted what was known as a "mammoth-plate" photograph, which measured around 18 by 22 inches, everything needed to be mammoth. This meant a negative also measuring 18 by 22 inches and a camera large enough to fit that negative. Everything else was sized up too: large negatives required larger developing trays, more chemicals, more water.

To shoot mammoth-plate views, particularly in a landscape as vast as Yosemite, you needed to be technically skilled, patient, and ambitious. Although we know little about the details of Carleton E. Watkins's life—much of his personal archive was destroyed in the San Francisco earthquake of 1906—to succeed as he did, he clearly had those qualities in abundance.

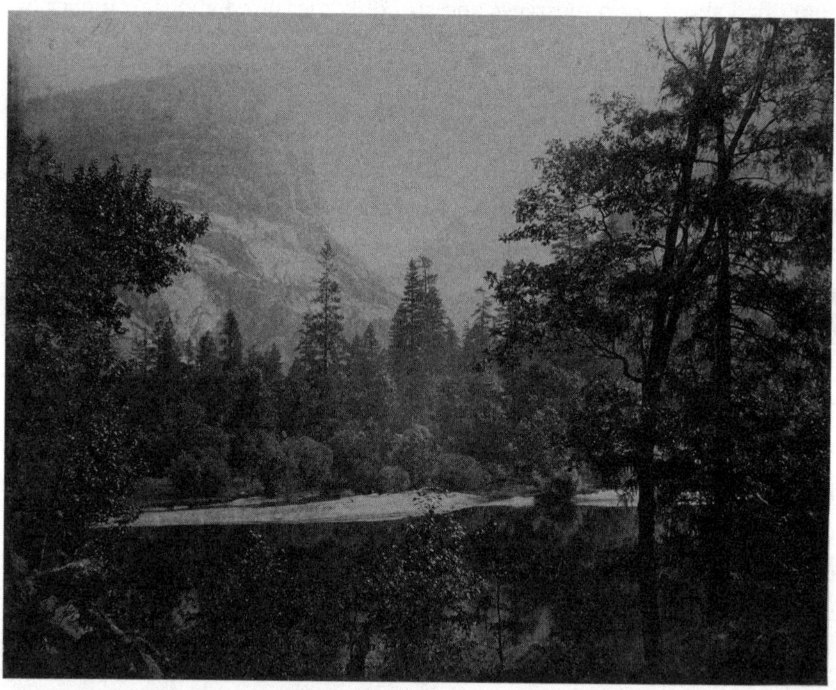

The Lake—Yo Semite Valley, *1861. Watkins preferred to shoot early in the morning, to avoid any wind moving the branches and leaves of the trees.*

Watkins was born in Oneonta, New York, and he died in California. We don't know for sure what the "E" indicates in his middle name. (According to various sources, it's either Emmons or Eugene.) We also can't confirm when he traveled to San Francisco; some historians have dated it to 1849, when he was nineteen, but it might have been 1851. He journeyed west with his friend Collis P. Huntington, a onetime tin peddler who reinvented his fortunes in California and cofounded, as one of the "Big Four," the Central Pacific Railroad. Whatever California promised these men, getting there was an arduous trip: they traveled by boat to Panama, crossed the isthmus, and took another boat to San Francisco.

We don't have an account from Watkins of his trip, but we do from a Scottish journalist, from 1851. In his book, he recounted miserable weather, rough seas, dolphins that frolicked off the boat's bow (and were later eaten for dinner), crowded hotels in Panama, and two deaths on board. He described a motley crew of passengers who toted knives and pistols, all of whom were assured of their soon-to-be-acquired California Gold Rush riches.

Once Watkins finally arrived in San Francisco—and one historian has estimated that, due to delays, it took Watkins nearly six months—he saw a harbor crowded with ships by an improvised city of staggering growth, at the far edge of a continent. He traveled with Huntington to Sacramento, where he worked as a store clerk and then a carpenter. At some point in the mid-1850s, back in San Francisco, Watkins learned photography.

We don't know with any certainty how Watkins came to practice photography. In a biography published after he died, and which has known inaccuracies, there's a story that Watkins was asked to help a local daguerreotypist who was short an operator. Once Watkins was shown the ropes, he proved himself a capable daguerreotypist, and became a portrait photographer. In this land of rich veins and lucky strikes, perhaps Watkins's photographic career did begin through some unexpected collision of events.

What we do know is that by the late 1850s, after working as a portrait photographer in San Jose, Watkins had turned his attention to landscape photography. He was also taking commissions. Watkins's earliest surviving landscape photograph, from around 1858, shows a large vista of the New Almaden quicksilver mine near San Jose. It was commissioned to provide evidence in a dispute over the mine, one of the many land grant cases that followed the Mexican-American War, in which the newly formed Land Commission adjudicated on the validity of Spanish and Mexican land grants. This was not a straightforward task when the original land grants often didn't include specific boundaries, and it was a highly fraught one when the land contained valuable resources, as it did with New Almaden.

Watkins's photo was submitted into evidence in the case of *U.S. vs. Fossat* in August 1858. (In his deposition, Watkins referred to himself as a "photographist.") It was a landscape view of the mining lands, 10 by 24 inches, which he produced by joining two prints together. Watkins was already trying to translate large vistas into large-format photographs, without having the necessarily large camera.

By this time, accounts of the Yosemite Valley had begun to appear in the press. A couple of years prior, in the summer of 1855, an article by gold-miner-turned-journalist James M. Hutchings was published in the *Mariposa Gazette*. He recounted, in awestruck wonder, a recent trip to Yosemite and its "wild sublimity, and magnificent scenery." Before this, the only other published accounts of Yosemite had come from members of a state-sanctioned local militia, the Mariposa Battalion, who had entered Yosemite in March 1851, in search of its Native American inhabitants, the Ahwahneechee, a subgroup of the Southern Miwok. The Battalion's purpose was clear: in January 1951, three US Government Indian Agents wrote that California had "but one alternative in relation to the remnants of once numerous and powerful tribes, viz: *extermination* or *domestication*." They added that the latter was preferable for it includes "all proper measures for their protection and gradual improvement," and also provides the state with "cheap labor."

In the incursions in March and May of 1851, the Ahwahneechee were forced to disperse: some escaped, some were taken to a reservation, and some managed to return to Yosemite. Soldiers also shot and killed the chief's son. The following year, after a skirmish at Yosemite, the US Army entered the valley and killed six Native Americans. After Hutchings made his first visit in 1855, he expressed regret, in the *Mariposa Gazette*, that such a tremendously beautiful landscape was only seen by "wild animals and Indians."

After a second visit in 1859, Hutchings published a series of four articles about Yosemite in his magazine. "It is beyond the power of language to describe the awe-inspiring majesty of the darkly frowning and overhanging mountain walls of solid granite that here hem you in on every side, as though they would threaten you with instantaneous destruction, if not total annihilation, did you attempt for a moment to deny their power," he wrote. "If man ever feels his utter insignificance at any time, it is when looking upon such a scene of appalling grandeur as that presented here." Such reverential wonderment was echoed by the editor of the *New York Tribune*, Horace Greeley, who wrote a book about his 1859 Yosemite visit, and the minister Thomas Starr King, whose accounts of his 1860 visit appeared in a Boston newspaper.

But descriptions, no matter how dramatic or God-fearing, could not have the same impact as photographs. Hutchings's articles had included engravings, based on the photographs of another San Francisco photographer, Charles Leander Weed. He had accompanied Hutchings into Yosemite and produced photographs measuring about 13½ inches by 10½ inches. This size was closer to what was known as "imperial plate" size, which encompassed negatives that were anywhere between 9 by 12 to 14 by 16 inches. It's likely that Watkins saw Weed's photographs, or that he read the accounts of Yosemite, or both. Shortly after this, Watkins received another commission, this time to photograph the Las Mariposas mining estate, which he captured in the "imperial" size of 12 by 18 inches. Watkins seemed to understand that certain subjects needed to be seen at a larger scale, at the exact time that one particularly dramatic landscape, Yosemite, was entering the consciousness of city dwellers. But to get to mammoth sizes, he would need a whole new camera.

The mammoth camera that Watkins brought to Yosemite for the first time in 1861 was a beast. It reportedly weighed 75 pounds and extended to around three feet long. Each glass plate—again, which measured 18 by 22 inches—weighed four pounds. Watkins's mammoth camera has not survived the intervening decades, and nor has information about its maker. Some historians have suggested that Watkins constructed it himself, possibly by adapting another camera; others have suggested that he commissioned a cabinetmaker, given there were no camera manufacturers in San Francisco at the time.

In addition to the enormous camera, Watkins traveled with thirty or more mammoth glass negatives, a stereoscopic camera and at least 100 glass stereoscope negatives, a portable darkroom tent (known as a dark tent), tripods, photo chemicals, and all his personal items for several months' stay, so it's unsurprising that he appears to have had two assistants. Just getting to Yosemite was several days via stagecoach and steamer.

Watkins probably also had a financial backer, given all the costs involved. But regardless of who he was with or who paid for it, Watkins arrived in Yosemite for the first time with the intention of taking mammoth photographs, and no doubt hoping to reap some future reward. Like so many in California, he was speculating.

The landscape he was speculating on was formidable and unforgiving. Trails were basic at best; Greeley recounted his descent on horseback into the Valley as "so steep, so rough, and so tortuous" that he decided it would be safer to dismount and attempt the trail on foot. Helen Hunt Jackson visited and wrote, "To an unaccustomed rider it is not pleasant to sit on a horse whose heels are much higher than his head." Hutchings clambered on his hands and knees up to what is now called El Capitan and came face to face with a rattlesnake. It was in this landscape that Watkins hauled all his equipment, moving it to every viewpoint, no matter how inaccessible, and setting up his dark tent to prepare his wet collodion plates. He had to transport glass bottles of chemicals and enough water for processing his negatives in large trays. The logistics of this process in Yosemite, with plates of mammoth size, are truly mind-boggling.

Aside from the risks from the terrain, the rattlesnakes, the heat, and the dust, Watkins also had to consider the wind. Movement in the leaves and branches translated, in the photograph, as blur. For this reason, Watkins photographed early in the morning, when the air was the most still. Of course, the downside was that the morning light was weaker, and exposure times were longer. With every photograph, Watkins had to weigh all the odds, and nothing was guaranteed to work.

One of Watkins's stereoscope views from 1861 shows his dark tent pitched underneath the Upper Yosemite Falls. Folds of fabric fall as a triangle in the left of the photo; in the center, an unidentified man—possibly Watkins, possibly an assistant—sits near a tripod. If you look closely, in

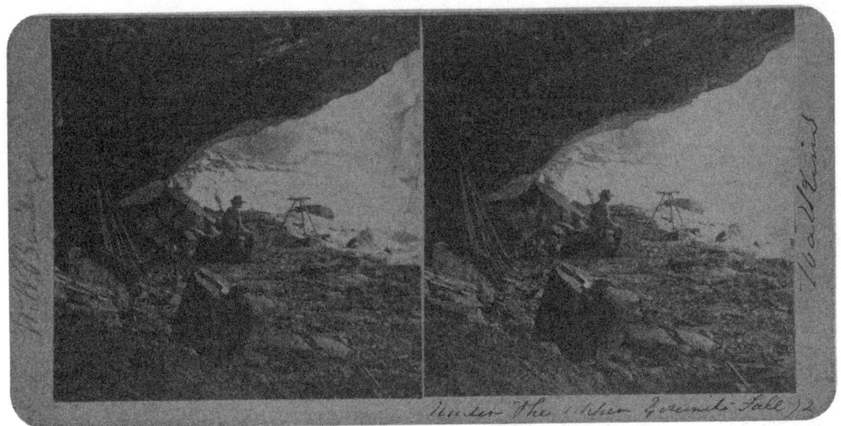

Under the Upper Yosemite Fall, *1861. Note the darkroom tent, tripod, and the anonymous figure looking out toward the falls. And, in the foreground, an axe.*

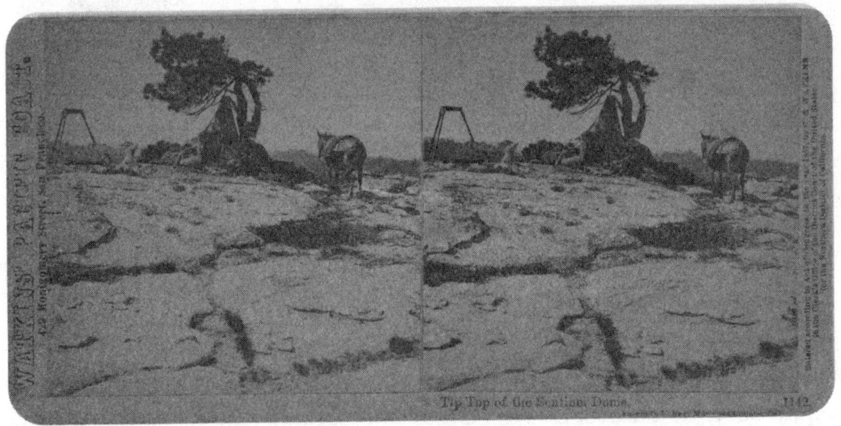

Tip Top of the Sentinel Dome, *a stereoscope view that Watkins produced around 1865.*

the foreground, you can see an axe. Another stereo view, from a subsequent trip, shows his dark tent hanging under a lone bent pine tree at the top of Sentinel Dome, some 8000 feet above sea level. To the left is the tripod; to the right, under the tree's meager shade, is an anonymous figure and a solitary, and presumably very tired, mule. Watkins called the view *Tip Top of the Sentinel Dome.*

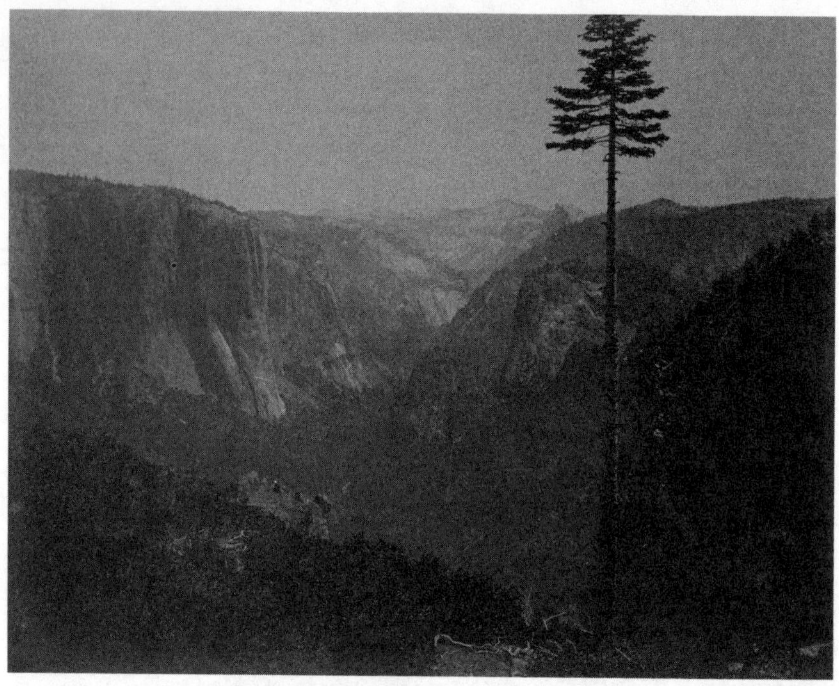

Yosemite Valley from the Best General View, *1866. Watkins, described as a photographer of "rare skill and indomitable energy," made multiple trips to photograph Yosemite with vast amounts of heavy and fragile equipment.*

By the time he left Yosemite in October 1861, Watkins had created at least thirty mammoth and one hundred stereoscope views of sights that would soon be on every visitor's and photographer's radar. (The stereoscope photos were trimmed to about three by three inches, pebbles to the mammoth plate mountains.) As Watkins took the long journey back to San Francisco, no doubt protecting his glass negatives every step of the way, perhaps he thought about the prospects for his photos—how he should display them, how he could sell them. Whatever Watkins's expectations were for his photographs, it is highly unlikely he was thinking about land conservation.

Once Watkins returned to San Francisco he began to produce prints and stereoscope views. In the spring of 1862, Watkins shared his prints with members of the California State Geological Survey and with Frederick Law Olmsted; by the end of that year, Watkins's mammoth photographs were on display some 3000 miles away in Manhattan, at Goupil's Gallery on Broadway. In its review of the exhibition, published on December 12, 1862, *The New York Times* had only the highest praise: "The views of lofty mountains, of gigantic trees, of falls of water which seem to descend from heights in the heavens and break into mists before they reach the ground, are indescribably unique and beautiful." Of Watkins's skill, the *Times* noted, "As specimens of the photographic art they are unequaled."

Watkins's mammoth Yosemite photographs also reached a California businessman named Israel Ward Raymond. In February 1864, after Raymond visited Yosemite, he wrote to California senator John Conness about protecting Yosemite. His letter included some of Watkins's mammoth photographs. Once again, Watkins's photographs were being used as evidence, and they were persuasive.

Shortly afterwards, Conness introduced a bill on the Senate floor stating that the land of Yosemite Valley "shall be held for public use, resort and recreation; shall be inalienable for all time." On June 30, 1864, President Lincoln signed the Yosemite Valley Grant Act. Some have suggested that Lincoln saw Watkins's photographs as well, given their reputation, and the fact that Lincoln was friends with John Conness. No records survive that can confirm this, but the fact that the act was passed was hugely significant. Though Yosemite was not the country's first national park, it was, in the words of art historian Hans Huth, "the point of departure from which a new idea began to gain momentum": to protect and conserve land for public enjoyment.

After the act was passed, the California governor appointed a Yosemite Board of Commissioners that included Raymond, California state geologist Josiah D. Whitney, and Frederick Law Olmsted, who became chairman. Olmsted had first seen Watkins's work in 1862, and he liked it so much he gave a set of mammoth prints to his family. He continued to hold Watkins in high regard: in 1865, Olmsted wrote to Watkins and to the painters Thomas Hill and Virgil Williams for their input on how best to maintain and enhance the scenery of Yosemite.

Watkins returned to Yosemite several times during his life. In 1865 and 1866, he visited with the State Geological Survey team. (According to

Whitney, Watkins's luggage weighed a massive 2000 pounds.) After the 1866 trip, *Humphrey's Journal* described Watkins as "one of the most persevering and successful" photographers in California. Persevering is correct: during his career as a photographer, Watkins produced over 1300 mammoth views, and not only of Yosemite. He photographed Mount Shasta, where he set up his dark tent on the ice, and the Pacific Northwest, where the dark tent caught fire from an engine spark. He documented, on assignment, the grand homes of San Francisco's elite. He took commissions from mining companies, where he composed images of industry as thoughtfully as he did of nature. Some of these photos capture the shadow of Watkins standing behind his tripod, visible yet entirely indistinct.

Watkins died in 1916. His last decades were desperate. He began to lose his eyesight in the 1890s; by 1906, when the earthquake struck San Francisco, Watkins was almost blind. The fires that followed the quake swept through his studio. Just a few days prior, a curator had visited Watkins. They had agreed to relocate his archive, including a trunk of his early

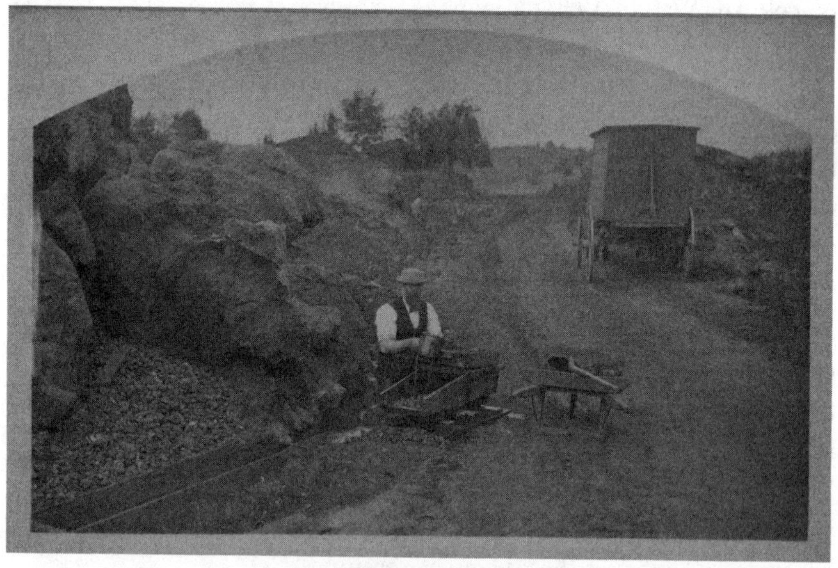

Watkins posed as a miner, circa 1883. Behind him is his photographic wagon, which he began to use in the 1870s to transport equipment and as a darkroom.

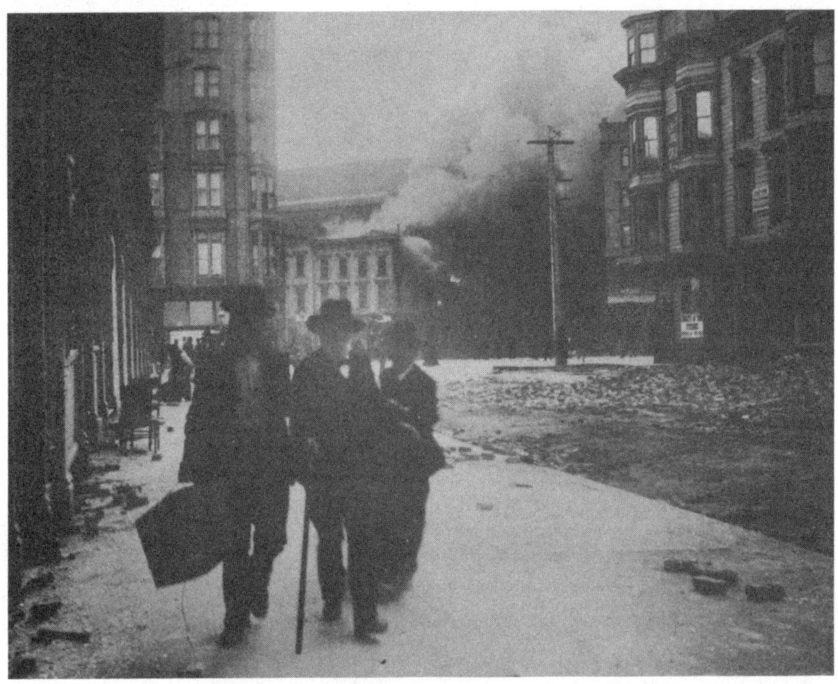

Watkins, with cane, after the 1906 earthquake, as his studio was engulfed in fire.

daguerreotypes, to a museum the following week. Instead, the daguerreotypes and many of his other works from his decades as a photographer were destroyed in the fire, along with any personal records he had kept.

Of the two confirmed photographs of Watkins, one, which he reportedly made for his children, shows him posing as a miner, his hat pulled low, bushy beard on display. The second shows him in 1906, being led away from his studio, white-haired, gripping a cane, as San Francisco crumbles around him.

The same year that Watkins died, 1916, Ansel Adams made his first trip to Yosemite, where he would go on to capture some of his most famous images. Even while Watkins was still alive, photographers like Eadweard Muybridge were following in his path. Yosemite became so popular that, in 1871, one visitor complained that the valley "abounds in photographers." But Watkins stands alone in his depiction of Yosemite—not only because he had the insight to photograph such an impressive landscape in mammoth format, and to withstand the arduous conditions, but that he did it so beautifully.

Trick or Truth

"Photography is truth embodied," wrote one anonymous commentator in 1867, which of course was a lie. Photography, the "mirror with a memory," as it was sometimes known, was also a funhouse mirror, one that could misrepresent, multiply, exaggerate, twist, and warp what it saw. The camera's lens distorted perspectives. Photographic chemicals only showed the world in monochrome. Exposures only captured what was still. One of the most famous early daguerreotypes, of the Boulevard du Temple in Paris, taken by Daguerre around 1838, did not show the whole truth: because of the long exposure time, the camera could not catch the restless life that bustled through the scene. Instead, it captured a ghost town, as devoid of life as it was of color. The only clearly discernible human figure was a wisp of a man who had stayed still for the exposure, through chance and not intention: he was having his shoes shined.

Photographers soon realized that the way to address the inherent shortcomings of this astonishing new process was to retouch the image after exposure. Some photographers also realized that the camera, far from being just a mechanical device that recorded facts, could be used as an artistic tool to create an image from imagination. And one photographer adopted what was known as combination printing—whereby multiple separate negatives were printed in sequence to create a single image—to construct a scene so controversial that it was banned from an exhibition, and its subject matter and technique were debated for years after it was produced. In one way or another, excessive retouching, faked photographs, and the notion of photographic truth have been contested ever since.

The members of the Birmingham Photographic Society who attended the meeting on April 28, 1857, were some of the first to see an astonishing new work by photographer Oscar Gustave Rejlander. Earlier that year, Rejlander had spent six weeks in his studio in Wolverhampton producing this single image, this dexterous feat of imagination and technical ability. That evening, dazzled Society members applauded Rejlander's artistry. Reporting on the meeting, *Photographic Notes* was effusive: "This magnificent picture, decidedly the finest photograph of its class ever produced, is intended to show of how much photography is capable, and how it may be made to assist the artist in historical and other paintings."

It was an enormous photograph, measuring about 31 by 16 inches, which Rejlander achieved by joining two prints together. At its center is a wise man flanked by two youths at a metaphorical crossroads. Sprawled out on either side of the men are figures representing the path of Virtue—knowledge, religion, industry, for example—and the path of Vice, as demonstrated by gambling, drinking, murder. The background and clothing (or lack thereof) are reminiscent of a fine art painting. Rejlander called it *Two Ways of Life.**

To produce the work, Rejlander first sketched out the composition. It wasn't possible at the time to achieve an image on this scale, with so many figures, in a single photograph. Instead, once he had selected his models, he photographed them separately, with careful consideration of lighting, poses, and proportions. He photographed the background separately, which he styled using some garden ornaments borrowed from a friend. Once he had all the components photographed, he needed to assemble them, one by one. He used a process known as masking, whereby for each glass negative, he covered the unwanted portion with paper, positioned

* An earlier title of the work was "Hope in Repentance."

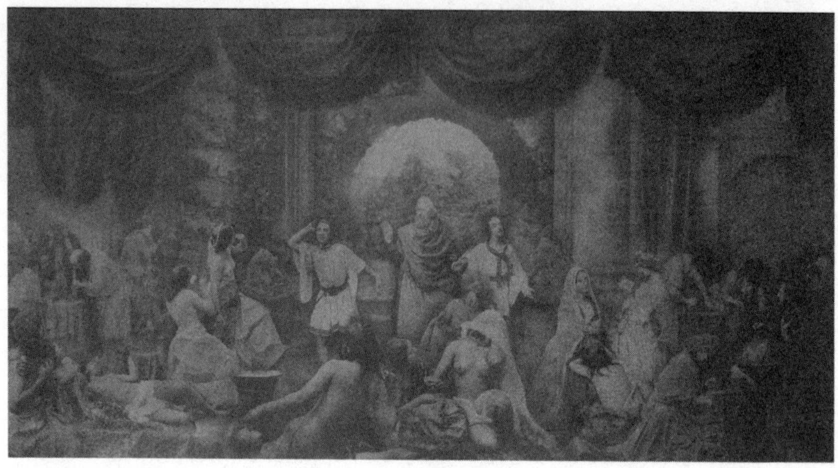

Oscar Gustave Rejlander's controversial Two Ways of Life, *1857.*

the negative on the photographic paper, and exposed it, slowly building a whole picture. Rejlander's final print had a combination of more than thirty separate negatives. Nothing on this scale, with this use of allegory, had been attempted previously.

Rejlander had trained as an artist in his native Sweden before moving to England in the late 1830s. Art remained a driving force in his photography. It was a commonly held perception that photography was purely mechanical, an act of chemistry and light, but to Rejlander this was entirely too limited. One of the reasons he produced *Two Ways of Life* was to show that photography could occupy the same realm as fine art. "I regard art as a means of making thought visible. If I can make a thought visible in a picture which people can understand, and be moved by it to laughter or tears, it is a work of art whether I produce it by the aid of the camera or of the pencil," he later wrote. To Rejlander, combination printing was a way to show what he called the "plasticity" of photography.

In that spring of 1857, Rejlander must have felt triumphant. After the applause of the Birmingham Photographic Society, *Two Ways of Life* was displayed at the Manchester Fine Arts Exhibition, which opened in early May. Prince Albert acquired three copies of Rejlander's work, reportedly for each of his royal homes, in Balmoral, Windsor, and Osborne House. In

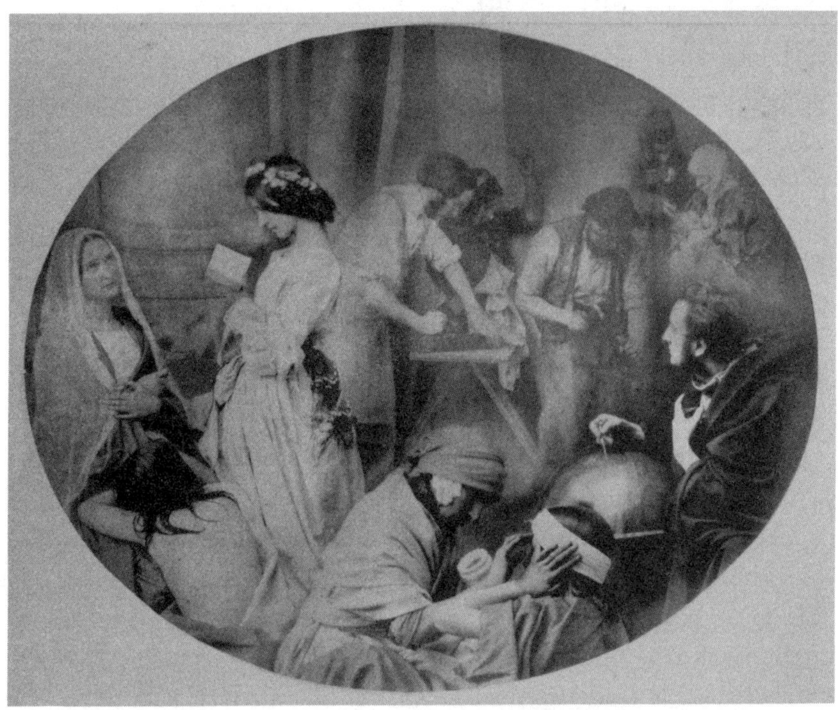

A detail from Two Ways of Life *provides clues as to where Rejlander possibly joined the photos. (Look, for example, around the bandaged head of the figure in the foreground, and along the back of the woman reading.)*

its review, the London journal *The Athenaeum* wrote that Rejlander's work "is in many points masterly—worthy to be painted as a fresco." But the accolades were short-lived.

For starters, the nudity was controversial. The Photographic Society of Scotland refused to display the print in its December 1857 exhibition when its more prudish members strongly objected to the naked figures. *The Edinburgh News and Literary Chronicle* followed the fuss and admonished those who judged the work "not by its merits as a work of art, but by some very tight-laced ideas of propriety." The furor became a farce the following year when the Photographic Society of Scotland agreed to exhibit the work but with the "vice" portion, and its naked bodies, covered up.

Other critics took aim at Rejlander's use of photography to create art, instead of to record. "Works of high art are not to be executed by a mechanical contrivance," declared *The Art-Journal* huffily. *The Photographic Journal* gave Rejlander the most backhanded of compliments: "The man who is capable of composing such a picture well enough to give satisfaction by a mechanical means, is entirely mistaken in his vocation, and should at once exchange the camera for the brush, and assume his proper position in the artistic world."

Two Ways of Life kick-started a debate about combining negatives that continued into the 1860s. This debate was further fueled by another controversial composite image, by the photographer Henry Peach Robinson, made in 1858, called *Fading Away*. It shows a somber family standing around a wan-looking girl dying of tuberculosis. To achieve the chiaroscuro drama of his death scene, Robinson had pieced together five separate negatives.

Like Rejlander, Robinson faced criticism for both his subject matter and his technique. "It would be difficult for him to have chosen a subject calculated to excite painful emotions in the minds of so many persons in anything like the same degree," wrote one correspondent in *The Photographic News*. In truth, that was Robinson's intention. He later wrote that he wanted to provoke strong feelings in the viewer to move away from the perception of photography as a "dull recorder of uninteresting facts."

Not surprisingly, Thomas Sutton, the peevish and polemical editor of *Photographic Notes*, had his opinions and did not hesitate to share them widely. Everything that he has seen of this nature from Rejlander and Robinson, Sutton said to an audience at the Photographic Society of Scotland in 1863, "has been a failure." A few months later he wrote that composite photography was "a mistake and failure" and that such works that have "been universally condemned." Just to drive the point home, he added that he would be pleased "when by common consent it is condemned to the limbo of experiments which have signally failed." Two weeks later, he followed up again—perhaps concerned that he hadn't been quite clear enough—to state that composite photographs are "productions" that are "no better than caricatures."

Sutton was also incensed by the nudity. Although initially he didn't agree with the Photographic Society of Scotland's ban of *Two Ways of Life*, he later changed his mind, saying they were right to prohibit a work "in which degraded females were exhibited in a state of nudity, with all the uncompromising truthfulness of photography." He also stated, inaccurately, that Rejlander's models were "nude prostitutes," whose photograph should not be exhibited to the public "in flesh-and-blood truthfulness and minuteness of detail." It's hard to tell what he hated more, nudity or combination printing.

Sutton wasn't alone. In the *British Journal of Photography* in 1860, the photographer Alfred H. Wall loudly and frequently criticized the "patchwork" nature of composite photography. His vocal response came after Robinson gave a talk at the Photographic Society of Scotland in which he described cutting negatives to compose his pictures, and staging a set to represent the countryside in his back garden. In response, Wall sputtered (in italics) against the use of *"scissors and paste-pot!"* In Wall's view, photography has "its highest and *only* value in its wondrous truthfulness."

Truth was at the heart of the debates about composite photography. If photography's defining characteristic was its truth, the argument went, wasn't creating a photograph from many negatives the antithesis of truth? But the meaning of "photographic truth" became malleable when photographers combined negatives for ostensibly honest reasons.

"The mass of chalky white sky is a great fault in a fine landscape, and I have brought a few specimens to illustrate the improvement easily obtained by the very simple operation of printing in a cloud sky from a second negative." This was photographer Samuel Fry, who spoke at a meeting of the South London Photographic Society in November 1860 about combination printing to remedy a well-documented and much complained about limitation in photography: the appearance of the sky.

The problem stemmed from the photo emulsions in use, which were far more sensitive to blue and blue-violet light than to the other colors on the spectrum, especially red. (This is why darkrooms have traditionally had a red safe light, the idea for which was first expressed in 1841.) That meant that when photographing a landscape, the exposure requirements for the sky were much shorter than for the rest of the scene. In practice, the result was a correctly exposed landscape and an overexposed, blank sky.

Photographers had other solutions to the sky problem, like painting in the clouds, but combining two negatives as Rejlander and Robinson had done gave the appearance of truth, particularly when the join was well hidden on the horizon line. Fry was not the first to use or even suggest using a different sky negative. Several years before Fry's talk, a French photographer named Gustave Le Gray was highly praised for his cloud-strewn seascapes, some of which were produced by more than one negative. Le Gray was decidedly less forthcoming than Rejlander and Robinson, who spoke openly about their techniques. But some of his photographs spoke for him—notably, the ones that contained identical clouds.*

In the same year as Sutton's ornery rants against composition printing, *The Photographic News* published an article, "On the Production and Use of Cloud Negatives," which advised photographers to expose the same scene twice in succession—once for the sky, once for the land—and join them together. Cloud negatives, the writer warned, "cannot be used indiscriminately without producing pictures open to very serious criticism." Another writer also advised caution against "indiscriminately printing-in clouds."

Their advice was indiscriminately ignored. Photographers combined skies from totally different landscapes with such gusto that by the late 1870s, they could purchase stock photographs of clouds to insert into their landscapes as they wished. (Printed-in skies even appeared in some of Carleton E. Watkins's landscapes.) "They produce charming effects, and must prove of inestimable value to the landscape artist," stated an "Unsolicited Testimonial" in an 1878 advertisement for "Perry's Cloud Negatives." Another advertisement stated the case more plainly: "Cloud Negatives. No More Inartistic Monotonous Background to Landscapes."

The ugly truth was that early photography had certain limitations that needed to be remedied, and nowhere was that truth better observed than in portraiture.

In 1853, the photographic firm of Ross and Thomson, "Photographers to Her Majesty," produced a short booklet they called *A Few Plain Answers to*

---

* As curator Sylvie Aubenas has pointed out, there's also evidence on one of Le Gray's negatives that he masked out a sign for a public toilet.

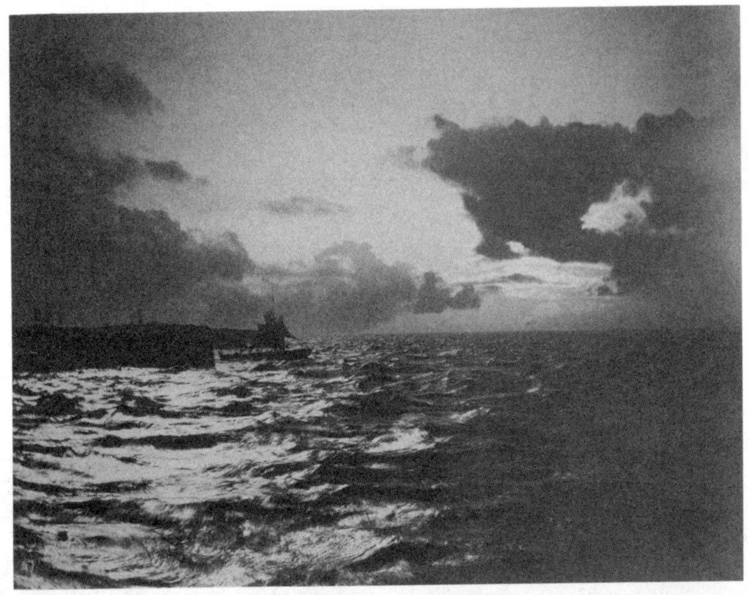

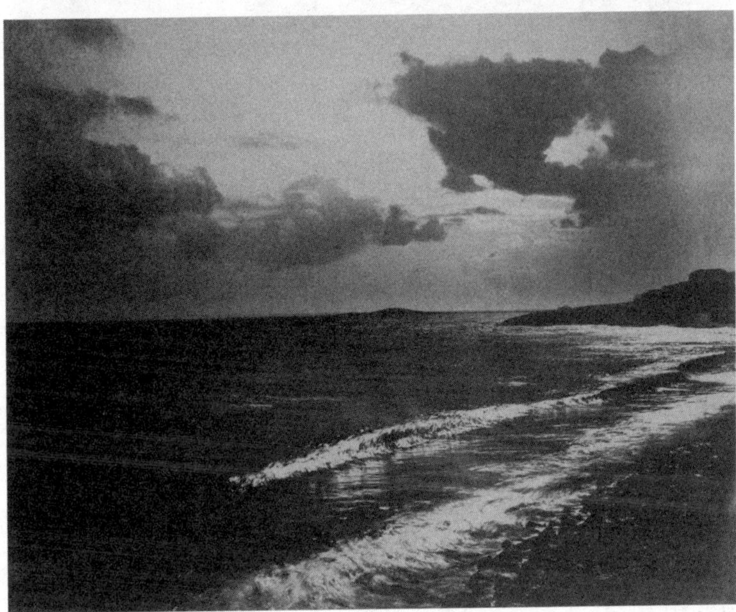

Two seascapes by Gustave Le Gray from 1857, Seascape with a Ship Leaving Port *and* Grande lame—Méditerranée—No. 19, *featuring identical skies. These photos were produced the same year, and using the same techniques, as Rejlander's* Two Ways of Life.

THE PHOTOGRAPHIC NEWS ALMANAC ADVERTISEMENTS   xlix

CLOUD NEGATIVES.

NO MORE INARTISTIC, MONOTONOUS BACK-GROUNDS TO LANDSCAPES, VIEWS, &c.

A New, Useful, and Necessary Adjunct in every Studio.

These splendid series of Cloud Studies, with varied effects to suit all subjects have been specially photographed from the grounds of our Studio at Durham Hall Bootle, which command an uninterrupted view of the open sea ; and, in order to render them perfect and suitable, they have been artistically worked by the eminent marine painter, Samuel Walters, Esq., and will be found better adapted for the purpose than any similar productions.

Nature and Art Combined.

The four effects supplied, which are distinct in character and subject, produce, by following the instructions sent, an endless variety of picturesque studies, and may be arranged to suit the lighting and contour of the view.

Four Sizes and Four Effects will be ready in February, viz.,
15×12, 12×10, $8\frac{1}{2} \times 6\frac{1}{2}$, $5\frac{1}{2} \times 4$.

Cloud negatives advertised in The Year-Book of Photography and Photographic News Almanac, *1877.*

Common Questions Regarding Photography. Their response to the question "Are they true likenesses?" would not have reassured a nervous potential client: "Yes; truth is stamped upon them all. Yet some of them shew how disagreeable truth may, in some cases, be made to appear. The veriest tyro in painting, that ever tried his prentice hand upon the human face divine, probably never produced greater monstrosities than have occasionally been perpetrated by this art. . . . In short, the good are very good, while the bad are very bad."

There was a reason people frequently likened a visit to the photographer to a trip to the dentist. Even the most enthusiastic of sitters (and many were deeply unenthused) had to contend with some combination of harsh lighting, an unfamiliar environment, an unnerving process, and the physical discomfort of a metal posing stand. None of this helped produce a relaxed, true-to-life portrait. Under these circumstances, "monstrosities" wasn't necessarily an exaggeration. As one commentator joked, "Honor the photographer. He can't help it."

Unhappy customers, with their sour expressions, occasionally sued photographers for bad portraits. In London, a client sat for his portrait over seventeen days, and the only result was a photo "that made him look as if he was going to be hung." He took the portrait photographer to court and was victorious. Not so for the case of *Windus v. Peat*, in 1876. This client, a "very awkward customer," sued the photographer over a portrait; a witness said the client was dissatisfied because the portrait "was not good-looking enough," at which point, the *British Journal of Photography* reported, the court laughed. The photographer won.

A photographer had tricks to flatter a client that weren't manipulation per se. An 1843 manual advised photographers to soften the focus and aim for a "vague likeness" when photographing someone with "wrinkles, or who is pitted with the small-pox, or one who has unpleasant features." But the obvious solution was to remedy the photo's inadequacies—whether they stemmed from the camera, the client, the photographer, or all three— after the portrait had been taken.

The camera couldn't capture color, for example, which was a problem from the very beginning. "It must want color," declared *Blackwood's Edinburgh Magazine*, in 1842, before continuing, "Its best likeness can be only that of a rigid bust, or a corpse." To address this criticism, photographers started to add color with a variety of materials, including crayon, watercolors, or oil paints. Photographers advertised for their portraits with color or without, as in one 1855 advertisement that the customer could opt for "tinted face and hands" or the pricier "Coloured Portraits, *in the best style.*"

An article from 1859 titled "Lessons on Colouring Photographs" included extensive notes about how to fix the "disagreeable expression" that came from a subject's eyes contracting in a bright studio. The article then enumerated how to correct it: "the upper eyelids will require raising and widening; the lower eyelids drooping; irises will require enlarging, eyebrows raising and opening (as they will probably be drawn together producing something like a frown), corners of the mouth raising, center of forehead making lighter; and the line down from the sides of the nose towards corners of mouth softening."

In addition to painting on color, photographers also applied charcoal or lead pencil to improve a print or negative. (In the wet collodion years, it was easier to retouch the print; with the dry plate process, retouching the negative was more common.) With all these options, retouching grew so rapidly that in 1857, Lady Elizabeth Eastlake wrote in *The London Quarterly Review* that "there is no photographic establishment of any note that does not employ artists at high salaries . . . in touching, and colouring, and finishing from nature those portraits for which the camera may be said to have laid the foundation."

Beyond their customers' vanity and their own business sense, photographers also retouched their images to combat the inconvenient fact that their prints were deteriorating. The London Photographic Society was so concerned that in 1855 it set up a "Fading Committee" to investigate this "all but universal" problem. The committee found several causes for the fading of prints, including certain developing processes and errors, and made recommendations to photographers to ensure that this supposedly permanent process was, in fact, permanent. Not that there weren't other hazards: an article in 1869 advised against hanging a photo on a green wall, as it was liable to fade due to arsenic in the green tint.

The advances in retouching continued as photo processes evolved. Retouchers used desks with a tilted surface and movable magnifying glass for greater precision. In 1884, a "Special Correspondent" for the *British Journal of Photography* reported on a contrivance he had seen at a photographers' convention in the United States. He described it as "a tool which, in educated hands, will produce effects of high finish upon a photograph with a great saving of labour." It was called the "Air Brush."

The widespread practice of retouching created a certain amount of hand-wringing in the photographic press. In 1855, *Humphrey's Journal* approvingly cited a French journal on retouching: "The essential value of photography, its indisputable truth, is wholly destroyed if retouching is to be allowed." But if a harshly lit portrait made the sitter look older than in real life, was it more truthful to restore some youth with retouching? Or was it less truthful because the portrait was accurate to what the camera saw?

Some argued that limited retouching was acceptable if it remedied a temporary defect in the sitter (like a pimple) or addressed an error in the

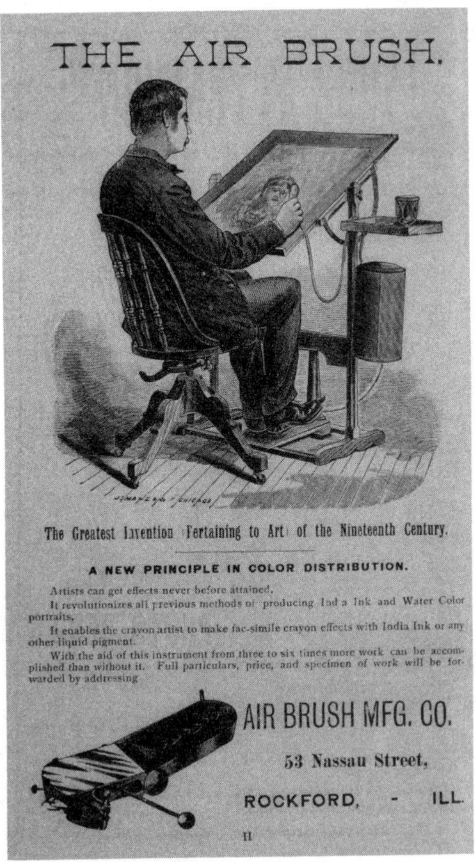

The Air Brush was, according to one reviewer, "a tool which, in educated hands, will produce effects of high finish upon a photograph with a great saving of labour." Advertisement from 1883.

photographer's technique (like lighting). In these cases (so the argument went), retouching was "not flattery, it is an act of simple justice." Beyond that, though, one commentator warned, a person's character is erased. "All life is gone; the suggestion of a breathing, feeling, living face is sacrificed to produce a pretty, smooth mask."

Retouching manuals and portrait photographers didn't always pay attention to such warnings. A photographer who wisely chose to remain anonymous for an 1886 interview with the *British Journal of Photography* stated plainly that his customers "desire to be made good-looking, and, secondly, young-looking, at any rate when past thirty." Another added weight to the list: "Do not on any account forget to touch ladies' waists in a specially hearty manner, if you want to keep on good terms with them," advised a British manual in 1891. For double chins, one article suggested, "considerable pruning may be effected," while a "clever retoucher" could remedy a child's portrait to produce a happier expression.

Once again, these debates revolved around the perception of, and the photographer's adherence to, the idea of truth. Too much retouching, warned *The Photographic News*, and it "becomes one of the most dangerous and degrading innovations to which photography could be subjected. The highest claim of photography is its unerring, its uncompromising TRUTH."

But even if a photographer largely avoided retouching, the truth (or rather, TRUTH) was finessed before the image was even exposed. For example, a photographer's studio might include any number of props and backgrounds to suggest that the subject was in a boat punting along a rural stream or enjoying a wintry alpine walk. The photographer might then daub in, on the final print, some gently falling snowflakes to complete the illusion. And while the camera might be a mechanical device, the photographer was not. As one article stated, "The rendering of the highest photographic truth is much dependent upon the photographer's perception of that truth, and skill in embodying by photography what he sees."

Upon his arrest for fraud in the late spring of 1875, the French spirit photographer Édouard Isidore Buguet wasted no time in confessing that his images were fakes. For several years he had produced photographs that purported to show, alongside a living subject, a ghost. Photos like Buguet's, which also appeared in the US and the UK, bolstered the Spiritualist belief that the dead could communicate from beyond the grave. But photographers could (and did) fabricate these pictures with photo manipulation, typically by either combining negatives or through double exposure, whereby two separate exposures were made on a single plate.

The full extent of Buguet's trickery was revealed during his trial. Buguet's cashier elicited information from customers about the spirit they hoped to see. In a back room, Buguet kept a mannequin shrouded in a veil

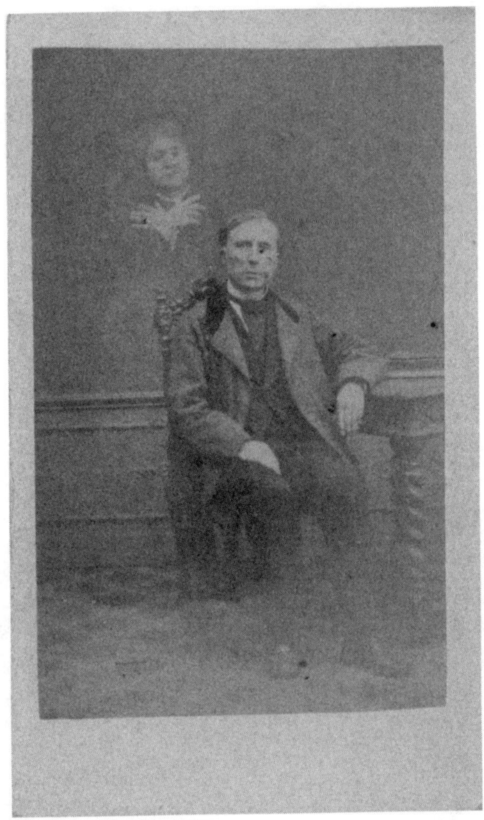

Édouard Isidore Buguet produced spirit photographs like this one, until he was tried and convicted for fraud.

and about 240 enlarged portraits pasted onto cardboard, "men, women, boys, and girls of all ages," as described by one newspaper. He matched the portrait to the description, clipped it onto the mannequin, made his first exposure, and reused the same plate for the client's portrait. During this exposure, Buguet enacted a charade of intensity. As one customer recalled, "He had me sit and, indicating the point on which I was to focus my eyes, said: 'let us pray.' At the same time, he turned his gaze heavenward; then, taking his head in his hands, he went to lean his elbows on a small table; he remained there for several seconds." He then presented the photograph to the amazed customer, who was perhaps a little gullible, and quite possibly

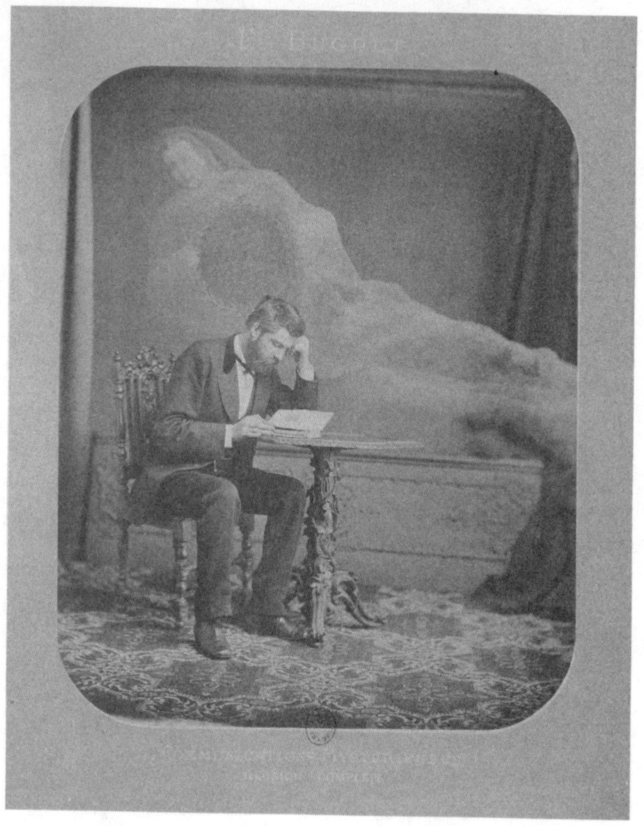

One of Buguet's spirit photographs produced after his trial, for entertainment, 1875. Part of the plaque reads, "The chosen ghost is guaranteed."

grieving. Unsurprisingly, the miraculous work of the supernatural had a satisfying effect on his profit margins. According to historian and curator Clément Chéroux, thanks to spirit photography Buguet increased his studio's annual turnover by about 20 percent.

Buguet's basket of tricks—his box of portraits—was produced at the trial, along with the mannequin. This didn't sway those witnesses who were die-hard Spiritualists, such as the Comte de Bullet, who saw the physical evidence and said, "That proves nothing." The court had a different view. Buguet was convicted and sentenced to a year in prison.

Buguet was shameless, but he was not alone. Other spirit photographers produced portraits with blurry, hazy figures that made it easy for the client to project the faces of their dear departed. As Oliver Wendell Holmes observed in 1863, "It is enough for the poor mother, whose eyes are blinded with tears, that she sees a print of drapery like an infant's dress, and rounded something, like a foggy dumpling, which will stand for a face: she accepts the spirit-portrait as a revelation from the world of shadows."

Buguet's brazenness continued after his trial. He produced another series of spirit photographs, with the same technique that he had used with his clients—except these were for entertainment, and he was the subject. In one photo, Buguet reads a book at a small table while a translucent, shrouded figure floats above him. She carries a plaque, part of which says, "The chosen ghost is guaranteed."

The crossover between photo manipulation and humor had come into play before Buguet's droll visuals. In 1856, David Brewster, who was so taken with microphotographs, had suggested using double exposures to create ghostly figures for humorous stereoscopes. Another popular visual joke was the double or triple portrait, in which a photographer's subject appeared repeatedly in the frame. Rejlander produced a photo he called *O.G.R. the Artist Introduces O.G.R. the Volunteer*, where he is featured twice, in different costumes. (It's not surprising Rejlander created this portrait: it was another display of the "plasticity" that he had spoken about, except this time it was for comedic effect.)

Along with the double portrait, photographers manipulated images to create all kinds of wacky effects: a decapitated man with his head stuck on the end of a sword, for example, or a woman made to look like a marble bust. Such effects and the means of their creation were explained, along with other goofy ideas, in the popular book *Photographic Amusements*, although the author warned against trick spirit photography for those "who practice on the gullibility of inexperienced persons."

Unlike Buguet, the American spirit photographer William H. Mumler never explained how he produced his spirit photographs. Mumler was also put on trial, in New York in 1869, after he was arrested for fraud and larceny. He had already been accused of fraud once, earlier in the decade, when a prominent spiritualist recognized one of the ghosts as someone who was alive and living in the same city as Mumler, Boston.

The fact that spirit photographs could be easily achieved through manip-

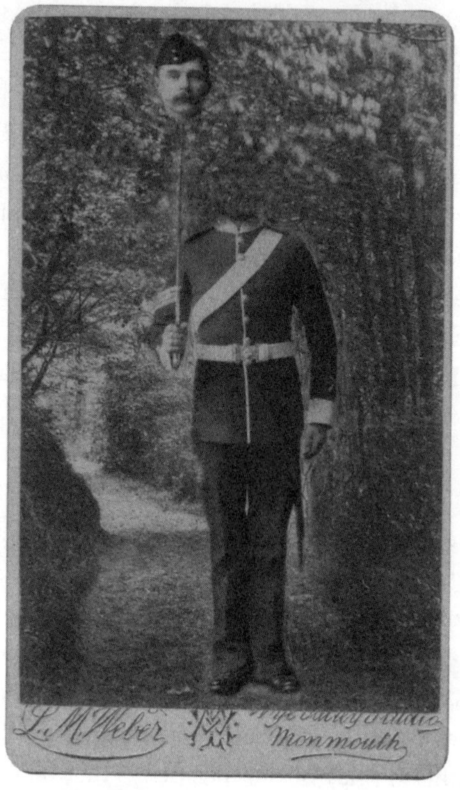

One of the many variations of humorous decapitation portraits made through photo manipulation.

ulation techniques was not lost on the photographic press. *Humphrey's Journal* ridiculed the idea that spirits "entertained the whimsical fancy of plaguing operators by skipping about photographic galleries." *The American Journal of Photography* described spirit photography as "a low swindle," before adding, "What a shame that all these things should come to pass in the nineteenth century and in America." The British publication *The Spectator* couldn't resist a jab at the United States either, not realizing that spirit photography's arrival in the UK was imminent: "The Americans, however unsuccessful they may be as politicians, certainly do contribute not a few cheering little stimulants to our social vivacity."

Mumler's trial was held at the Tombs Police Court in Lower Manhat-

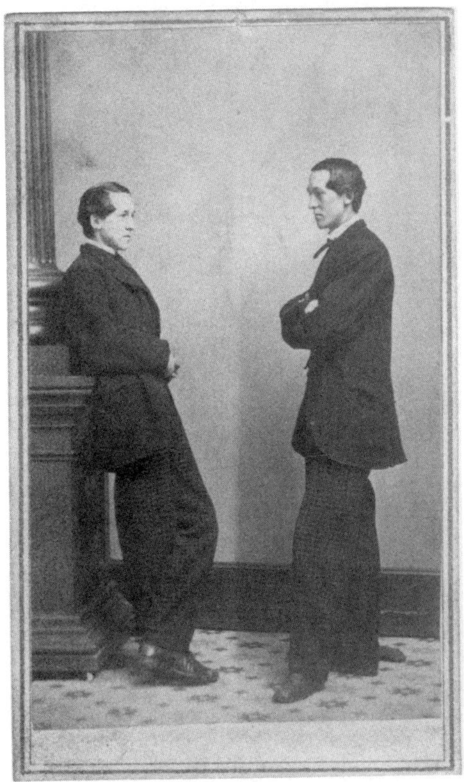

A double portrait.

tan, with the rapt attention of the press. *The New York Times* published no fewer than seven articles covering the trial. *Harper's Weekly* dedicated a front page to Mumler, featuring nine engravings of his spirit photographs. In court, Mumler asserted that he "never used any trick or device" to produce his spirit photographs.

At the end of the trial, the judge noted that even though he was "morally convinced" that a deception had taken place, the prosecution had not proved its case. Mumler was acquitted. He moved back to Boston and continued to produce spirit photographs, including one showing the ghost of Abraham Lincoln. The customer, who reportedly wore a veil and used an assumed name, and whom Mumler claimed he didn't recognize, was Mary Todd Lincoln.

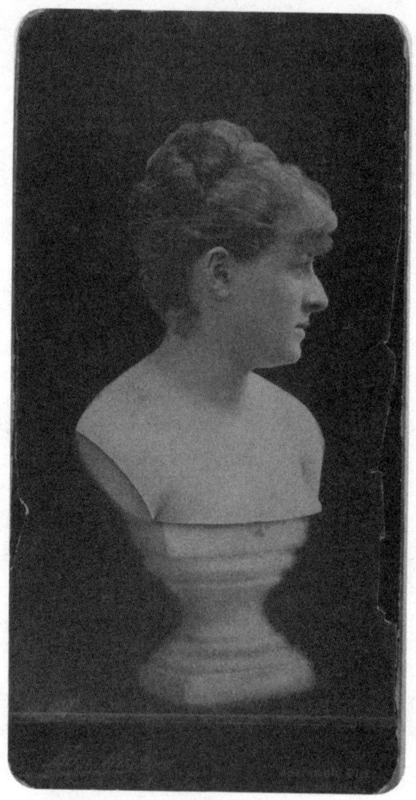

Another comic photo manipulation technique was to make a portrait look like a marble bust.

As for Buguet, he later renounced his confession, declared bankruptcy, worked under an assumed name, and then disappeared from historical records sometime in the early 1890s. Whatever Buguet's notion was of "photographic truth," he was speaking the truth when he said, in court, "I am just a photographer with more or less skillful tricks."

In 1894, a photo appeared in a London publication that purported to show Prime Minister Gladstone standing outside a pub in Covent Garden. In fact, it was produced with combination printing for an article about how a photograph cannot be relied on as evidence of a fact—at least, not without some wary interrogation. The other photos in the article included a portrait of a man pointing a gun at his double, a spirit photograph, and

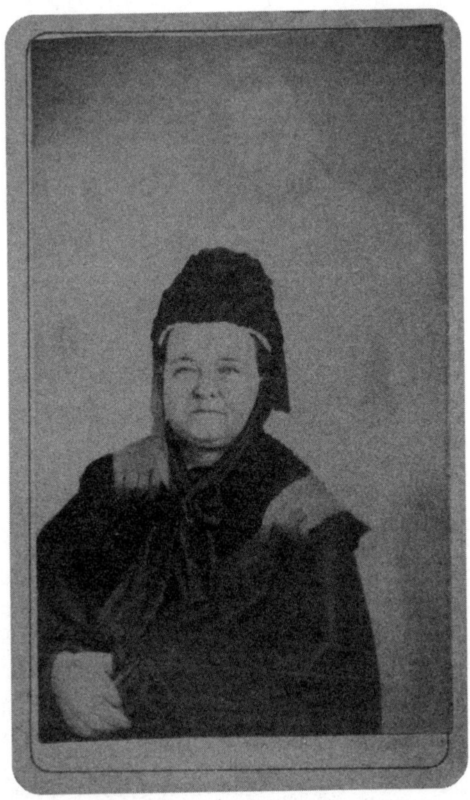

Mary Todd Lincoln, photographed by William Mumler in 1872, after his fraud trial. With her in the photo is another figure, which some claimed to be the spirit of Abraham Lincoln.

the stage actor Henry Irving in a leg-kicking comic dance with the music hall singer Lottie Collins. The improbability of these situations perhaps undermined the article's central argument: that the viewer needs to exercise caution before assuming that a photograph always shows a fact. But perhaps nineteenth-century audiences were more credulous than us, in the twenty-first century; we would know better than to believe, say, a fake, AI-produced photograph of the pope wearing a white puffer jacket. Wouldn't we?

Chapter 7

In Color

A photographic process that could capture color was as desirable as it was elusive. For most of the nineteenth century, color photography existed as a series of false starts and empty promises. By the 1890s, some five decades after the daguerreotype was first introduced, photographers finally started to make some headway—but these color photography processes were either impractical, not easy to commercialize, or both. The search for a workable color process even prompted *Punch* to mock, in 1904, "Never Despair. The secret of photographing in color has again been discovered. We were getting afraid that this year was going to be an exceptional one."

If anyone could crack color photography, it was the Lumière brothers, Auguste and Louis. The Lumières had a long association with photography. Their father, Antoine, was a portrait photographer in Lyon; he had also designed the red gaslit sign that blazed across the exterior of Nadar's studio. In the early 1880s, while he was in his late teens, Louis, the younger brother, began to experiment with a new gelatin dry plate formula. In 1884, the family formed the Société Antoine Lumière et ses fils; by the early 1890s, they were producing some 350,000 boxes of dry plates per year.

In 1894, during a trip to Paris, Antoine saw Edison's Kinetoscope, a box-shaped device that a single viewer peered into to see a moving picture. Antoine told his sons about it and they started to experiment with a projected moving picture device. In 1895, they introduced a camera that could both record and project short moving pictures, which they called the *cinématographe*, and which is generally regarded as the start of motion pictures. One of their earliest films shows Lumière employees, mostly women, leaving the factory in Lyon.

But color photography took much longer. It was over a decade from the time the Lumière brothers first tried out a color process, devised by Gabriel Lippmann, to when their own final product, the autochrome, went on the market in 1907.

The process by which they (eventually) created the autochrome began with tiny grains of potato starch. Using levigation, a system of grinding the starch while wet, the Lumières extracted grains of starch that measured around 12–16 microns each. (A micron is one-thousandth of a millimeter. The thickness of the single page of a typical book is about 90 microns. The starch grains were *minute*.) The grains were dyed red-orange, blue-violet, and green, then scattered on a varnished glass plate, to which carbon black was added to fill any gaps. Each plate was then laminated, which flattened the shape of the grains to improve transparency and absorption. This step required the Lumières to invent a new machine, because every existing press cracked the glass. The final steps were to add a watertight varnish and the panchromatic emulsion.

Theirs was a practical, in-camera color process: photographers could use the plates with their normal cameras to produce a glorious color positive transparency. The viewer either held the transparency up to a light source such as a window or used a special stand with a mirror called a diascope.

"The spectator was met by such a wealth of color in these pictures that one could scarce find words to express his admiration," wrote one commentator in the *Bulletin of Photography*. The Lumières began to sell the plates in France in the summer of 1907. The demand was so high that availability was delayed in England. Photographer R. Child Bayley described his impatience, under the headline "Colour Photography: A Revolution," of trying autochromes for the first time. "Never since I developed my first plate, twenty-three years ago, did I feel the interest and impatience that filled me as I carried that slide into the dark room," he wrote. The results, he declared, "must be seen to be believed. I did not believe them til I saw them."

Autochromes made the news in the daily press as well. "Color-Photography Made Possible at Last: The Lumière Process," wrote *The Illustrated London News* in September 1907, and devoted a full page with elegant illustrations to show how it worked. *The Times* of London reported that the Lumière process was "an entire revolution in colour photography." In *The New York Times*, under the headline "Nature Reproduced by Color Plates," the writer stated the results were "scarcely less than astonishing." Upon seeing autochromes produced by Alfred Stieglitz, the writer added, "To the layman it seemed that he was simply looking through a window at the person or thing reproduced."

Autochromes had their downsides. They were expensive. They could not be reproduced. The exposure times were much longer than for black-and-white film. To be exhibited in public, they needed a magic lantern. Many photographers who had embraced autochromes, including Stieglitz, eventually returned to black-and-white. But autochromes stayed in production until the 1930s, and those that still exist are a precious glimpse into history in color. They include some 73,000 autochromes from around the world that make up the Albert Kahn collection, autochromes of World War One, and even the first underwater color photograph, taken in 1926. Stieglitz was not wrong when he wrote, in 1907, "Soon the world will be color-mad, and Lumière will be responsible."

In the spring of 1889, after a long search, a Scottish book collector named William Lang Jr. finally acquired a copy of a title that was, at that point, already some thirty years old. The book contained 377 plates of bright blue prints, known as cyanotypes, showing delicate pale outlines of algae. The author was simply listed as "A.A.," whom Lang dubbed "Anonymous Amateur" in a journal article. But a botanist from London's Natural History Museum recognized the author. "A.A." was not an "anonymous amateur," but the initials of an early photographer and author of what was subsequently identified as the world's first photographically illustrated book: Anna Atkins.

In her introductory text, Atkins wrote that it was "the difficulty of making accurate drawings of objects so minute as many of the Algae" that led her to cyanotypes. It's hard to think of a better application than images of algae. Their irregular, wispy tendrils appear cloudy white against the gloriously rich blue, suggesting visions of swaying seagrass. (If Louis Boutan had seen Atkins's book, he would have been a fan.) The ocean blues of the cyanotype must have seemed startling in comparison to the monochrome tones of photographs.

The cyanotype was invented by Sir John Herschel in 1842, around the same time that he was carrying out experiments with color photography. Herschel was a highly respected scientist who had a significant influence on the early history of photography. He suggested to Daguerre's rival and inventor of the calotype process William Henry Fox Talbot, for example, the use of sodium thiosulfate (hypo) to fix an image. He also coined the photographic use of the terms "negative" and "positive." Herschel's experiments with color photography were so thorough that he even tried extracts from flower petals as well as from his pet boa constrictor.

Herschel also experimented with light-sensitive iron salts on paper. He discovered that the paper, if dried in the dark and then exposed to the sun while in contact with a drawing or an object, transformed. When he washed the paper with water, the exposed areas took on a vivid blue tone. This was a cyanotype, and Herschel used this process to copy his handwritten notes. Cyanotypes are better known as blueprints, which is how they were used well into the twentieth century.

Herschel published a paper on cyanotypes in 1842, and sent a copy to friend and fellow scientist John George Children, who was Anna Atkins's father. Atkins's mother had died the year she was born; growing up, Atkins had received an education in science, which served her well in her photographic endeavors. By the time she started making cyanotypes in the autumn of 1843, she was forty-four, married, and living in Halstead, Kent, in a home that her husband had inherited after the death of his father, John Atkins, in 1838. In addition to the Halstead home, Atkins's husband also inherited land holdings in Jamaica and Bermuda; John Atkins had been a slave owner. With the abolition of the slave trade in 1807, and of slavery in the British colonies in 1833, John Atkins was entitled to claim compensation, which he did, successfully.

Atkins issued her cyanotype prints to her "botanical friends" in sets, for the recipient to compile and bind into books. She rendered the title page to look like algae (either typographically or with algae) to create a distinct and cohesive visual style. Atkins spent ten years making cyanotypes for her algae book, during which time she produced, according to photo historian Larry Schaaf, more than 6,000 cyanotypes. Atkins would have had servants to assist, and a childhood friend, Anne Dixon, also appears to have worked with her. Anna Atkins died in 1871 and had already been forgotten by the time of Lang's rediscovery of the "Anonymous Amateur," less than twenty years later.

Schaaf researched Atkins extensively in the 1970s after her name and her work had again disappeared into obscurity. It was Schaaf who determined that her book on British algae was the first to be photographically printed and illustrated. Atkins's cyanotypes, writes Schaaf, "are powerful early examples of the expressive potential of photography."

Atkins also took photographs, but none have yet been found. Photography was considered an eminently suitable pastime for a woman of education and means. "This last summer, being in the country, I amused myself Daguerreotyping, and a more delightful amusement never was invented; it is perfectly bewitching!" wrote an anonymous female amateur in *The Art-Union* in 1840. "And I wonder every lady in the land has not an apparatus, for it is particularly fitted for a lady's recreation." Queen Victoria set the tone: together with Prince Albert, she was a patron of the Photographic Society of London, collected photographs, and even had a darkroom installed at Windsor Castle. In the early decades of photography, both Constance Talbot (married to William Henry Fox Talbot) and Cecilia Glaisher (married to James Glaisher, who had ventured so high in a balloon that he passed out) took photographs as amateurs.

Women also worked as professional photographers in Great Britain from the very beginning. In 1846, a Mrs. Ann Cooke, who took over her

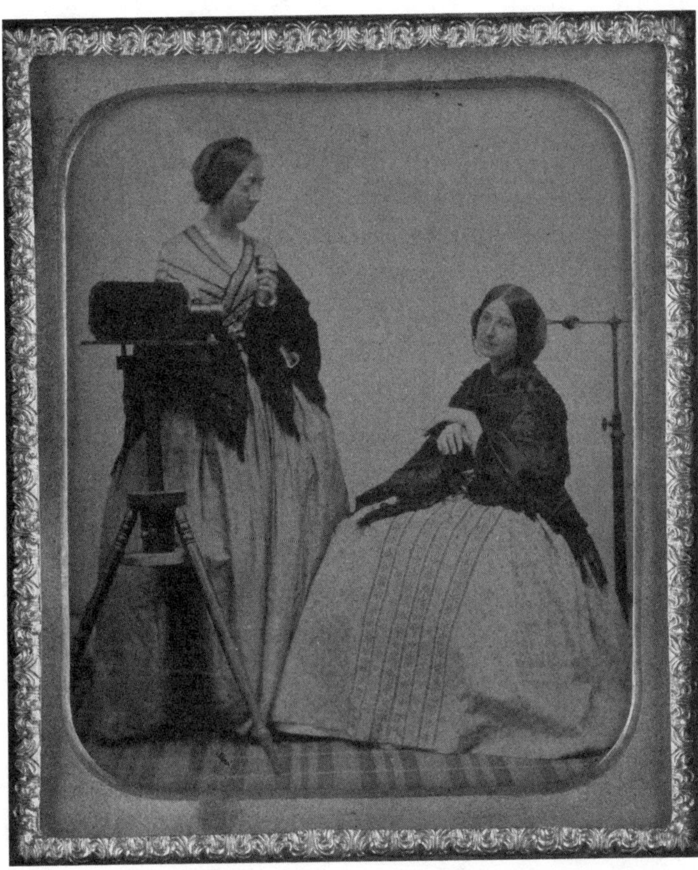

An unidentified daguerreotypist and her subject, circa 1855. The identities of many women who worked in early photography remain unknown.

husband's photography studio after he died, advertised her daguerreotype studio with the promise that "an unerring Likeness will be taken, with or without Colours." (Colors were painted on, of course.) Another photographer working around this time, whose name also appeared in advertisements for her studio, was a Miss Jane Nina Wigley.

It's interesting to note that Wigley was single, and that Cooke was widowed. From a legal standpoint, in mid-nineteenth-century England, any wages earned by a married woman belonged to her husband. In fact, once

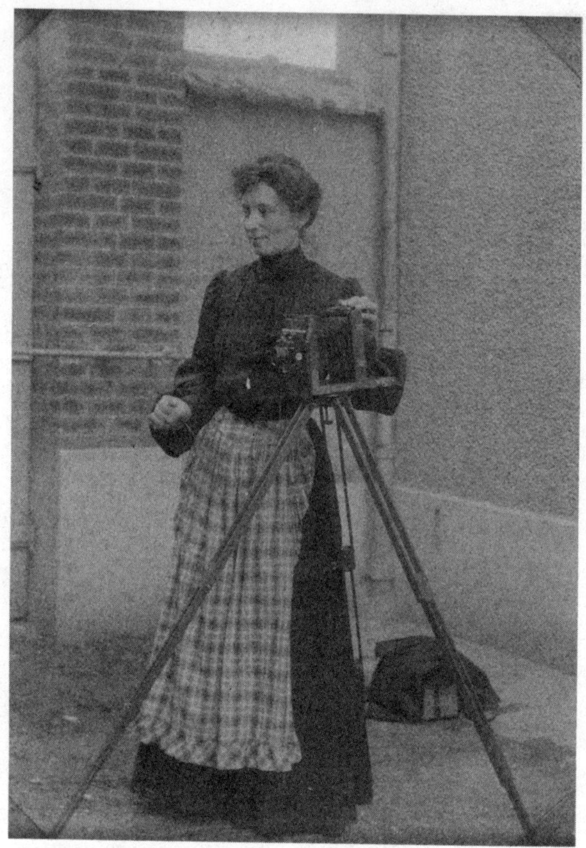

An unidentified French photographer, circa 1900. Women were involved in photography, as both professionals and amateurs, from the very beginning.

married, a woman's legal identity was entirely subsumed into her husband's. She couldn't own property or sign a contract. Not until 1870 could married women control the wages they earned, and even then with limitations. This legal situation was also reflected in naming conventions. Women in photography often used their husband's name or just an initial, which has made the study of early women photographers more challenging for historians.

Women were also involved in all aspects of photo studios in behind-the-scenes roles, as printers, attendants, receptionists, retouchers, and colorists. In London, almost every photo studio employed women in some capacity, according to the 1867 book *Young Englishwoman: A Volume of*

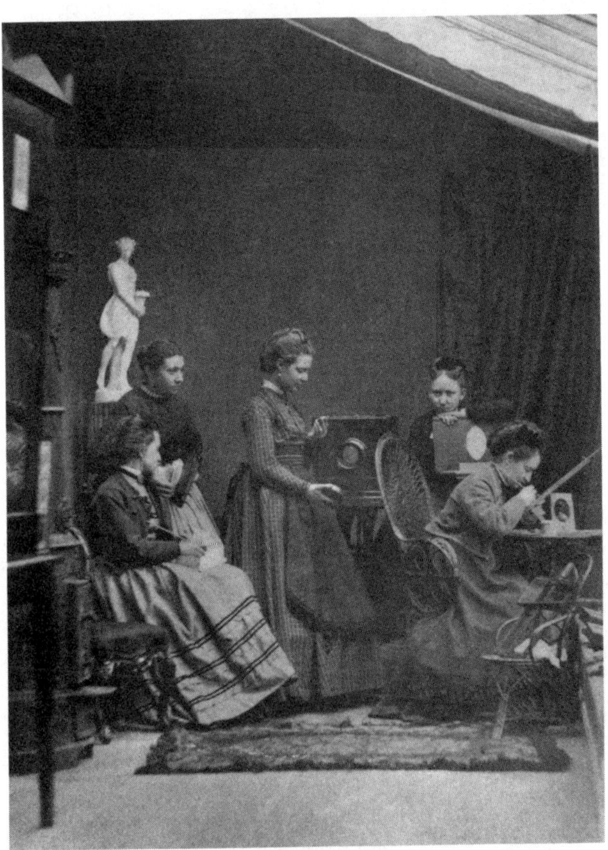

The studio of Swedish photographer Rosalie Sjöman, circa 1860s. The woman on the right appears to be retouching or adding color to an image. (Note, too, the skylight above.)

Pure Literature, New Fashions, and Pretty Needlework Designs. The author also went on to note that, despite their value to the studio, it was difficult for women to advance: "In photography, as in almost every other trade or profession, men will thankfully employ women as their subordinates, while they are jealously careful and anxious to exclude them from the knowledge which might permit them to practice the art independently."

Applying color to photographs, like other aspects of retouching, was for the most part uncredited. One notable exception was Elizabeth Bond: the name "Miss Bond of Southsea" was incorporated on the verso of her

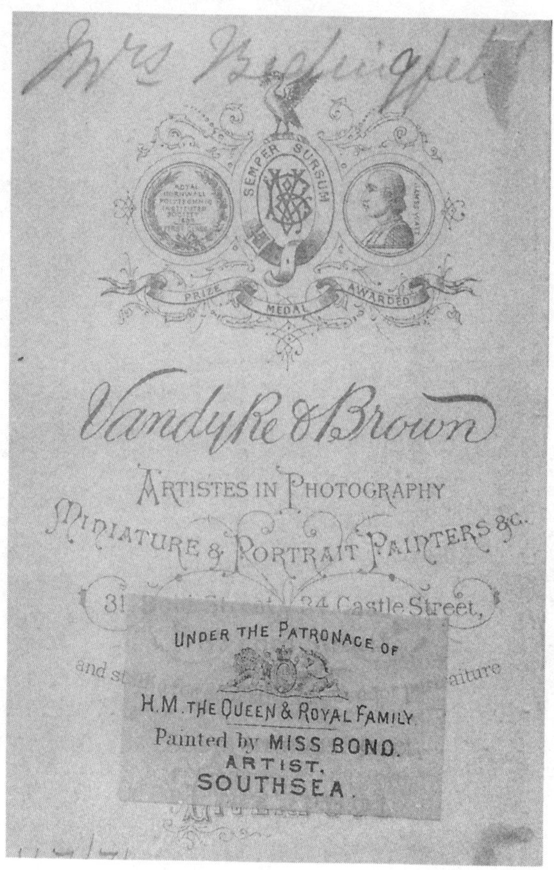

One photo colorist whose name does feature on her work appears here—Elizabeth Bond, or "Miss Bond," of Southsea.

photographs. Another was Amélie Guillot-Saguez, who appears to have worked with her husband as both a photographer and a colorist.

Coloring photographs, reported *The Photographic News* in 1873, was "essentially well suited to young ladies of liberal education and gentle nurture desiring to make an income, whilst it would possess the further advantage that it need not necessarily be given up on marriage, as it might often be continued at home without necessarily interfering with home duties."

Una Howard, a British photo colorist, had also suggested lessons for "respectable, yet bereaved" women in need of work. In a letter to the *Lon-*

don Evening Standard in 1865, Howard wrote, "the art of colouring photographs, to make them lifelike, is one that has already been attempted with great success, and is calculated to afford employment to many young ladies who, being left in straitened circumstances, have yet, while their natural protectors were living and could afford it, received lessons in painting." Howard's system was to render the photograph slightly transparent and then paint both the front and back, work that was praised for its "delicacy." She also founded the "Self-Help Institute for Distressed Gentlewomen," the leading focus of which was the coloring of photographs.

But others sniffed at the idea that women were bringing valuable skills to photography. Photographer Cornelius Jabez Hughes opined, in 1873, that coloring photographs had an "enormous and lucrative scope for female skill," adding there was "no more obvious occupation for a woman, who is without the higher genius to lift her into the legitimate sphere of art, and yet has ability enough to use her pencil skilfully, than in painting photographs." Then there was the question of wages. Hughes was doubtless sharing the view of many when he wrote that the chief advantage of employing women to do basic studio work like printing was "not that they do their work better, but that they can be got cheaper than men."

The perceived lack of "higher genius," and the fact that women were paid less, influenced views of photo coloring as a profession, argues historian Nicole Hudgins. As she writes, "the acceptability of coloring as a manly profession declined in direct proportion to the visibility of its women practitioners."

Aside from the cyanotype—which was not color photography, it was *a* color, and a camera-less process—and applied color—which was also not color photography but a post-production method of adding the appearance of color—the first leap toward color photography as a process that might, someday, be captured with a camera occurred on a Friday evening in May 1861.

James Clerk Maxwell was a precociously talented and influential Scottish physicist who started working on color theory in 1849 while attending the University of Edinburgh and later, at Cambridge.* (He'd had his first sci-

---

*   Maxwell's wife Katherine also contributed to his color theory experiments.

entific paper published at age fourteen, but because he was too young to present it to the Royal Society of Edinburgh, a more senior member had to do it for him.) Throughout his scientific career, which continued up until his death in 1879, he made crucial discoveries in astronomy, thermodynamics, and electromagnetism; but on that Friday evening, at age twenty-nine, in front of the scientifically minded members of the Royal Institution of Great Britain in Central London, he presented the world's first color photograph.

Maxwell's experiments in color photography were based on the work of Thomas Young, who had proposed that the human eye is only receptive to red, blue, and green. Ahead of the presentation, Maxwell had asked Thomas Sutton, who was so frequently bashing out his opinions for *Photographic Notes*, to photograph a piece of knotted tartan ribbon. Sutton used the wet collodion process to photograph the ribbon against a black velvet background, outside in sunlight, three times. As requested by Maxwell, one exposure was made through a red filter, one with blue, and one green. For the color filters, Sutton used solutions of metallic salts. During the lecture, using a magic lantern, Maxwell projected each positive photograph as a transparency through its correlating filter. When he superimposed the three images together, they showed a color photograph.

But there's a strange anomaly to this experiment, which was only resolved one hundred years later. As with all the photo emulsions at this time, the one Sutton used was highly sensitive to the blue, violet, and ultraviolet end of the spectrum, far less to green, and not to red and yellow. According to Sutton's notes, exposure time for the red filter was eight minutes, twelve minutes for the green; the blue exposure took six seconds. For later photographers and photo historians, how any image showed up for the red was a mystery until 1961, when Ralph Evans, an employee at the Kodak Research Laboratory's Color Technology Division, decided to recreate the experiment.

Evans discovered that Sutton's red filter had also transmitted ultraviolet light—invisible to the human eye, but not to the plate—as did the red dye from the tartan ribbon. The green had also transmitted a small amount of blue. Evans described the presence of the ultraviolet light as a "happy acci-

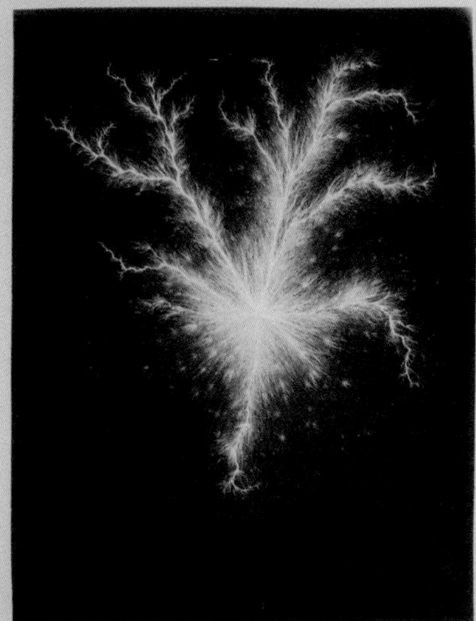

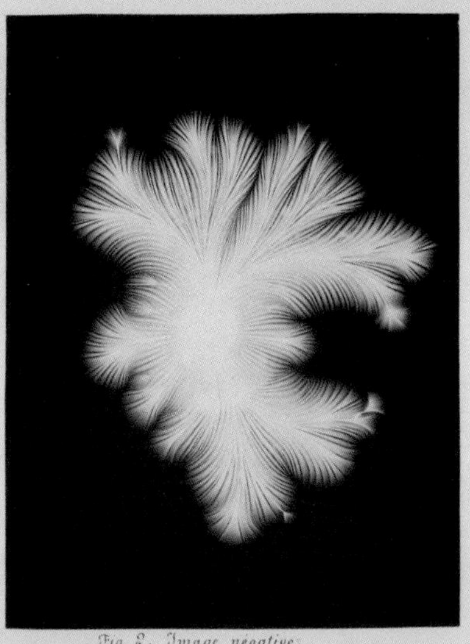

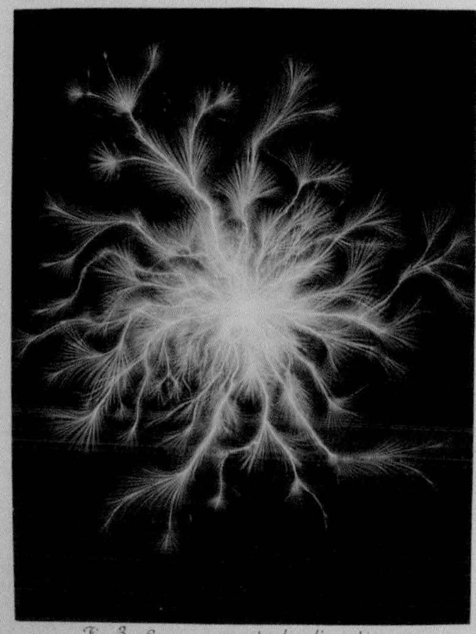

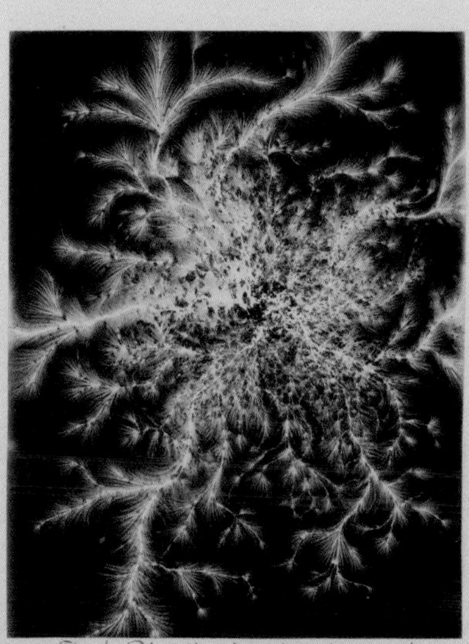

An explosion of light: Étienne Trouvelot's electric sparks, circa 1885. Trouvelot likened their appearance to meteors, fireworks, and the curved branches of a palm tree.

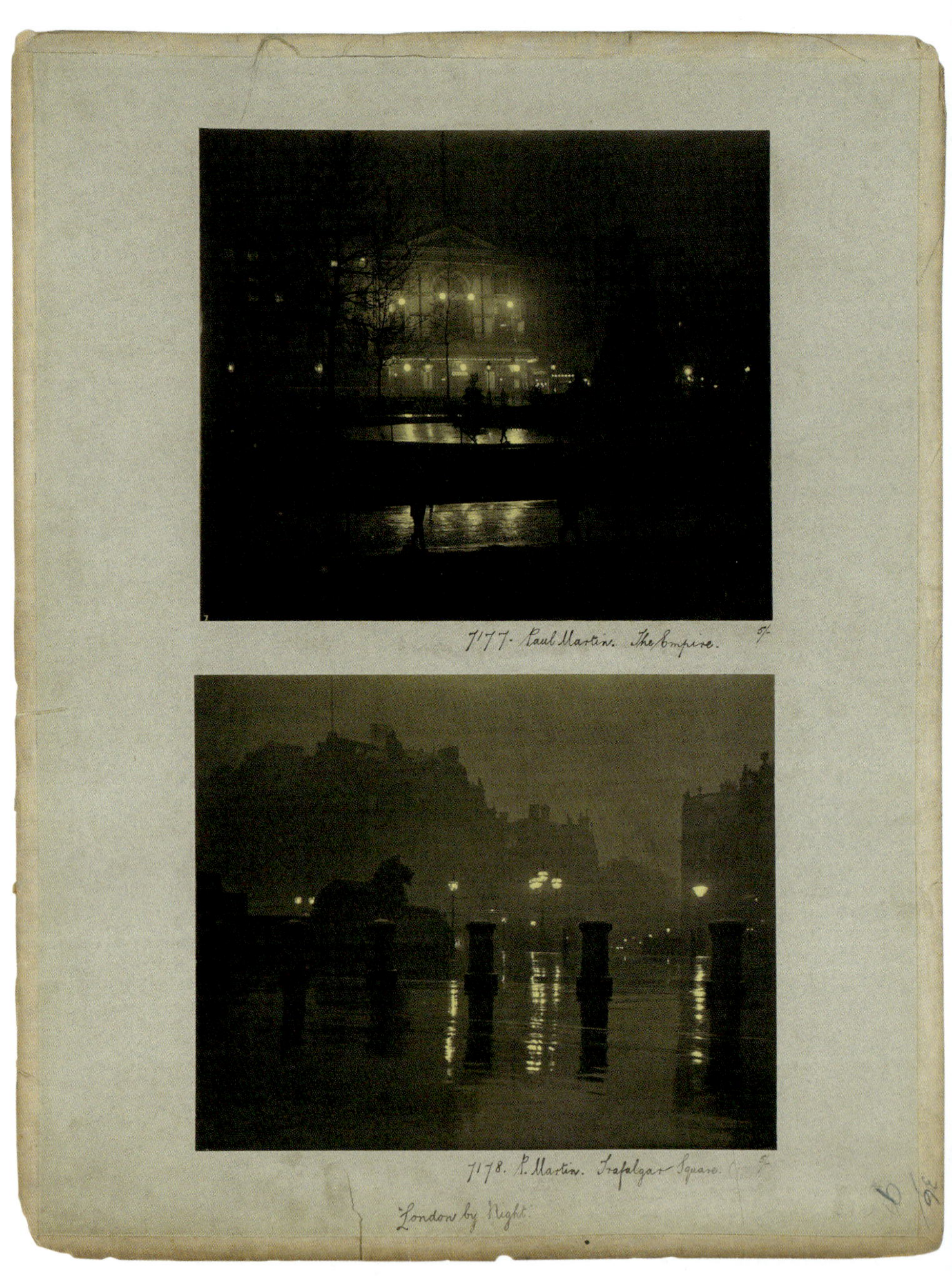

The moody nighttime scenes of the Empire Theatre and Trafalgar Square, photographed by Paul Martin, 1896.

daguerreotype of the moon by John Adams Whip-
e, 1852, similar to the one Warren De La Rue
uld have gazed upon at the Great Exhibition.

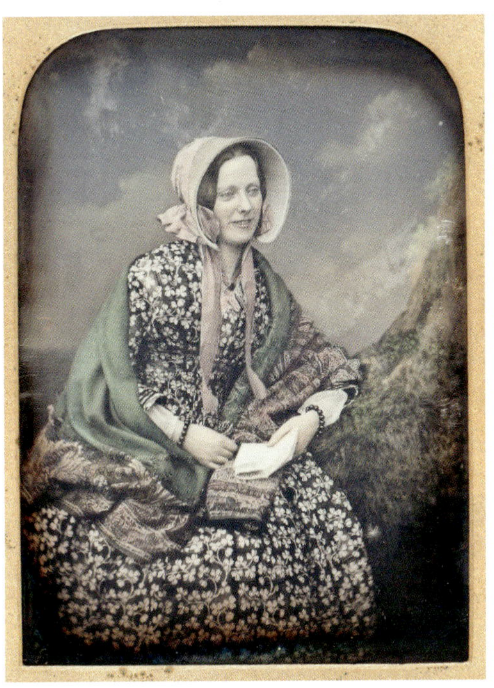

A beautifully tinted daguerreotype from Beard's
Photographic Institutions of London and
Liverpool, circa 1852–54.

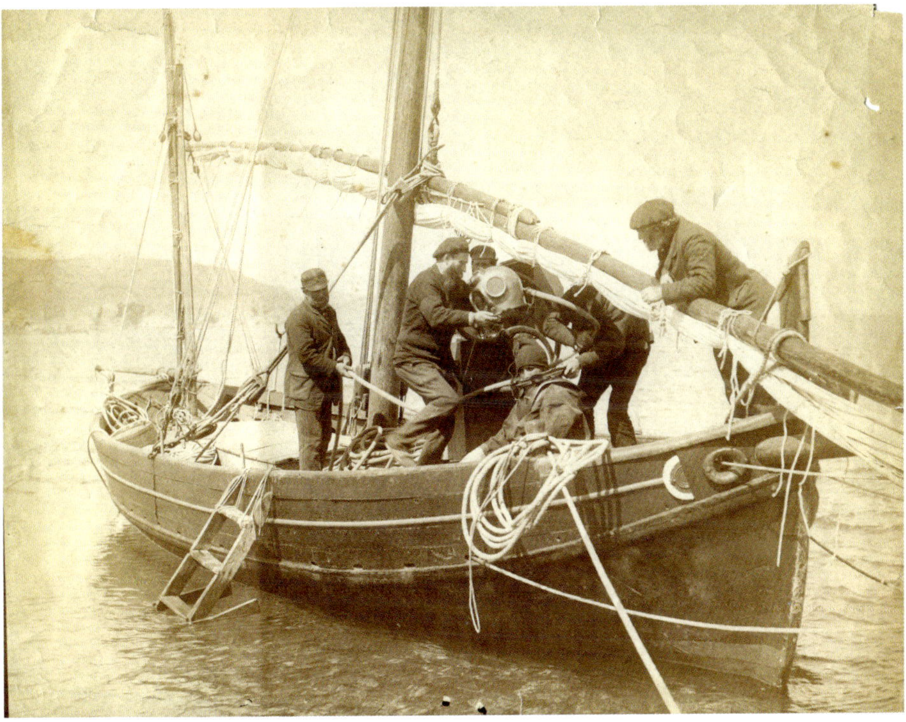

Joseph David prepares to enter the water around Banyuls-sur-Mer. The average weight of
a helmet diving suit was some 170 pounds.

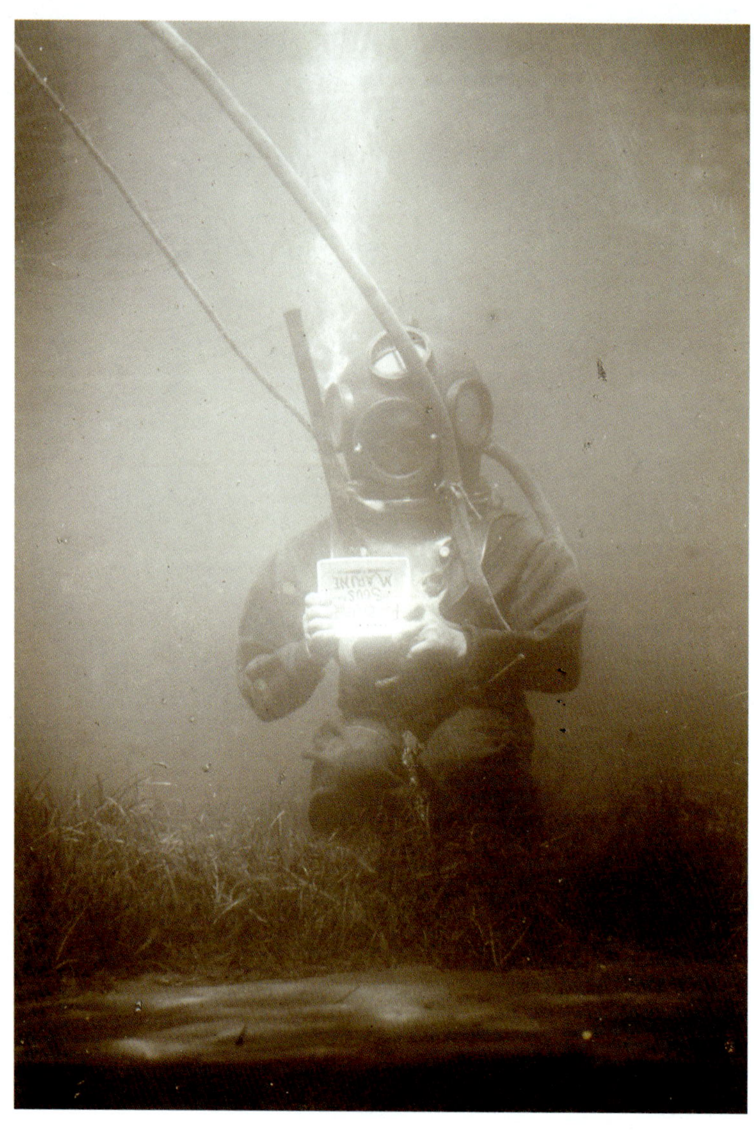

An underwater portrait of Joseph David captured at around 11 am on September 22, 1898, holding an upside-down sign. Louis Boutan operated the camera from the boat above.

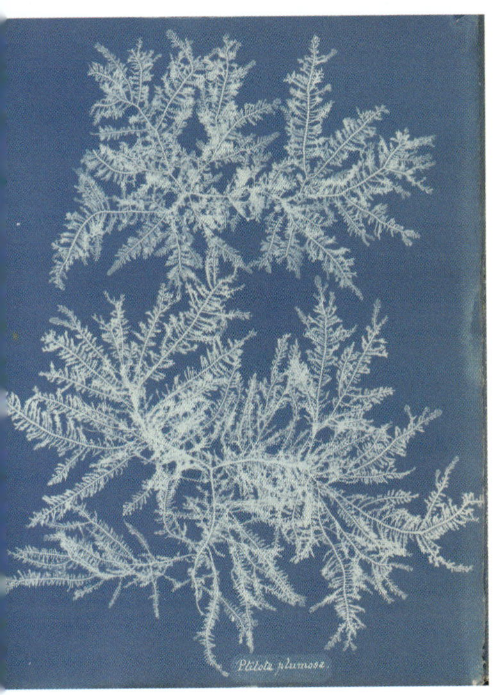

Ptilota plumosa, *rendered in cyanotype by Anna Atkins.*

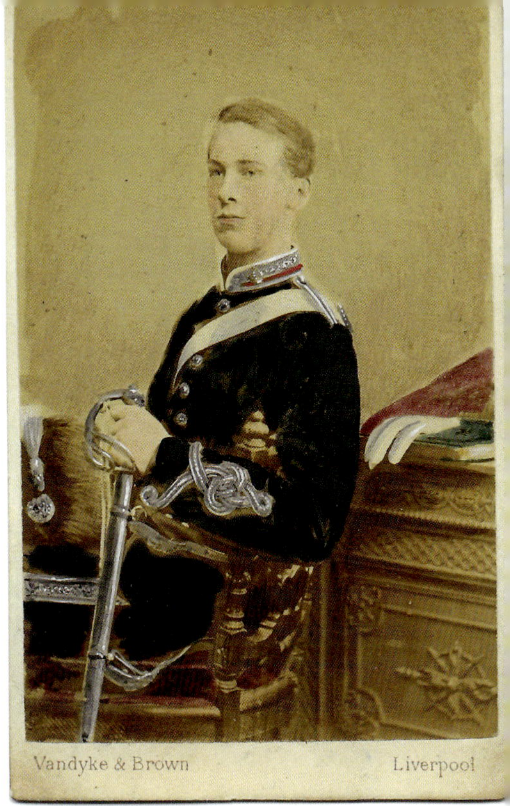

A photo with painted-on color by "Miss Bond of Southsea."

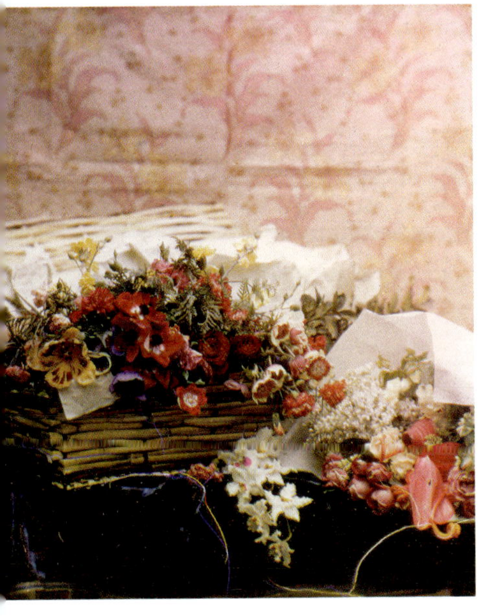

n autochrome by the Lumière brothers. pon seeing autochromes for the first time, e writer marveled, "The spectator was met such a wealth of color in these pictures at one could scarce find words to express his dmiration."

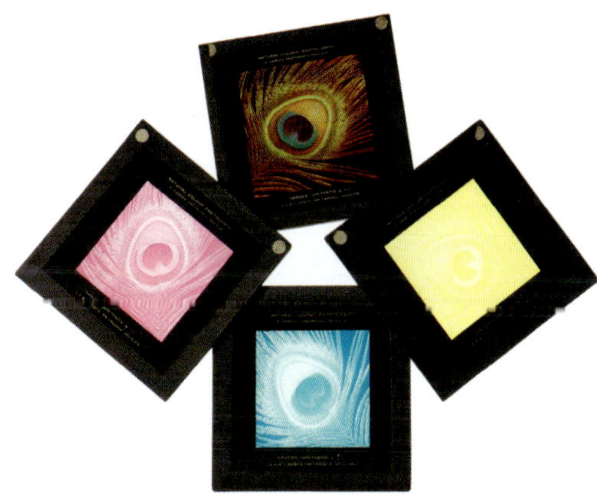

Slides that illustrate the Sanger Shepherd process, circa 1905.

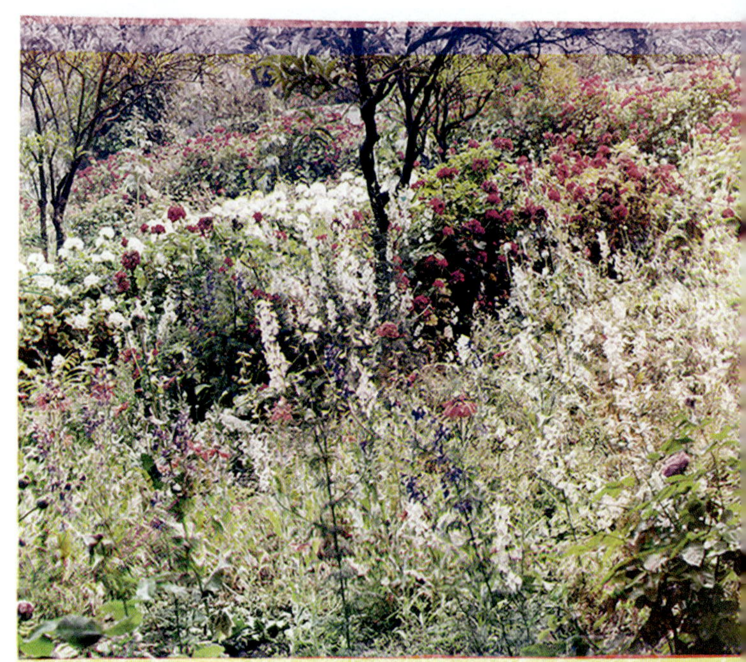

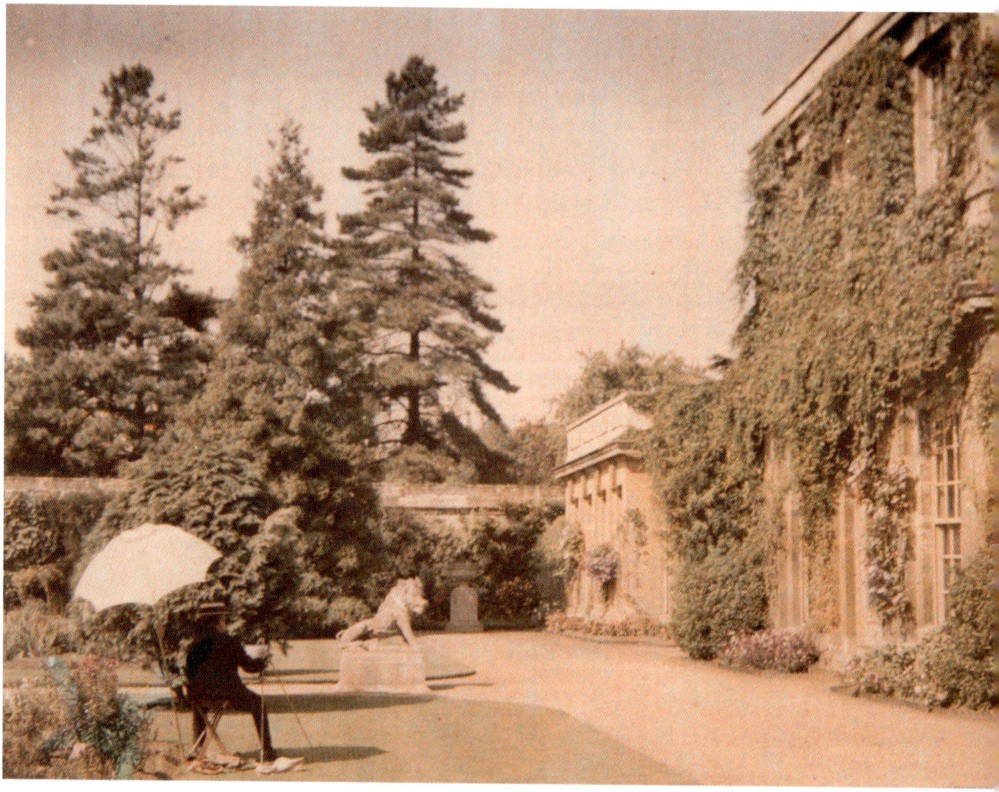

An autochrome of the Botanic Garden, Oxford, by Sarah Angelina Acland, circa 1908.

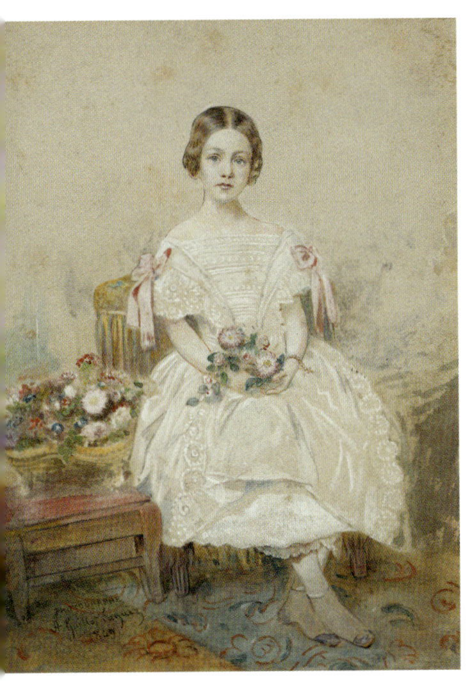

...rtrait of a Girl, a salted paper print with ...ded color, by Amélie Guillot-Saguez, 1849.

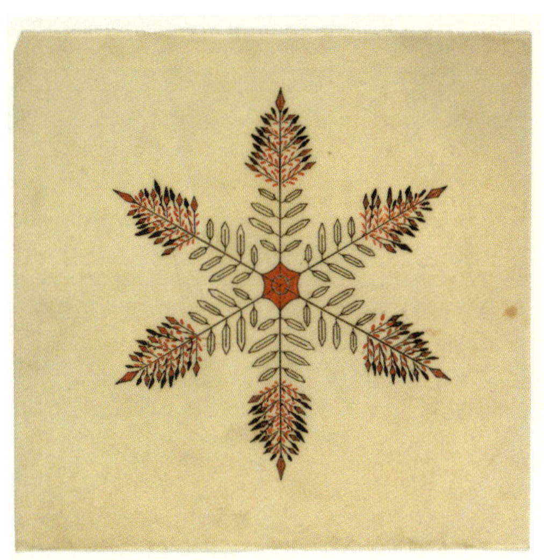

A snow crystal, illustrated by Cecilia Glaisher, mid-1850s.

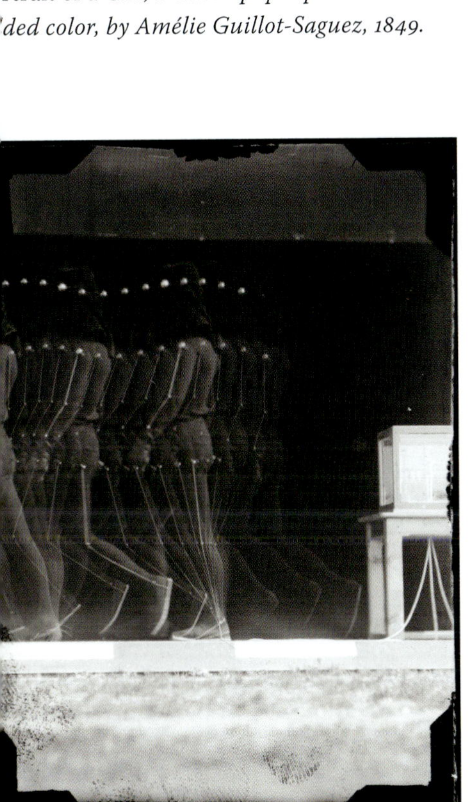

...chronophotograph of Étienne-Jules Marey's ...an in black, circa 1884.

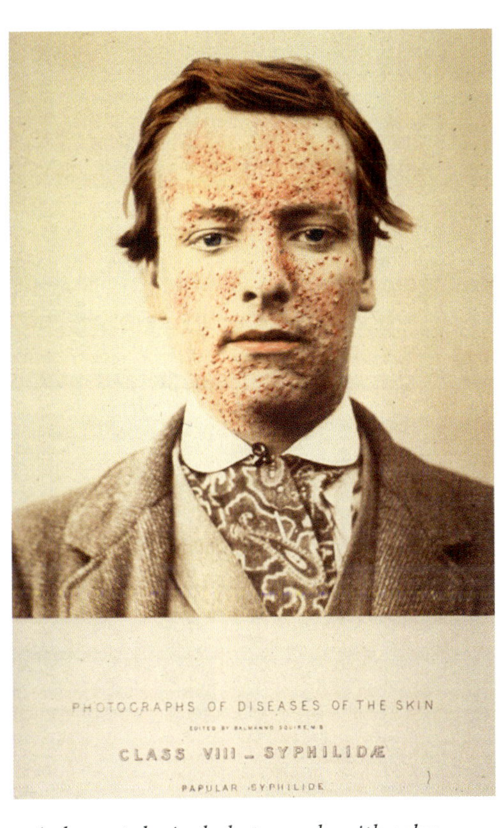

A dermatological photograph, with color added, of a patient with syphilis, from A. J. Balmanno Squire's book, mid-1860s.

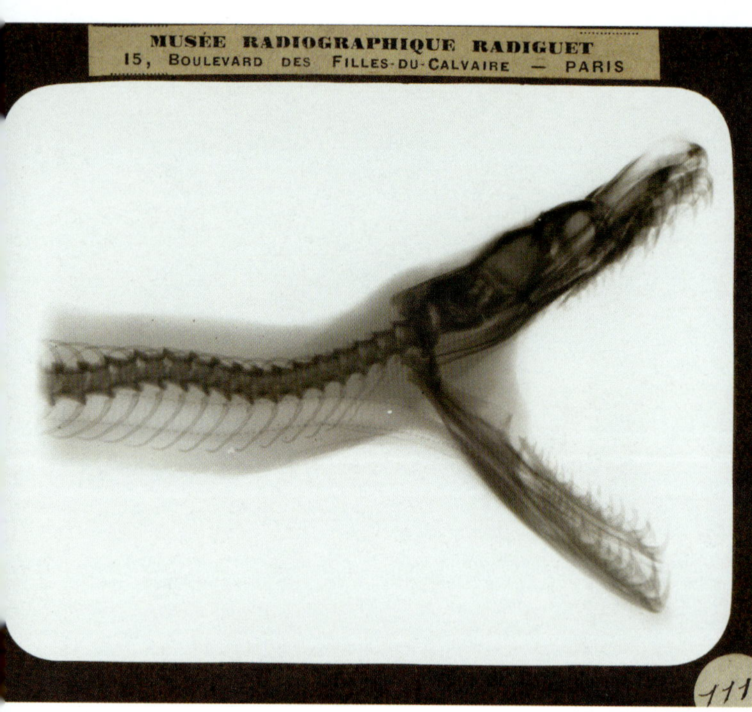

Arthur Radiguet, X-Ray o[f]
Snake Head in Profile wit[h]
Mouth Open, *1896–98.*
Like many other early X-r[ay]
practitioners, Radiguet d[ied]
from radiation exposure.

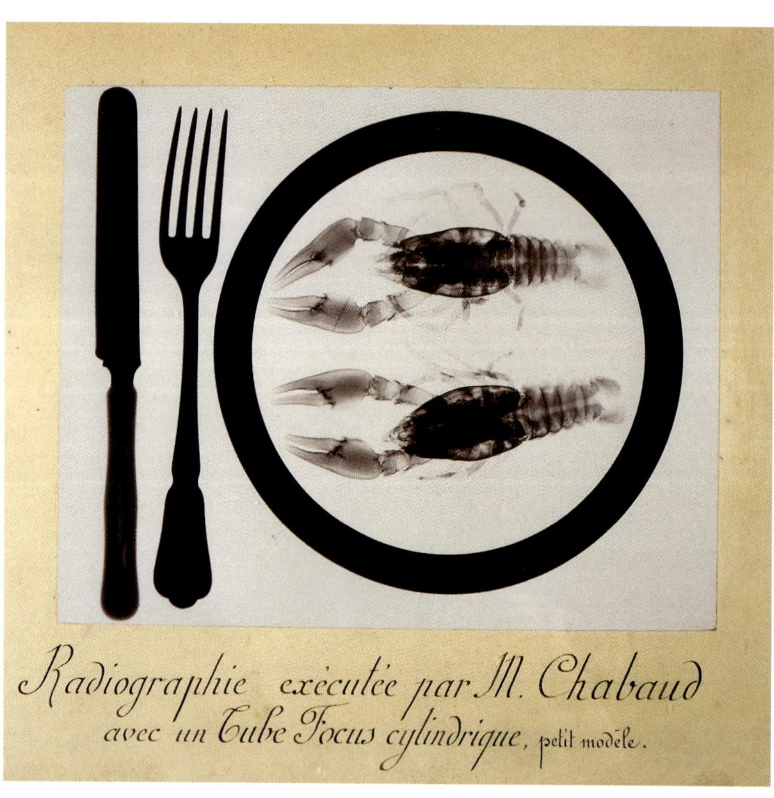

Victor Chabaud's weird and wonderful X-ray of crayfish on a dinner plate,
circa 1898.

dent," and it's true: without it, the experiment would have failed. Maxwell, with Sutton's assistance and a smidgen of good luck, had proved the theory of trichromatic color and, in doing so, provided the basis for color photography.

Later the same decade, Louis Ducos du Hauron and Charles Cros proposed three-color photographic theories, but these didn't result in an immediate practical color photography process either. Before that could happen, the photo emulsions needed to have a broader sensitivity. Thanks to experiments by Hermann Wilhelm Vogel, in the 1880s photographic plates began to be produced that were sensitive to both blue and green. They were known as "isochromatic" or "orthochromatic" plates. "Panchromatic" plates—sensitive to the full spectrum of colors, including red—were first made commercially available in 1906, right in time for the Lumière brothers.

On February 27, 1899, Sarah Angelina Acland gave her first public lecture to members of the Oxford Camera Club. She had taken up photography in 1891 when she was given a camera for her forty-second birthday, and had since focused on domestic interiors, regarded as an eminently suitable subject for the Victorian woman photographer, and portraiture. Her sitters were high-profile visitors to her family home near Oxford University, where her father held the position of Regius Professor of Medicine. Like Atkins before her, Acland was wealthy and didn't need to work; she pursued photography as a dedicated and highly proficient amateur. She was also her father's caregiver after her mother died; Acland herself had a leg condition since childhood that affected her mobility.

Acland had joined the Oxford Camera Club in 1894 as its first woman member. The subject of her first lecture was "home portraiture," an influential subcategory of portraits taken within the less formal confines of home. The *British Journal of Photography* reported that Acland's "beautiful portraits" had made the members "anxious to learn her methods, and there was a large attendance." Acland's experience in her subject was clear; she advised her audience to get "rid of all ideas of colour" and think, instead, about the "distribution of light and shade." The meeting closed with a presentation of a color portrait made by Frederic Ives, one of the several color photography innovators of the 1890s.

When another color process became available around 1900, called the "Sanger Shepherd Process of Natural Colour Photography," Acland grabbed the opportunity to work with color. She also had the means and the time to work with this new method—and it was time-consuming. For this process, the photographer had to capture the same image with three separate color filters (red, green, and blue); print them as positives (with cyan, magenta, and yellow, on either glass or celluloid); and then combine them with Canada balsam (a kind of glue that Dagron had also used to adhere his microscopic photos into Stanhopes). At first, it took Acland two weeks to produce a single image.

But by 1902, Acland was exhibiting her color work at that year's UK Photographic Convention. During the president's address on color photography, photographer and editor of *The Photogram* Catharine Weed Barnes Ward noted, "I am proud to say that several of the best slides were by a lady, Miss Acland, of Oxford."

Acland then commenced what is now a genre unto itself: color travel photography. In 1903 and 1904, she traveled to Gibraltar, where she used the Sanger Shepherd process to capture sun-drenched landscapes. One extant slide has deteriorated with age but is nevertheless glorious: a mass of wildflowers, rendered in shades of purple, blush pink, magenta, lime green, and sage. She presented her work to the Oxford Camera Club and on four separate occasions to the Royal Photographic Society in London. At one talk in 1905, she said that color slides from a trip showed "a more vivid and correct idea of the places described, than could be the case if only slides in monochrome were exhibited."

Praise for her work was effusive: she was described by *Photography* as being the "most untiring and successful exponent" in three-color photography work, and again by Catharine Weed Barnes Ward, who said her work was "quite the best I have ever seen." Acland began to work with the autochrome almost as soon as it became available, and produced more color photographs from a trip abroad to Madeira, which she had made for health reasons.

In his monograph on Acland, photo curator Giles Hudson argues that the significance of her work with the Sanger Shepherd process "is difficult to overstate. Her slides made her a household name among photographers, created an appetite for colour in the lecture-going public, and remained a point of reference for years to come." Unfortunately, much of Acland's work in the Sanger Shepherd process was accidentally destroyed by one of the lantern presentations at the Royal Photographic Society, but her work with autochromes, along with some other color processes, has been conserved.

Autochromes, as with other early color processes, suffer with age. Almost every component of an autochrome is liable to deteriorate: the glass cracks, the dyes can fade and change, and the tape can seals break and let in humidity, which damages the emulsion. In 2011, when the New York Metropolitan Museum of Art wanted to exhibit five autochromes by Alfred Stieglitz and Edward Steichen, it did so by using a special low-oxygen display. (Even then, the autochromes were only exhibited for less than week, and were then replaced with copies.)

Despite the problems with conservation, early color photography is evidence of the years of effort and experimentation to depict the world beyond monochrome. And autochromes in particular allowed photographers to capture what had been there all along, and what could now, at long last, be seen.

Chapter 8

Caught!

For several decades, photographers endured the woes of wet collodion. They rushed to expose and develop their photos while the plates were still wet. They handled chemicals that smelled and stained their fingers and that, from time to time, exploded. The few attempts to extend the collodion's sensitivity resulted in much longer exposure times. But in the late 1870s, after almost three decades of wet collodion, a new process appeared that set off an utterly seismic shift in photographic practices. This faster and more convenient process, which used gelatin rather than collodion, changed more than just exposure times and darkroom practices: it changed camera design, and with it, what—and who—was photographed. And it was all kick-started by a letter published in the *British Journal of Photography* on September 8, 1871, from an amateur photographer and physician named Richard Leach Maddox.

His letter detailed an experiment with gelatin that he hoped "may perhaps not be uninteresting" to readers, "though little more can be stated than the result of somewhat careless experiments tried at first on an exceedingly dull afternoon." It's fortunate for the history of photography that this is how he chose to fill a boring afternoon. Maddox's experiment set in motion the process for creating a dry gelatin photographic plate, one that didn't require carefully timed on-the-spot preparation. Maddox himself didn't pursue the process further, but, over the course of the 1870s, other photographers did. By the end of the decade, after nearly thirty years of wet collodion, photographers could buy ready-to-go dry plates.

The adoption of dry plates was swift and decisive. In an eighteen-month period between the spring of 1878 and late in 1879, the number of dry plate manufacturers in Great Britain jumped from four to fourteen. In 1879, the editor of the *British Journal of Photography* wrote, with no overstatement, that the year would be regarded "as one of the most noteworthy epochs in the history of photography." Dry plates weren't just a new industry; they were the start of a whole a new era.

Exposure times were less than a second. Tripods were no longer a necessity, and neither were portable darkrooms to prepare plates. This drove changes in camera design, including the production of smaller, easy-to-use handheld cameras, known as hand cameras. (Within this category was the detective camera, the hand camera mentioned earlier that was designed to look as unobtrusive as possible.) Cumulatively, these changes were the difference between climbing six flights of stairs on a hot day to reach an apartment and being whisked up in an air-conditioned elevator.

All of this gave photography mass appeal. "The noble army of amateurs has already assumed colossal proportions and new recruits are being added daily to its ranks," reported the *British Journal Photographic Almanac*, in 1885. "The amateurs are insatiable and ubiquitous," reported the *San Francisco Chronicle*. A firsthand account from a freshly minted amateur photographer in *The Photographic News* confirmed the appeal of the new process: "The ease with which photographs can be taken by the modern dry plate process leads the tyro to roam about always on the look-out for whomsoever he may focus."

Catering to the swelling masses of eager amateur photographers, *The Amateur Photographer* began weekly publication in 1884. Its remit was clear: "Above all things, the *popularity* of the journal will be its leading feature. No dry, scientific details will here be set down, but our pages will ever brim over with information put in the clearest and most readable form."* *The Amateur Photographer* was filled with advice, letters, photo competi-

---

* *Amateur Photographer* is still published today in print and online.

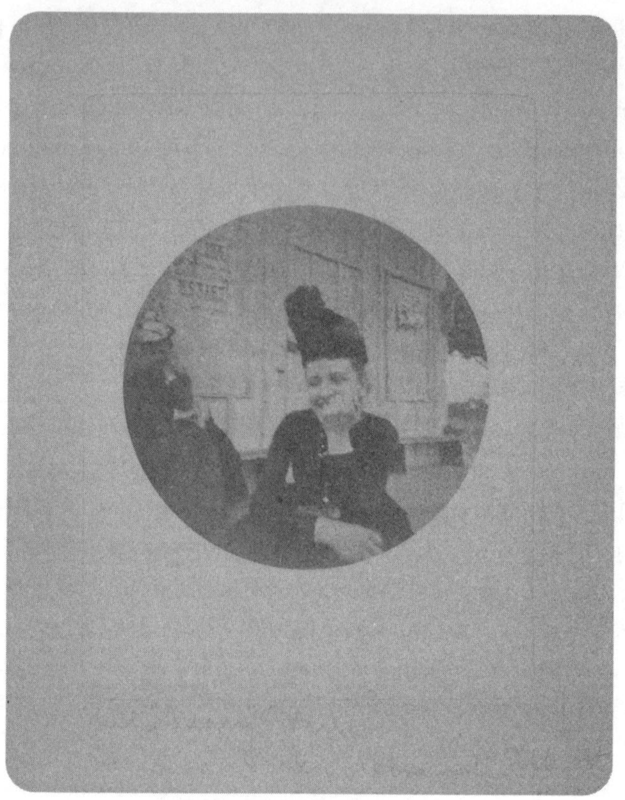

The technological developments in the 1880s made candid photographs like this a reality—and a revelation. Miss "Mary" Winn, Laughing, *1888.*

tions, reviews of photographic manuals, equipment sales, and accounts of snap-happy photo jaunts through Derbyshire or Wales or Normandy.

With this influx of photographers, amateur societies sprung up with abandon: between 1880 and 1885, the number of photographic clubs in Britain surged from fourteen to forty; by 1895, that number reached 250. Amateur photographic societies proliferated in the United States too, jumping from 20 in 1885 to 109 in 1895. The enthusiasm was such that in 1893 *The Amateur Photographer* politely suggested that smaller clubs should combine with the larger ones, given the "enormous increase" in amateur clubs in limited areas.

Not all amateur photographers were content to just snap away with the new, simpler tools. George Eastman, a bank clerk and amateur photographer living in Rochester, New York, first read about dry plates in the *British Journal of Photography*. He was not short on ambition or ingenuity, but he was short on funding: in the early years, he kept his job and used his mother's kitchen stove to experiment. Eventually, he rented a space, found some startup capital, quit his job, and devoted himself to his new company, the Eastman Dry Plate and Film Company. In addition to producing his own dry plates, he devised a machine to coat dry plates, and experimented with roll film and roll holders. He also hired a chemist, Henry

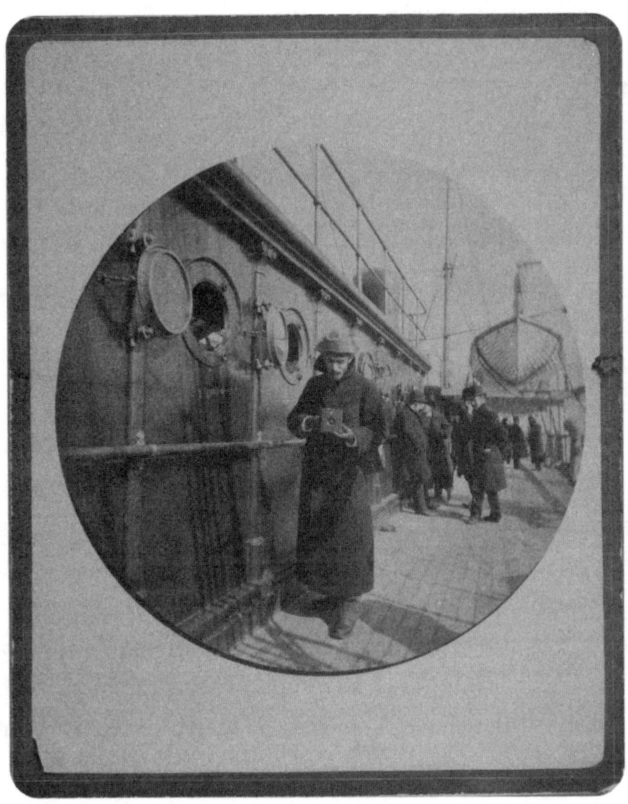

George Eastman holding one of his cameras on board the SS Gallia, *February 1890. The photo was taken with a Kodak No. 2 camera.*

Reichenbach, to find a new non-paper material for film backing, which resulted in celluloid roll film. But most crucially for the history of amateur photography, in 1888 Eastman introduced a new camera, the No. 1 Kodak.

The camera was compact, 3¼ by 3¾ by 6½ inches, and cost $25, at a time when an annual subscription to a photo journal was $3. It arrived preloaded with 100 exposures on a roll of paper-backed circular negatives, about 2½ inches in diameter. Once the film had been used up, the photographer sent the camera back to Rochester to be developed. The prints, along with a reloaded camera, were returned to the photographer for a $10 charge. Thanks to Eastman, for the first time in history people could send their film away to be processed.

Eastman reportedly came up with the name "Kodak" while playing anagrams with his mother; it turned into a brand name so recognizable that it has become synonymous with photography. With it came one of the most innovative and enduring slogans in the history of advertising: "You press the button, we do the rest." In 1892, Eastman renamed the company Eastman Kodak; the following year, he launched the long-running "Kodak Girl" advertising campaign, which depicted women as the photographers of their families and their adventures, ably assisted by their trusty hand-held Kodak cameras.

By 1898, Kodak had reportedly sold a staggering 1.5 million cameras. Five years later, with even more cameras on the market, the company was making some $3 million in profits. With this enormous financial success, Eastman donated more than $100 million to various groups and charities in his lifetime. He also built himself a 37-room mansion in Rochester; in its conservatory he installed a pipe organ and had musicians play it to him as he ate his breakfast.

One Kodak owner lived in Eastman's home state, and appeared, at first glance, as just the type of snapshooter who might grace Eastman's "Kodak Girl" advertisements. She liked to travel, play tennis, and attend parties. But she owned several cameras, and photographed constantly and methodically. In so doing, she created an extraordinarily valuable archive, and one that was nearly lost.

Alice Austen became fascinated by photography after an uncle brought home a camera in 1876, when she was about ten years old, and showed her the basics. She was an amateur photographer in the sense that she did not need to work, but her dedication and her skill set her apart. For most of her life, Austen lived in the comfort of her family home on Staten Island overlooking the Narrows. From here, she photographed passing ships, and her family on the wisteria-shrouded porch.

She also photographed her women friends in candid scenes that were humorous and intimate: dressed up as men, faking drunkenness, pretending to smoke cigarettes in their undergarments. One quieter but still striking photo is an interior of a friend's bedroom: as Austen historian Ann Novotny pointed out in her book *Alice's World*, this was a space that was inaccessible to a male photographer. Austen's photographs reveal a rarely visualized perspective by and of nineteenth-century women: one that shows spontaneity, playfulness, affection, friendship, and love.

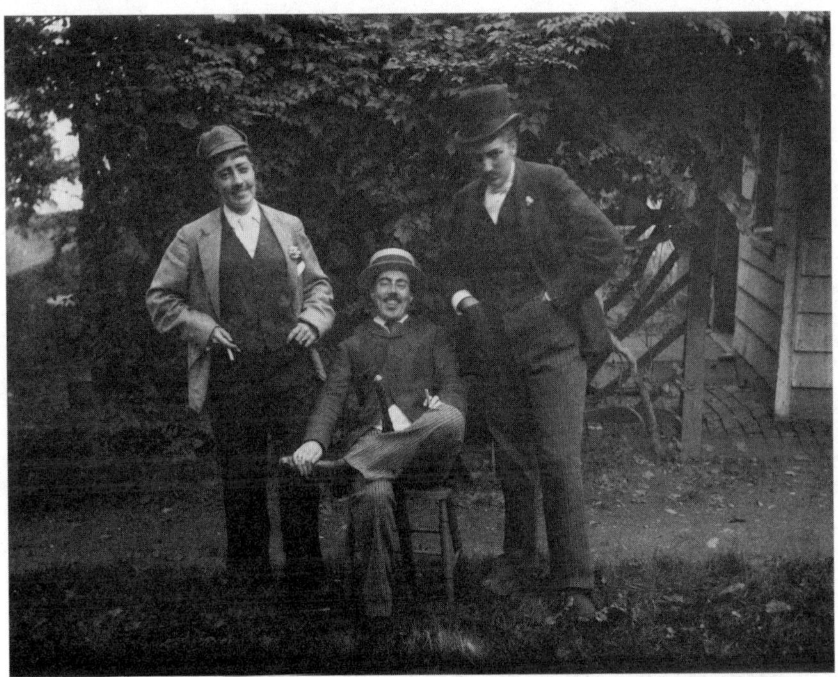

Alice Austen, left, with Julia Martin and Julia Bredt, in men's clothes, with a strategically placed umbrella handle, October 1891.

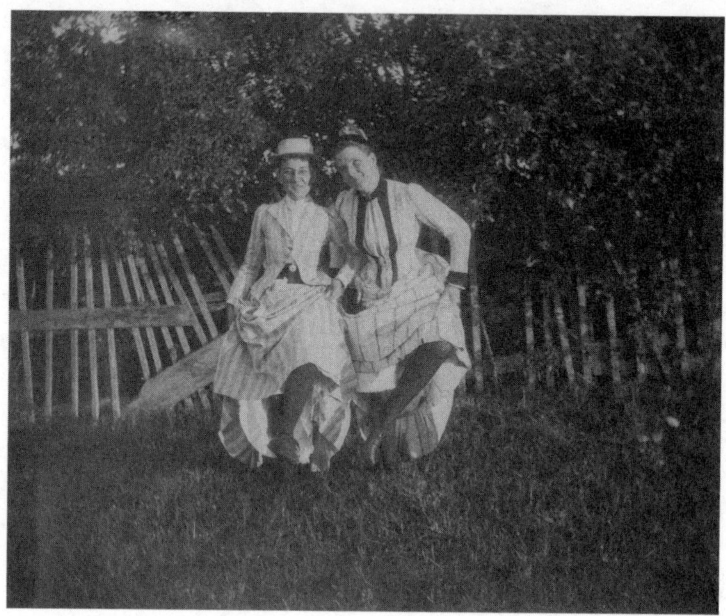

Miss Sanford & Mrs. Snively, *photographed by Alice Austen in Vermont, August 1890.*

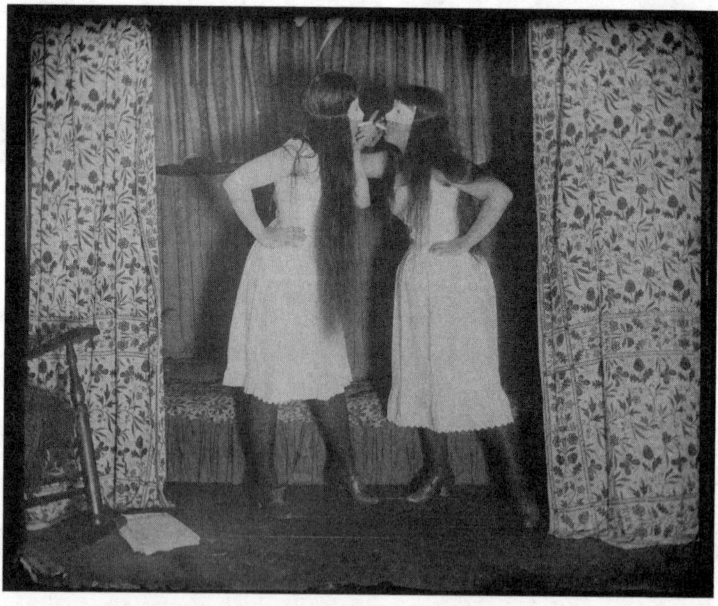

Alice Austen and Trude Eccleston, in masks and pretending to smoke cigarettes, 1891.

For a book on cycling, written by her friend Violet Ward, Austen took photographs of Daisy Elliott, and the two women became romantically involved. As their relationship wound down, Austen took a trip to the Catskills, where she met Gertrude Tate. Austen took photos of this trip too, of course, and inscribed an album as a gift to Tate. It was the start of a loving relationship that lasted fifty-six years. They traveled together and, from 1917, lived together, when Tate moved into the Austen home. (Tate's family disapproved of what they termed her "wrong devotion" to Austen.) Today, Austen's family home is a museum dedicated to her life and work, and a designated LGBTQ+ history site.

In the mid-1890s, Austen took her bicycle and camera on the ferry from Staten Island to Manhattan and photographed street scenes that she copyrighted in an album called *Street Types of New York.* Just like New York street photography of today, Austen's focus is not on architecture or infrastructure but on the people she encountered. There's no account from Austen of her experiences as a lone woman photographer on the streets of New York, but there is from Catharine Weed Barnes Ward, the same photographer who had earlier praised Acland. During a lecture in 1905, Ward said, "Men can, as a rule, do street work more easily than women, for although the camera is a common sight in all large cities, and little attention is paid to a man, if a woman stops to photograph, passers-by more often than not stop also, generally where they block the view, and as a lady hates to make a scene the chance is often lost."

If Austen encountered a similar experience—and her subjects are nearly exclusively looking at the camera, aware of her presence—her portraits still feel spontaneous. This project was not social documentary photography, like Jacob Riis's; it was early street photography. Austen's photos have immediacy and atmosphere. They draw you into their subjects. They also tell you something about the person behind the camera, someone who was bold and decisive, who approached strangers on the streets of Manhattan to take their portrait. And Austen was methodical, jotting down the details of her exposures on her glass plate envelopes: brand of plate, lens, exposure time, focal distance, light conditions, subject, and date.

Of an estimated 8000 images, around 5000 have survived, but they too

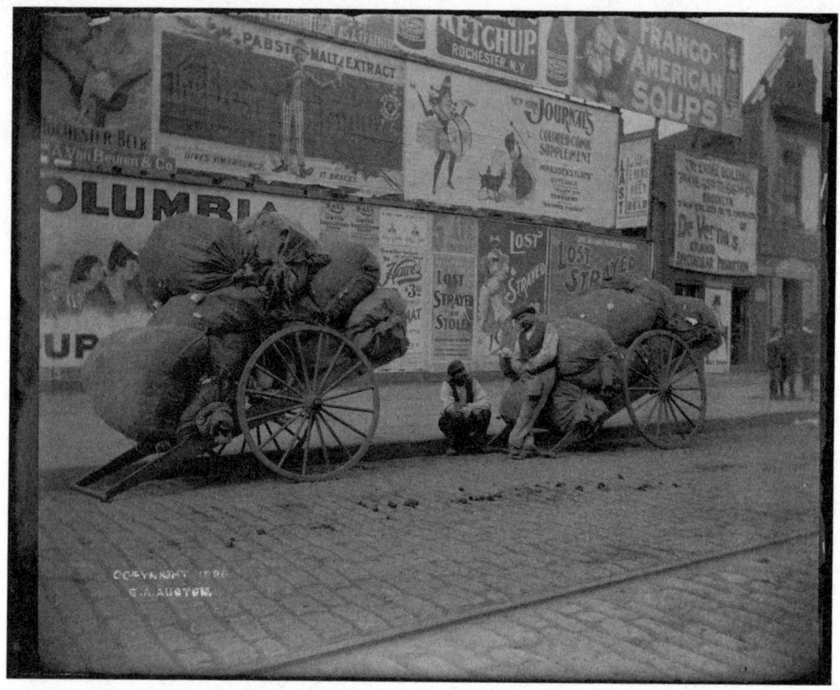

One of Austen's images from her trips into Manhattan, taken in October 1896. Austen copyrighted a selection of these photos, and created an album she called Street Types of New York.

were nearly lost.* In the mid-1940s, Austen and Tate were forced to leave their Staten Island home after more than a decade of financial difficulties stemming from the stock market crash of 1929. Some of Austen's negatives went to an auction dealer; some were lost; the rest ended up at the Staten Island Historical Society. They were rediscovered in 1951, when the writer Oliver Jensen contacted the Historical Society about work by women photographers. By then, Austen was destitute and living at the Staten Island Farm Colony; Tate was with her sister in Brooklyn.

Jensen, with Tate's help, licensed some of Austen's photos to *Life* magazine, which ran an eight-page photo essay, headlined "The Newly Discovered Picture World of Alice Austen: Great Woman Photographer Steps

---

* Austen also reportedly experimented with autochromes, but none have yet been found.

Out of the Past." Jensen licensed photographs to other publications and made enough money to transfer Austen into a private nursing home. Austen died the following year, age eighty-six; Tate survived her by another ten years. Despite their wishes, their families didn't bury them together.

For its essay, *Life* sent Alfred Eisenstaedt to take Austen's portrait. He photographed her in her wheelchair with his Leica camera. At some point during their meeting, she borrowed it and turned the lens on Eisenstaedt. It's fitting that this must have been the last camera she ever held; the small, discreet Leica had revolutionized photojournalism just as the hand camera had so profoundly altered amateur photography. These two worlds momentarily collided as Austen snapped that candid portrait.

As amateur clubs proliferated, so too did a new kind of amateur photographer, one who committed far worse sins than just poorly composed snapshots of trips to the Welsh seaside. This category of amateur even had its own moniker, which described any obnoxious, overeager photographer who possessed few skills and even fewer manners: the camera fiend.

"This fiend in human shape delights in portraying his acquaintances in all sorts of ridiculous situations," griped the *British Journal Photographic Almanac* in 1885, as it described the "colossal" army of amateur photographic types. For the fiend, nothing was off-limits. One report complained about a camera fiend stopping a fire truck, mid-emergency, to take a photo. In an anonymously written article from 1893 titled "Confessions of a Snap-Shot Fiend," the author proudly described his candid photos of Otto von Bismarck, the former German chancellor, tying his shoe, and one that showed a duchess chewing on a large piece of pear.

The camera fiend's prying photographs were regarded as unwelcome, unflattering, and embarrassing. Amateur photography was embraced by both men and women, but complaints about the male fiend's activities around women were numerous. "Many estimable young women gave up bathing in the surf for fear their lower limbs would adorn somebody's photograph album," reported *The American Journal of Photography*. The *British Journal of Photography* said that the conduct of amateur photographers seeking a seaside snapshot "is highly reprehensible, and tends to bring pho-

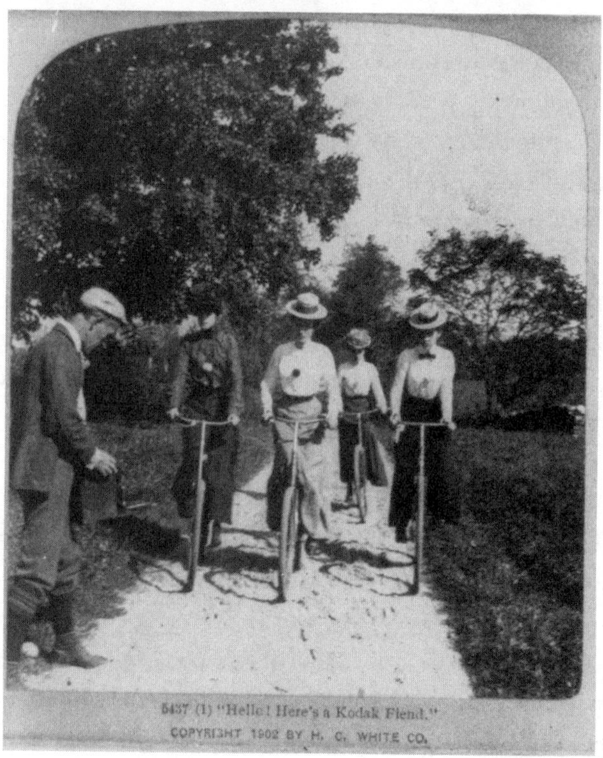

A camera fiend at work. One journal described the fiend as someone who "delights in portraying his acquaintances in all sorts of ridiculous situations."

tography into contempt," adding that women "have a very strong objection to being photographed, surreptitiously, when they are in a state of *deshabille*." Another writer reported, in 1889, a "brotherhood of Peeping Toms" in the Berkshires who waited for women to climb out of their vehicle and snap "a glimpse of feminine draperies and hose and record it for future use."

Some fiends used their photos for genuinely devilish purposes. The scheme was simple: photograph someone—either a woman or a man and a woman together—in a pose that was either embarrassing or suggestive, then make her buy the negatives. In 1891 in New York City, a camera fiend was "haunting" Fifth Avenue, "making snap shots of every pretty woman he meets. Then in a few days he appears at the house of the lady and offers to sell the photograph." Later the same year, *The Photographic Times and*

American Photographer warned of an amateur "doing a blackmailing business among certain Philadelphia society belles at the shore."

The rich and famous were also popular targets for the camera fiend. In 1887, a photographer captured an unprecedented candid photograph of Queen Victoria laughing in her carriage during her golden jubilee celebrations, which he published and displayed in a London shop window. The *British Journal of Photography* commiserated with Her Majesty for the "unfortunate incident," although the *Edinburgh Evening News* was decidedly less empathetic: "it is said to be the best portrait that has been taken of the Queen since 1861." In the United States, Newport society was plagued by camera fiends, resulting in one incident where William K. Vanderbilt II assaulted a photographer and grabbed his camera. In 1901, President Roosevelt blocked off public access to the rear grounds of the White House because of "ill-mannered camera carriers."

If the fiend had an enabler, it was the detective camera. Hand cameras were easy to use and portable; a detective camera was all this, but in disguise. At its simplest, it looked like a plain box or parcel. "It requires no skill other than to get within range of focus of the unsuspecting victim," boasted an advertisement for Scovill Detective Cameras, from 1887. One tricky design was the "Deceptive Angle Graphic," a camera with fake stereograph lenses in the front, so the photographer could snap his subjects at a right angle while he was seemingly looking the other way. In the advertisement, a man was shown using the camera on two women in bathing costumes.

At the Henry Slater Detective Agency in London, the detective camera lived up to its name in the most literal sense: the agency advertised "instantaneous photography" services "for secretly securing photographs of persons when together or separately, for identification and corroboration." Similarly, in 1888, a newspaper reported—under the heading "Midnight Hunt for Divorce"—that a man had used a detective camera (and magnesium flash) to barge into a hotel and catch his wife having an affair.

Disguised cameras soon began to appear on the market with the sole purpose of taking someone's photo on the sly. These cameras were typically of poor quality, but highly inventive in how they were concealed. "Most Deceptive in Appearance," stated an advertisement for a camera disguised as a picnic basket. Other hidden camera disguises included a hat, tie pin,

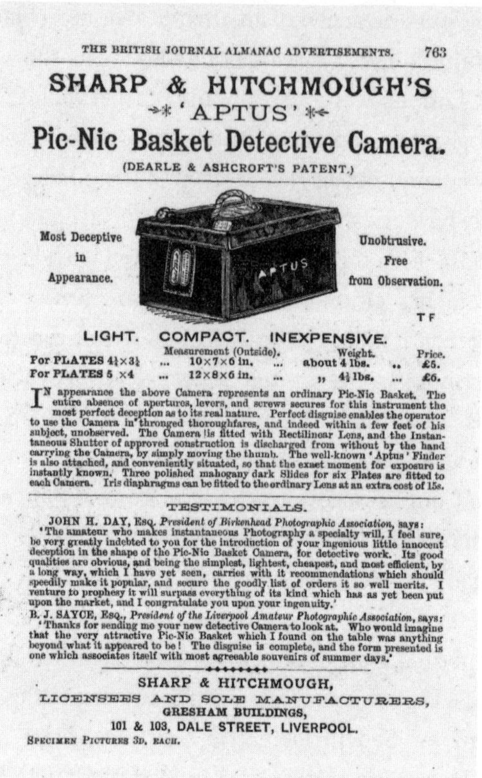

From 1889, the "most deceptive in appearance" picnic basket detective camera.

scarf, briefcase, book, watch, handbag, walking stick, and binoculars. One lawyer even had a bespoke camera made to resemble volumes of law books, possibly to capture all those fascinating candid photographs of client meetings. One extraordinary archive of secret photographs still survives in a museum in Norway, shot with a camera hidden behind a buttonhole.

Carl Størmer was a young university student in the 1890s when he began to take secret photographs of street scenes in Oslo. The camera, which was patented in 1886, was disk-shaped and worn around the neck on a chain and positioned under a waistcoat. The lens poked through a buttonhole. On each plate there were six circular negatives, which Størmer rotated with a small knob. When he pulled a concealed string, the camera made an exposure. Størmer had initially bought the camera to take a pho-

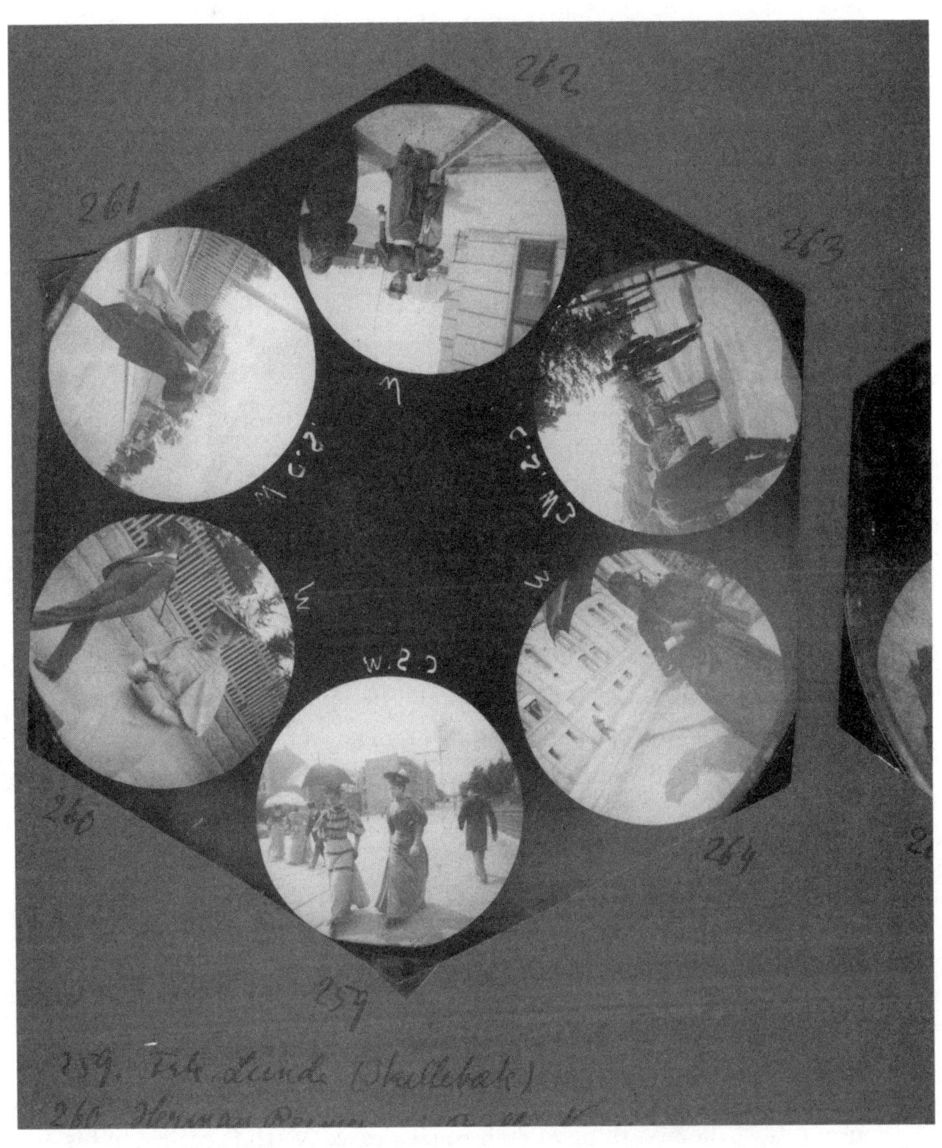

A contact sheet from Carl Størmer's buttonhole camera, 1890s.

tograph of a woman whom he had a crush on. (He also took photographs of women swimming.)

Størmer's archive of circular glass negatives is held at the Norwegian Museum of Science and Technology, along with some of his scientific photographs. (Størmer became a well-known mathematician and astrophysicist, and also took photos of the aurora borealis.) His son donated his buttonhole images nearly one hundred years after they were taken, and they are an important record of how people used concealed cameras. Størmer's hidden photographs have all the movement of the street: women pass by, sometimes slightly blurred, their skirts swaying, and men doff their hats to the face above the unseen, but all-seeing, lens. Each photo is circular, reminiscent of a spyhole, and just as furtive.

Paul Martin, who captured the moody, rain-slicked streets of London by gaslight, also secretly photographed London streets using his detective camera. It was disguised as a brown paper parcel, as though it was "one or two books of the same size neatly wrapped up by a shopkeeper," according to a review of the camera. To protect it from rain, Martin decided to give it a black leather case along with a few other modifications, and went out into London's streets. "It is impossible to describe the thrill which taking the first snaps without being noticed gave one, and the relief at not being followed about by urchins, who just as one is going to take a photograph stand right in front shouting, "Take me, guv'nor!," he later recalled.

Martin's goal was to show something new—not the aesthetic of "careful composition, superimposed clouds, soft printing" but, rather, "people and things as the man in the street sees them." His photos are about as close as we can get to a sensory slice of Victorian London. In one photo, a Billingsgate market fishmonger walks casually across cobblestones with a tall basket balanced on his head. Another photo shows a traffic accident on High Holborn, where a policeman helps a passenger down from a tipped-over hansom cab. "The first thing to do in such a case," Martin recalled in his 1939 book, "is to sit on the horse's head." Martin captured more: street vendors, a knife grinder, a fair in Battersea Park, a crowd waiting for a policeman's funeral, and yes, "urchins," three barefoot children sitting on a street curb. In Martin's photos, London street life—messy, unfiltered, vibrant—

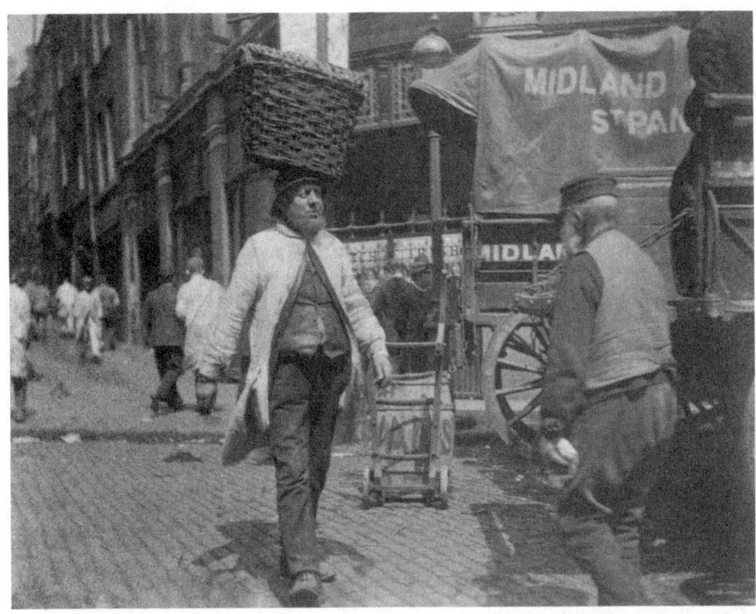

Paul Martin, Porter Carrying a Basket of Shrimps, Billingsgate, *1893.*

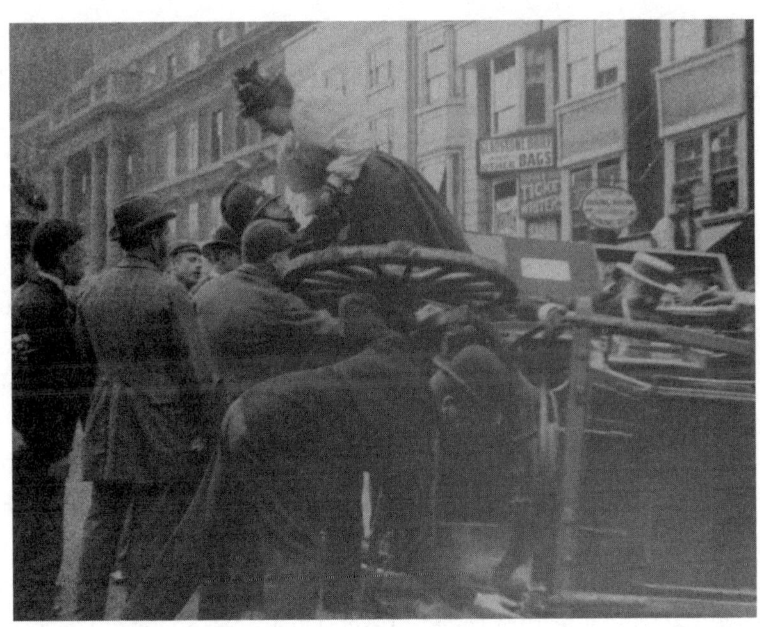

Paul Martin, Cab Accident, High Holborn, London, *1893–96.*

does not so much cry out as roars. It demands to be seen as a record of the ordinary yet extraordinary activities of people who once lived.

He also photographed public holiday festivities in Hampstead Heath with a Kodak provided to him by the Eastman company, although he preferred his own detective camera. (Eastman Kodak also offered Martin a job that he turned down, a decision he later regretted.) Martin took his camera outside of London, where he photographed beach scenes and seaside activities: children watching a Punch and Judy show; a crowd behind a barrier, listening to a concert. Martin also surreptitiously snapped couples canoodling in the sand, and women tentatively stepping into shallow, foaming water, their long skirts raised and billowing at mid-calf.

Had Martin been caught taking photos of women at the beach, he might have faced some retribution. In articles and letters in both British and American photographic journals, many lamented the camera fiend's activities and suggested various forms of punishment, particularly if the fiend was caught photographing women at the beach. They ranged from demanding that the photographer hand over negatives, to destroying the camera (with a brick, suggested one reader helpfully), to a "good horse whipping" of the fiend.

Regardless of the near-universal hatred of the camera fiend, the attitude of the amateur photographer journals ranged from oblivious to obliging. In 1893, for example, *The Amateur Photographer* encouraged its readers to go to Brighton Beach to photograph "fat men" for some "funny plates, or some laughter-moving lantern-slides." Some years earlier, the same journal had expressed regret that no one had photographed the actress Sarah Bernhardt falling down some stairs. Upon learning that some readers were deterred from entering competitions because their names would be published, the editors readily suggested pseudonyms. Eventually, *The Amateur Photographer* did publish guidelines for photographic etiquette, which included not using a camera "for a thunderingly good practical joke."

In the 1890s—as amateur photography and its devious brigade of camera fiends boomed—newspapers and magazines began to reproduce photographs in print using the halftone process. Candid snaps of well-known figures soon became photographic fodder. In 1899, *The Penny Pictorial Magazine* started a feature called "Taken Unawares: Snaps of Celebrities Taken Without Their Knowledge," essentially a nineteenth-century precursor to the twenty-first century's "stars without makeup" magazine editions. The August issue published the photograph of Queen Victoria smiling, the Belgian king standing awkwardly by the sea at Ostend, and the German emperor meeting some clerics. The page invited contributions from photographers to be sent to the "Snapshot Editor." Similarly, in 1910 *The Graphic* published "Unconventional Portraits of Celebrities," in which it praised the "snapshottist" who supplies a "constant round of pictures of the great taken unawares."

One place seemingly safe from a prying camera on the hunt for news was an English courtroom. But in October 1908, a press photographer named Arthur Barrett entered Bow Street Magistrates' Court with a camera squirreled away in his top hat. He'd cut a small hole for the lens and, once inside the courtroom, he placed the hat on a ledge and directed it toward the three highly newsworthy defendants in the dock. He later recalled how he coughed to hide the sound of the exposure, as he captured the faces of the alleged criminals Emmeline Pankhurst, her daughter Christabel Pankhurst, and Flora Drummond.

The three suffragists had been arrested for conduct likely to breach the peace after encouraging the public to "rush," or invade, the House of Commons at the opening session of Parliament. Christabel Pankhurst acted as the defense attorney. (Despite having a law degree from the University of Manchester, as a woman she could not practice law professionally.) The interest in the trial was such that "a stranger passing the police court earlier in the day might have mistaken it for a theatre at which a matinée performance was to be given, so large was the number of women waiting for the opening of the doors," reported *The Daily Mirror*.

Despite Christabel's questioning of Chancellor of the Exchequer David Lloyd George and Home Secretary Herbert Gladstone, whom she had audaciously called as witnesses, and her two-hour closing argument, the judge issued an order against the defendants. When they refused the order—to pay a fine and keep the peace for twelve months—the judge sentenced the three women to Holloway Prison (Christabel for ten weeks, Emmeline Pankhurst and Drummond for three months). As they left the dock, *The Times* of London reported, "a cheer was raised in the crowded Court" with "hysterical shouting and the waving of handkerchiefs." Amid the ruckus,

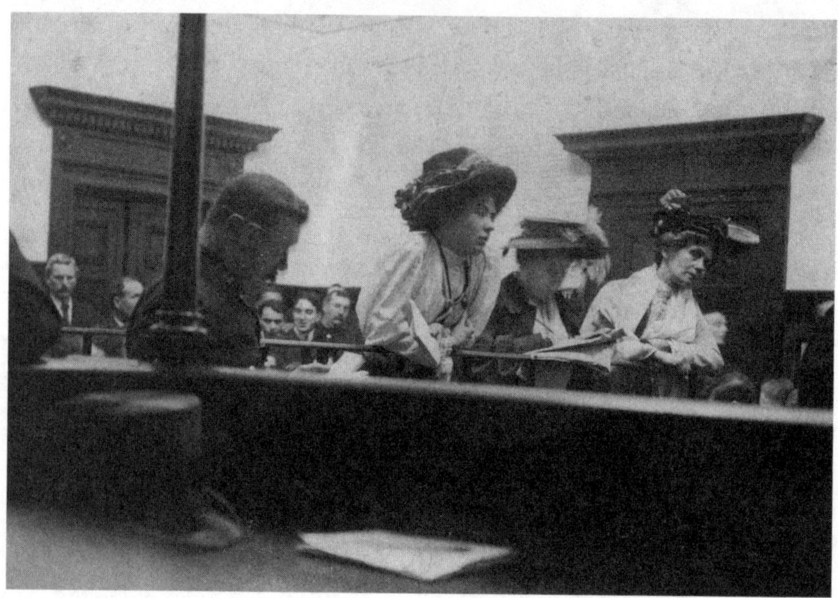

Arthur Barrett's secret hat photo from inside Bow Street Magistrates' Court, 1908, showing, from left, Christabel Pankhurst, Flora Drummond, and Emmeline Pankhurst.

Emmeline Pankhurst shouted, "I congratulate the Government on having another servant who does what he is told to do."

Barrett's photographs were a coup. Pictures of the suffragists in the dock ran in *The Daily Mirror*, *The Daily News*, and the *Sheffield Daily Telegraph*, and one was also reproduced as an illustration in *Lloyd's Weekly Newspaper* accompanied by the hysterical headline "Suffragette Attack on Parliament." The press coverage of the trial was described by one suffragist as a "suffrage meeting attended by millions." It was a scoop for Barrett; it was also not the last time he would photograph suffragists without their knowledge.

By this point, anyone who was arrested in England and Wales had to have their photograph taken, a requirement that had been in place since the 1871 Prevention of Crimes Act. But not all prisoners complied, including the suffragists, who strongly resisted any attempts to get a clear photograph of their faces. One example is a photo of suffragist Evelyn Manesta, who was photographed at Strangeways Prison in Manchester with her eyes tightly shut and her lips pressed together. She was in prison after using a hammer to smash the glass on thirteen artworks at the Manchester Art Gallery in early April 1913, along with two other suffragists. In his sentencing, the judge made it clear he would prefer a harsher publishment, stating, "If the law allowed me I would send you round the world in a sailing ship."

Except that there are two versions of Manesta's photograph. In one, she wears a scarf and is shown against a blank background. In another, the original version, she is pictured in what looks like a prison doorway. The scarf is partially covered by an arm—the arm of the prison warden who was restraining her around her neck. The prison governor evidently realized that circulating a photo that depicted violence against a suffragist prisoner was unwise. "The photographer informs me that he can easily

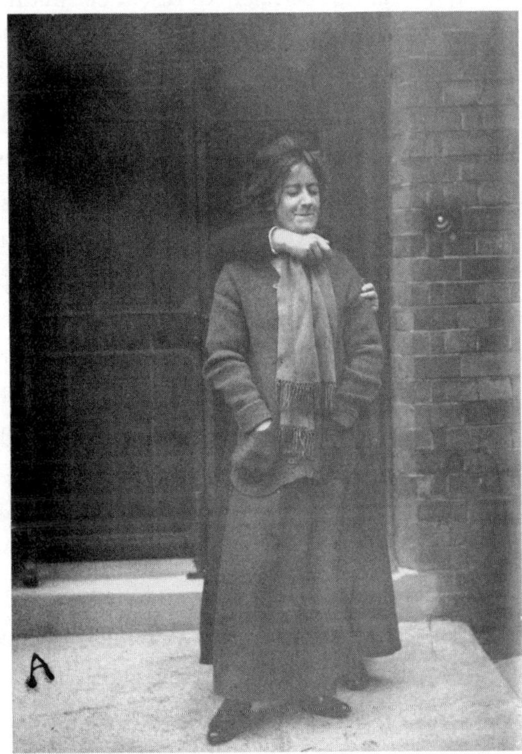

The original, unmanipulated portrait of Evelyn Manesta.

print out the matron's arm around Manesta's neck, should it be considered desirable to do so," he wrote, after the photo was taken, adding that it was "required in order to prevent her holding her head down." Only the manipulated version was circulated.

In that same month, April 1913, Parliament passed what was known as the "Cat and Mouse" Act, officially titled the Prisoners (Temporary Discharge for Ill-Health) Act. Under this new law, an imprisoned suffragist whose health was in danger from a hunger strike could be released from prison, and rearrested once she recovered—provided she could be identified. (Manesta was released from prison early under the Cat and Mouse Act; *The Suffragette* reported that she was "in a state of collapse, after a forty-eight hours hunger-strike.") So on April 26, the order came from the Home Office to photograph and fingerprint all suffragist prisoners without fail. "As, however, photographs cannot be taken by force, you should, if the prisoner resists or is thought to be likely to resist, communicate by letter or telephone with the Special Branch, New Scotland Yard, when an expert photographer would be sent to the Prison and would endeavour to take a photograph without the prisoner's knowledge." The expert photographer in the form of Arthur Barrett was now taking surveillance photographs of suffragists in the yard at Holloway Prison. (Fingerprints, the same document noted, should be taken "in every case," and by force "if necessary.")

Barrett attempted some photographs from what we'd now consider a classic stakeout position, snapping clandestine photos from inside a van. But he struggled to focus the lens on his subjects from such a distance. The police were perhaps also not as subtle as they had hoped. At the end of May 1913, *The Suffragette* reported that imprisoned suffragists had been taken outside to exercise one by one, not as a group; and after being led by a prison warden to a particular area, they described hearing a "click" and seeing a camera hidden behind a metal gate. The article noted, with some satisfaction, "evidently the Government is finding the enforcing of the Cat and Mouse Bill a difficult task."

The police were also concerned about the expense of maintaining a freelance photographer, so in August 1913, New Scotland Yard purchased an eleven-inch telecentric Ross lens, an early version of a telephoto lens. Testimonials from photographers around this time praised the lens as being especially useful for sports photography, news photography, and for photographing wild animals such as birds and big game. When photographing captive animals, wrote one, the lens offered a "decided advantage from a point of personal safety while making exposures of savage and bad-tempered animals." Another photographer, after using it during the previous year's Olympic Games, called the long lens "absolutely epoch-making." But now, long lenses were being used for a new and much more sinister kind of sport, surveillance photography, in which the British government had become a leading player.

Such was the covert nature of the photo operation that the Home Office designated code names for its telephone conversations with Barrett and Holloway Prison. The suffragists were code-named Wild Cats; Barrett was Banjo. Secret communications were also part of the suffragists' operations. The Women's Social and Political Union (WSPU), founded by Emmeline and Christabel Pankhurst, had, by this point, developed code names to use in secret messages. Individual cabinet members were code-named after plants: Lloyd George and Gladstone were known as Nettles and Thorns.

The following year, on March 10, 1914, Mary Richardson entered London's National Gallery with a butcher's knife and slashed Diego Velázquez's painting *The Toilet of Venus*, also known as the *Rokeby Venus*.* After Richardson's attack, the National Gallery closed for two weeks; the British Museum stayed open but permitted women to enter only if they had a letter of recommendation from, or were accompanied by, a man. The Home Office then released its mugshots of suffragists to galleries. One set, still held by the National Portrait Gallery and attributed to the Criminal Record Office, shows ten suffragists, some of whom are clearly being seen through the end of a long lens. In the bottom right corner is the manipulated image of Evelyn Manesta.

In 2003, one hundred years after the WSPU was founded, the suffragist surveillance photographs were included in the exhibition "The March of the Women: Suffragettes and the State 1906–1918," at the National Archives in London. Up until this point, the surveillance records had not been widely seen, and the press reaction to the exhibition conveyed a deep

---

* Mary Richardson was arrested, tried, convicted, and sentenced to six months at Holloway Prison. While on hunger strike, she was force-fed multiple times. In the 1930s, she joined the British Union of Fascists, but left in 1934 because of its policies toward women.

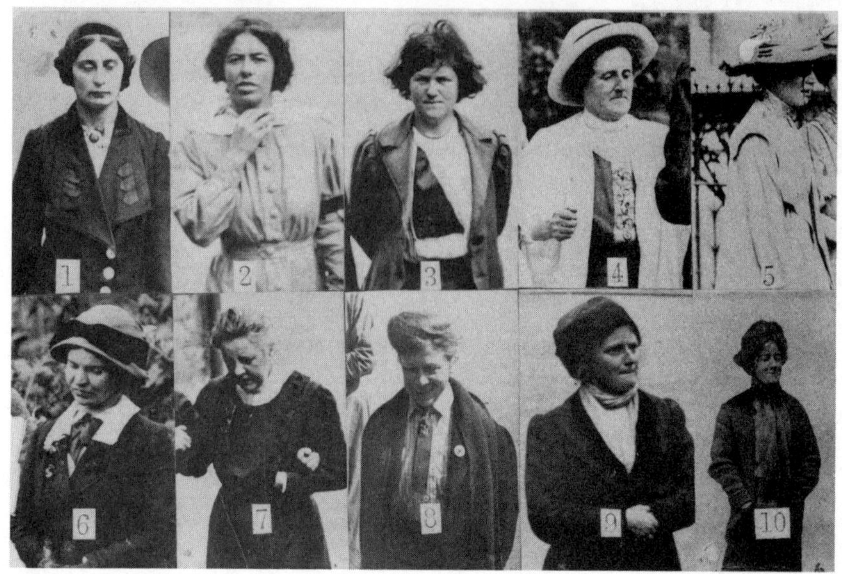

Suffragist mugshots. The manipulated photo of Evelyn Manesta is at bottom right.

unease at what the archives had revealed. Looking at these photos now, the feeling remains. As historian Jill Liddington has noted, it was evidence of how the British government "secretly spied on women imprisoned merely for demanding the basic democratic right to vote."

Barrett, for his part, went on to join the Royal Naval Air Service as an aerial photographer during World War One. But his involvement with the suffrage movement hadn't quite ended. In 1955, Barrett joined a reunion of British suffragists and brought with him his top hat and secret camera. He showed some of the suffragists how it worked, which was captured on a British Pathé newsreel, available on YouTube.

The hand camera, in all its guises, changed photography. It changed what people photographed, and how they photographed it. It brought out some photographers' worst instincts; for others, it created a crucial record of everyday life. The candid shots it produced marked the start of long-running debates about privacy and consent, and precipitated sleazy paparazzi shots and silent surveillance images. The hand camera truly was, in the words of one Kodak advertisement, "the camera that takes the world!"

Chapter 9

Cartomania

Amateur fiends and their detective cameras were not the only all-consuming photographic craze in early photography. In the 1860s, a new trend in photography took hold with such force it had its own nickname: cartomania.

It referred to what were known as *cartes de visite*, portraits mounted on a board about the size of a playing card that people exchanged and collected in albums. Cartes de visite were affordable, collectable, and so massively popular that in England, at the height of the cartes de visite craze in the 1860s, 300–400 million cartes de visite were sold every year. As noted in the weekly journal *All the Year Round*—edited by Charles Dickens, who also sat for multiple cartes de visite—they offered "the opportunity of distributing yourself among your friends, and letting them see you in your favourite attitude, and with your favourite expression. And then you get into those wonderful books which everybody possesses, and strangers see you there in good society, and ask who that very striking-looking person is?"

Who invented cartes de visite was subject to some debate, but the person who patented and popularized (and certainly profited) from them was the Parisian photographer André Adolphe-Eugène Disdéri. In 1854, he created a more economical process of portrait production, by using a multi-lens camera to create several small portraits (at first ten, and then eight) on a single

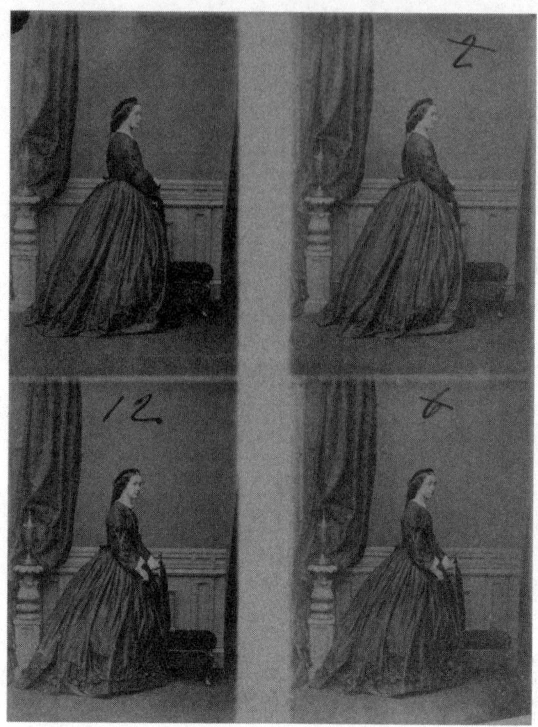

A proof sheet of carte de visite photographs, 1863.

plate. Because of their size, the portraits didn't need retouching, which saved even more time and money. Once developed, the prints were cut and glued onto a cardboard mount measuring about 2.5 by 4 inches. Cartes de visite were initially embraced by the wealthy and the fashionable of Paris, and only became widely popular at the end of the decade. Around this same time, cartes de visite arrived in the UK and the United States, and it would not be long before cartomania reached feverish proportions. This was all good news for Disdéri: by 1861 he was said to be the richest photographer in the world.

For the first time in photographic history, people shared, exchanged, and collected images in large quantities. The intimacy of holding a carte de visite must have felt revelatory: portraits were no longer stuck behind glass in a gallery wall, or interpreted through an engraving in the newspaper, or viewed through a stereoscope device. Cartes de visite could fit into a pocket or purse or handbag. The card portrait, wrote Oliver Wendell Holmes, is "the cheapest, the most portable, requires no machine to look at it with, can be seen by several persons at the same time,—in short, has all the popular elements. Many care little for the wonders of the world brought before their eyes by the stereoscope; all love to see the faces of their friends."

The subject of a carte de visite was typically a studio portrait, although it could feature virtually anything (including "obscene" photographs). For the portrait, the sitter usually appeared in full length, standing or sitting, with props, often in front of a painted background to add a sense of place, although which place exactly is sometimes unclear. The portrait might be solo, or it might be a pair—a married couple, for example. One journal advocated for men to pose next to a column, to represent an exterior view, while women's portraits should include a chair or curtain, to convey their appropriate nineteenth-century domain, the interior. In many cartes de visite of married couples, the woman stood while her husband sat, suggesting, as historian Andrea Volpe has pointed out, her deference. (In the cartes de visite of Queen Victoria and Prince Albert together, she is depicted both standing and sitting.) The subjects of cartes de visite were not limited to humans either: Disdéri opened a studio in Paris that specialized in cartes de visite of horses, which led to some inevitable jokes, in the English and American photographic press, about putting the cartes before the horse.

If you were in London in 1862 and wanted a carte de visite portrait of your own, you might visit Regent Street, which at that point had no fewer than thirty-five photography studios. Once you had chosen your photographer, you needed to carefully consider how you wanted to present yourself. Depending on the photo studio, props could range from a simple table and chair to the elaborately artificial. You might decide to lean against a faux balcony, gate, balustrade, or pedestal. You might stand next to some drapes, a column, a piano, a bridge, or a well. You might sit on a swing or a rock, or perch in a rowboat. You might even command the deck of a steamship. The background scenery was painted on panels reminiscent of amateur theatrical productions although possibly less convincing. All of which meant that if you wanted to pose on the fake balcony of a pretend villa in front of a painted background of some vaguely European-looking location, despite never having set foot in a villa, nor ever having seen the Continent, you could jolly well do so.

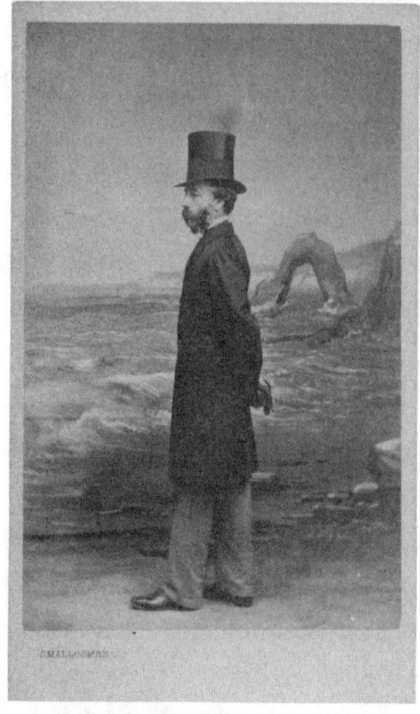
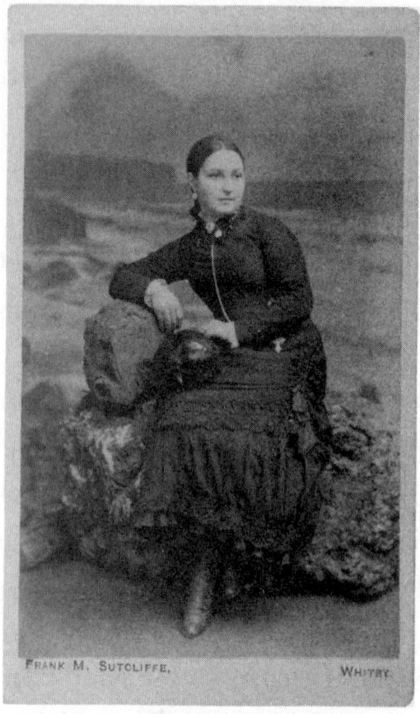
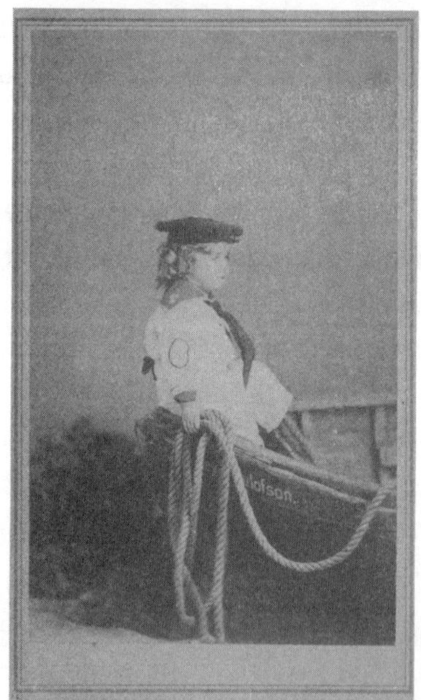
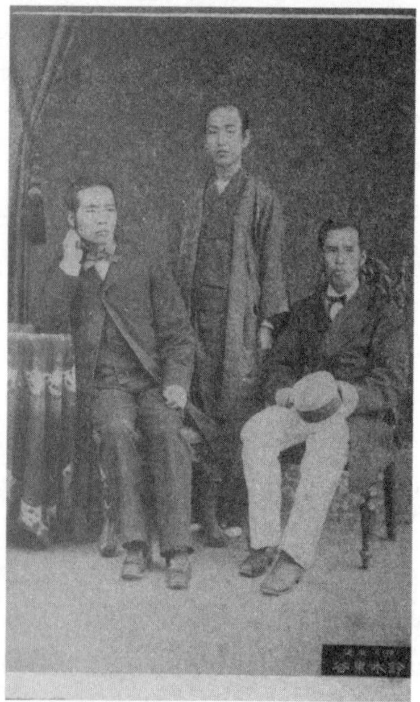

A sampling of cartes de visite.

The excessive use of outlandish props and unnatural scenery did not go uncommented upon. "Nothing can be more unpleasing than to see a portrait of a person surrounded with a crowd of incongruous objects, and it becomes almost painful to see the same things monotonously repeated in the portraits of a number of friends," wrote *The Photographic News*, in 1861. *All the Year Round* made the point more humorously: "What is the profession of that unhappy and misguided wretch who is supposed to pass his life in a perpetual environment of pillar and curtain?"

The more snobbish disapproved of portraits that, they felt, elevated their subjects beyond their class. "It will be observed that the lower the neighbourhood the more varied the amount of properties or scenic decorations to be found in the studio," wrote Andrew Wynter in the British journal *Good Words*, noting with some disapproval that "servant maids" are seated in "splendid boudoirs, and respectable tradesmen are placed in extensive libraries, whereas the only books they feel at home with are their day-books and ledgers."

In addition to this clutter of accoutrements, there was one other prop that you would inevitably face at the cartes de visite studio, and that was the posing stand. This was a metal rod anchored to a heavy base with two clamps, one for the neck and one for the base of the spine. (Over time, manufacturers began to make padded posing chairs instead.) Here the props came in handy, as they could sometimes be made to obscure the base of the stand. But the physical restraint was visible in other ways: of a carte de visite of Lord Clyde, one commenter wrote, "Head-rest is written in every line of his features."

Children also sat for cartes de visite, although it was such a nightmarish process that some photographers refused to work with children at all. The result was highly dependent on the photographer and the parents' abilities to keep the child still for the exposure, for which there were a few methods. Sometimes the child was simply held in place by one of their parents, who was disguised (not very convincingly) with a draped cloth. In other instances, the photographer used a sash or rope to tie their young sitters to a chair.

As millions of cartes de visite changed hands, a new industry sprung up: the cartes de visite album. Albums ranged in size and style, and could feature gilt edging, chunky clasps, and leather binding; one of the more ridic-

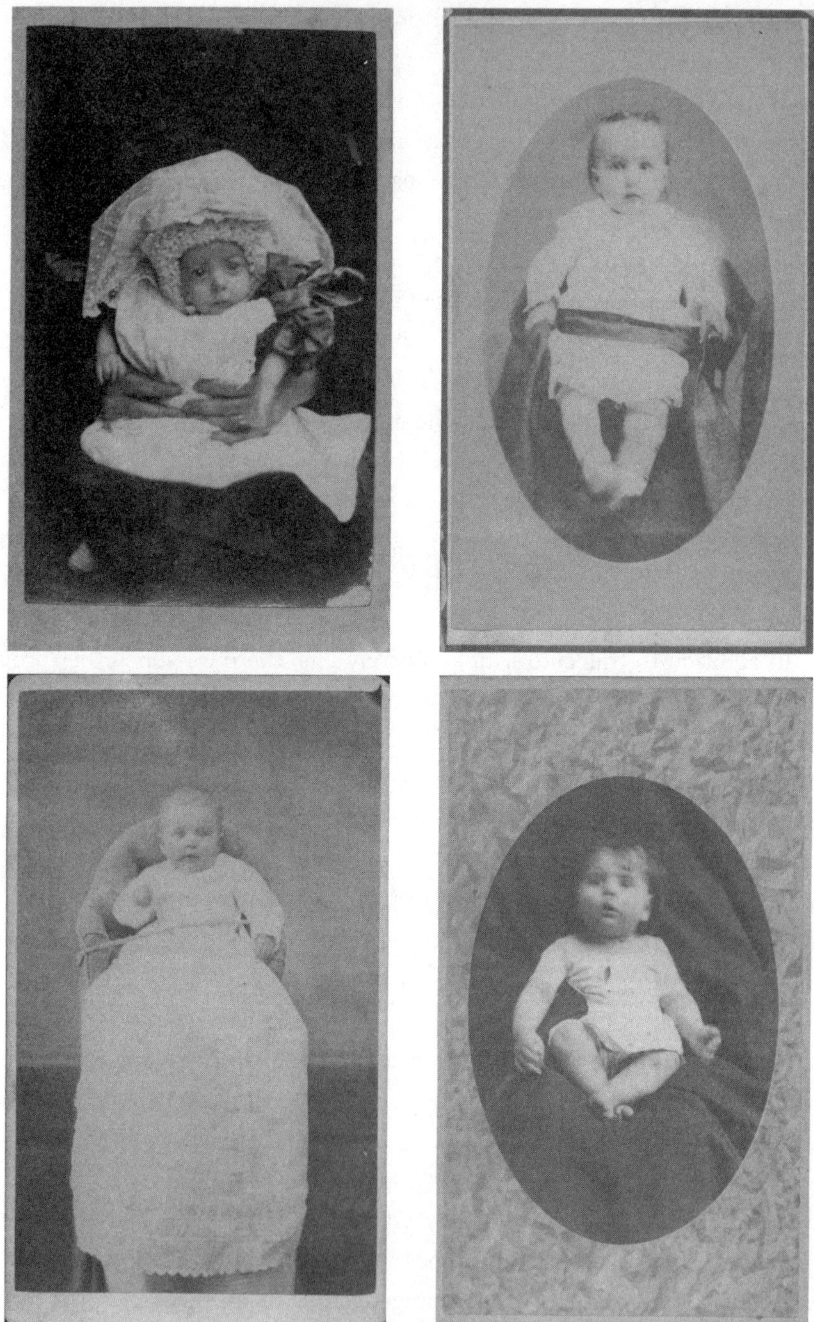

Various methods of keeping babies still for a portrait.

ulously ornate incorporated a music box on the cover. "No drawing-room table of the day can be considered furnished without its 'Photographic Album.' . . . It is at once a mild form of hero-worship and an illustrated book of genealogy," declared *The Saturday Review*, before further praising the album as the ultimate pre-dinner conversation starter, with which "even the dullest mind in the dullest half-hour of the day may find something to talk about." For album manufacturers, it was boom time: in 1867, one British firm claimed to have sold almost a million albums.

The enormous appeal of cartes de visite was not only that you could collect photographs of your friends, family, acquaintances, and that person you met last week, but you could also collect photographs of public figures. Consider, for a moment, that now–perpetually popular photographic subject, the British royal family. The first time a photographic portrait of the royal family was officially exhibited in public was at the 1857 Manchester Art Treasures Exhibition. Just three years later, the royal family became a photographic commodity when Queen Victoria approved the sale of a series of cartes de visite portraits by John Jabez Edwin Mayall. That same year, 1860, she also approved an entire "Royal Album" of cartes de visite for general sale. For monarchy-loving cartes de visite collectors, this was a splendiferous visual feast—"exquisite studies," gushed *The Athenaeum*—with fourteen cartes de visite of the royal family. Anyone who could afford it could, for the first time in history, own portraits of the royals, to keep in one's home at one's fingertips, to scrutinize and admire, alongside family portraits of Great-Aunt Gertrude and Cousin William. In two years, Mayall reportedly sold between three and four million copies of the "Royal Album."

Queen Victoria herself owned thirty-six cartes de visite albums, which contained cartes of British aristocrats and European royalty. Evidently one of her ladies-in-waiting was not enamored of the queen's new hobby, noting, in the autumn of 1860, "I have been writing to all the fine ladies in London, for their or their husband's photographs, for the Queen; . . . I believe Miss Skerrett is right when she says 'the Queen could be bought and sold for a Photograph'!"

The desire for cartes de visite of the British royal family was insatiable. In the week after the unexpected death of Prince Albert, in 1861, the largest cartes de visite retailer in England, Marion & Company, received 70,000

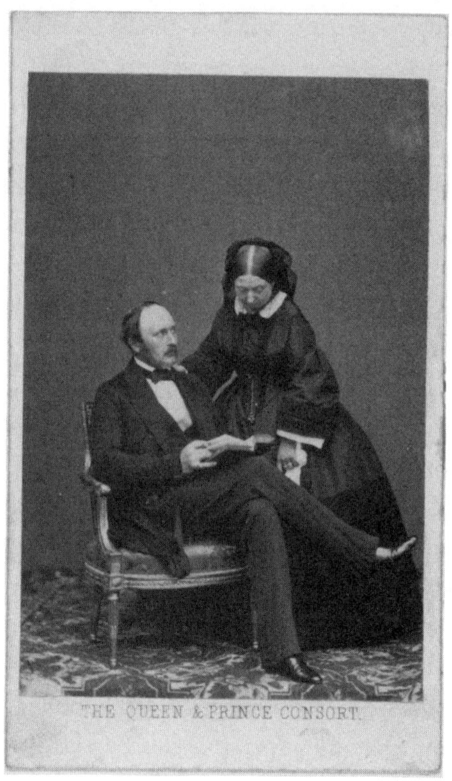

One of Mayall's hugely popular cartes de visite of Queen Victoria and Prince Albert, 1860.

orders for his cartes de visite. London diarist A. J. Munby recounted his attempts to purchase one in Regent Street, with "crowds around the photograph shops, looking at the few portraits of the Prince which are still unsold. I went to Meclin's to buy one: every one in the shop was doing the same. They had none left: would put my name down, but could not promise even then." One of the most popular cartes de visite of all time was a portrait of the Princess of Wales carrying a young Princess Louise on her back, which sold an astonishing 300,000 copies.

The popularity of celebrity cartes de visite extended beyond the royal family. In 1862, *Living Age* reported, "The commercial value of the human face was never tested to such an extent as it is at the present moment in these handy photographs." The windows of cartes de visite stores contained the faces of whoever was popular at that moment: politicians, royalty, actors and actresses, clergymen, scientists, writers, sportsmen, "ballet girls," and any other figure who had suddenly found themselves famous. A Londoner returning to the city from a long absence, reported *All the Year Round*, "might say with certainty from an inspection of the cartes-de-visite in the shop windows what would be the prominent subjects of conversation at his first dinner party." At least one public figure was known to pop into stores to check the sales of his cartes de visite portrait, the same way that today a celebrity might track their social media followers.

For celebrities, according to *The Saturday Review*, a full-length carte de visite was appealing because its small size made it less easy for viewers to discern awkward features. It was also a chance to make some extra cash: some public figures entered into financial arrangements with photographers for their cartes de visite portrait. The photographer either would offer the celebrity a fee to sit for the portrait, or would agree on a royalty split for the sales. During his trip to the United States in 1867, for example, only after he was paid did Charles Dickens sit for Jeremiah Gurney, the New York photographer who nearly died from exposure to mercury. Even so, photographers and publishers regularly copied other photographers' cartes de visite, particularly as copyright legislation did not extend to photographs in England until 1862, and not in the US until 1865. If a photo studio wanted to sell an image of a celebrity but didn't possess the negative, they just copied an existing carte de visite.

For the larger publishers that churned out vast numbers of cartes de visite, the process was factory-like. "The workmen in large establishments, where labor is greatly subdivided, became wonderfully adroit in doing a fraction of something," wrote Oliver Wendell Holmes in 1863, after his visit to E. & H. T. Anthony, a photo supply company in New York, continuing, "One forlorn individual will perhaps pass his days in the single work of cleaning the glass plates for negatives." In 1862, the Anthony's catalog listed about 2000 different carte de visite portraits available to purchase. That same year, wholesaler Marion & Company in London reportedly supplied 50,000 cartes de visite to retailers every month.

This also created a vast subindustry for all the materials needed to supply the eager demand for cartes de visite. It was the wet collodion era, so all those millions of photographs needed chemicals and albumen paper for printing, and the mounts needed cardboard. One manufacturer in England used half a million eggs for its albumen paper per year at the height of cartomania. If a customer wanted reprints at some later date, the photographer needed to store the glass negative; eventually, this became such a hassle that the photographers set a time limit for storage before they cleaned and reused the glass. Manufacturers expanded their offerings of backgrounds and props in ways that barely seem possible, and for decades after the cartes de visite craze first arrived. One company in the United States advertised its "new style papier mache rock"; another sold no fewer than eight different varieties of rocks, one of which was constructed so that "a baby cannot fall out. It has a hole in back of seat so that a child can be held from behind the rock if desired."

With cartes de visite, portraits were time efficient and mass produced. One photographer claimed he took 97 negatives in eight hours. Those who aspired to photographic originality deplored cartes de visite, including Nadar, but he was canny enough to produce them as well. (Nadar held a low opinion of Disdéri, who was also his competitor, believing him to be a sellout.) But some photographers managed to produce quality cartes de visite portraits. Camille Silvy, one of London's most renowned cartes de visite photographers, often photographed stage performers and fashionable London society. "That sense of beauty and instinctive art of catching the best momentary *pose* of the body is a gift which cannot be picked up as a mechanical trade can be. This gift M. Silvy possesses in an eminent degree," wrote Andrew Wynter, in 1862, of a visit to Silvy's studios in Porchester Terrace. Silvy's output was nevertheless prodigious: in a single month of 1861—May, and the summers were busiest, because of the light—Silvy photographed 806 customers. This intensity also took its toll: Silvy's later ill health was, he believed, caused by cyanide exposure.

As the craze progressed, the uses for cartes de visite ranged from the romantic to the transactional. The *British Journal of Photography* reported that one couple had fallen in love through the exchange of cartes de visite, and only saw each other in person on the wedding day. At a New York brothel, patrons perused an album of the sex workers' cartes de visite. A

US journal, in 1864, suggested sending cartes de visite when introducing a governess or maid to a potential employer.

Cartes de visite flourished during the American Civil War. "The soldier, though physically absent, will be photographically present until his lineaments fade from paper," wrote *The New York Times* in June 1861. In addition to portraits of their absent loved ones, people collected war-related cartes de visite, such as landscapes of battlefields or portraits of military leaders. One popular carte de visite was a portrait of Colonel Elmer E. Ellsworth, an early casualty in the war; he was killed on May 24, 1861, just after he had removed a Confederate flag flying from the roof of a tavern in Virginia.

It was also at the start of the Civil War that a public figure took a singularly astute approach to her carte de visite portraits. Abolitionist and suffragist Sojourner Truth had around fourteen portrait sittings in carte de visite format, and a larger style known as a cabinet card, during her lifetime. Truth's earliest known carte de visite dates from around 1861, when she was about sixty-four years old. It's impossible to know Truth's exact age, as the year of her birth into slavery in upstate New York was not recorded; historians date the year to around 1797. In 1826, as slavery was coming to an end in New York, after her master refused to free her as he had promised, she ran away, emancipating herself and her baby daughter. In 1850, she self-published her autobiography, *The Narrative of Sojourner Truth*, which she had dictated; as an enslaved child, she was not taught to read or write. But in photography, the sitter can speak directly to the viewer.

By creating and selling carte de visite portraits, Truth could both fundraise and represent herself as she wanted to be seen, without the need for words. In her carte de visite portraits, she sometimes appears next to a table with a vase of flowers, or with a book. Sometimes she holds knitting, which historians have linked either to the wartime drive to knit for soldiers, or as a symbol of gentility. Often, her right hand, which had been maimed in an accident while she was enslaved, is obscured. In one series of portraits, she displays in her lap a portrait of her grandson James Caldwell, who had joined the all-Black infantry force, the 54th Massachusetts Infantry Regiment. Truth sold her cartes de visite at her lectures and through the mail, made possible through their small size. (One report noted that Truth kept her cartes de visite in her bag when at speaking conventions.)

But how Truth engaged with the carte de visite format itself was unprece-dented. Although she was the sitter and not the photographer, in 1864 Truth copyrighted her cartes de visite under her own name. Photographs were not explicitly covered by copyright law in the United States until 1865, but there was an old law that was opaque enough to be used by some photographers—and Truth—for copyright protection. Truth's copyright ownership appears in type, on the back of her cartes de visite. "Against law, convention, and general practice, Truth was claiming legal rights to her image and the right to sell her body's shadow as property," writes art historian Darcy Grimaldo Grigsby in her book *Enduring Truths: Sojourner's Shadows and Substance.*

Truth's name also appears on the front of almost all of her cartes de vis-

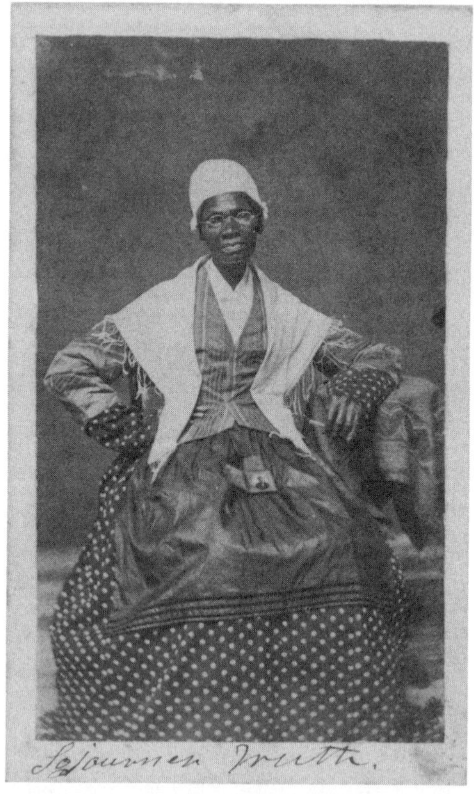

A carte de visite of Sojourner Truth with a photograph of her grandson in her lap, circa 1863–65.

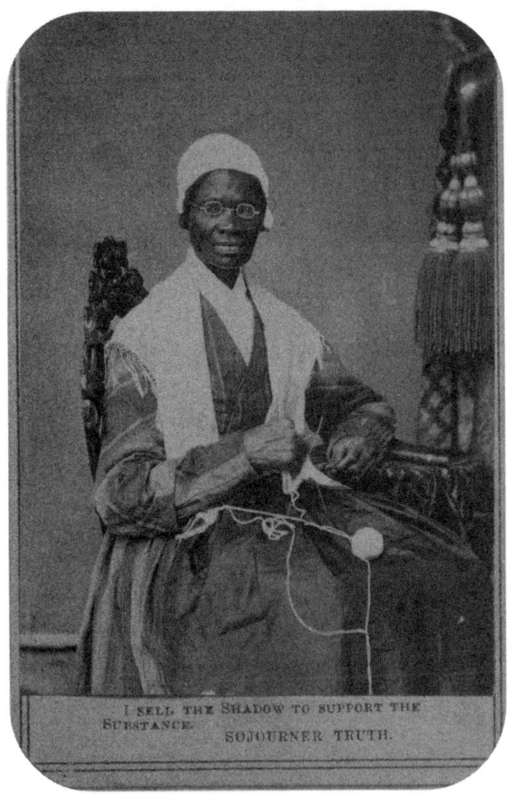

Sojourner Truth, 1864.

ite, where typically the photographer's name would appear. This was highly unusual; so too was her decision to include the phrase "I sell the shadow to support the substance." Although "shadows" and "substance" had previously been invoked in relation to photography, the phrase has since become closely associated with Truth's portraits, and with her singular, self defining approach to portraiture. Its power comes from the placement next to Truth's name, under her portrait, where it seems less a caption and more a declaration.

Truth's biographer Nell Irvin Painter notes that images like Truth's "were largely missing from American culture, even from the feminist and antislavery subcultures." Truth's cartes de visite portraits were circulating in an era of derogatory, racist portrayals of Black Americans in illustrations and cartoons, and in images made by ethnologists that promoted vir-

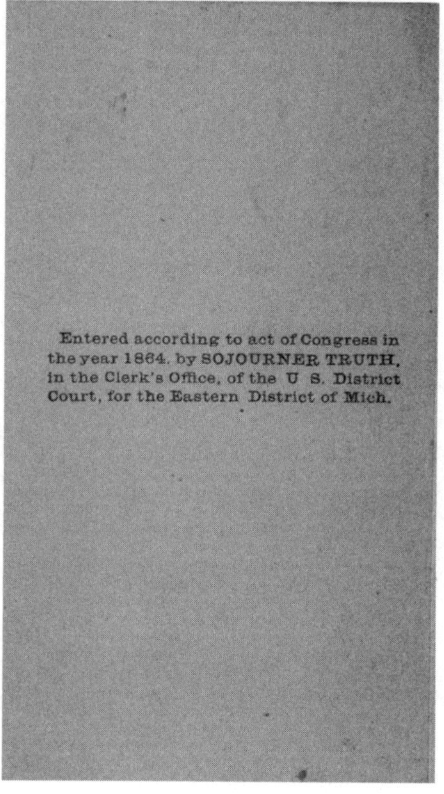

The verso of one of Sojourner Truth's cartes de visite indicating her copyright ownership.

ulent notions about racial hierarchies. Frederick Douglass addressed this in a speech to the literary societies of Western Reserve College in 1854, where he spoke critically of the degrading imagery of Black Americans that were presented by ethnologists. "The importance of this criticism may not be apparent to all;—to the *black* man it is very apparent. He sees the injustice, and writhes under its sting."

By this time, Douglass had already started his own long and purposeful alliance with photography, which he saw as a way to counter the era's pervasive racist imagery that misrepresented and mocked Black Americans. Douglass was keenly aware of how photography could shape perceptions. "It is evident that the great cheapness and universality of pictures must exert a powerful, though silent, influence upon the ideas and sentiment of present and future generations," he said, in one of several speeches he made in the 1860s about photography. Through portraiture, Douglass could present his humanity and exert his own powerful influence.

Douglass's first known portrait, a daguerreotype, was taken in around 1841, some three years after he escaped from slavery. Douglass continued to sit for photographic portraits until the end of his life in 1895, for an estimated 168 separate portraits, making him the most photographed man in nineteenth-century America. Historians Maurice O. Wallace and Shawn Michelle Smith have argued that Douglass and Truth, along with other leading Black figures like Ida B. Wells and W. E. B. Du Bois, were "practitioners" of photography: "Although they might never have focused the lens

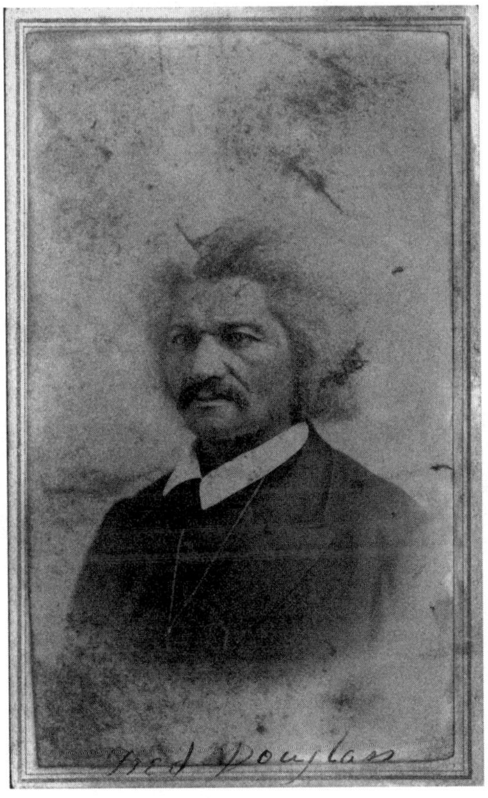

Frederick Douglass, photographed by James Presley Ball, 1867.

or released the shutter on a camera, they put photographs to striking use in their varied quests for social and political justice." Douglass made use of almost every photographic medium, including cartes de visite, and he worked to share his photographs widely.

One of Douglass's cartes de visite has since been found in an album kept by an African American schoolteacher by the name of Arabella Chapman. The album is one of two from Chapman, held in the Clements Library at the University of Michigan, which contain both cartes de visite and another highly popular portrait format, tintypes, which were photos produced on a thin sheet of metal. How the albums came to this university remains unknown; Chapman was born in New Jersey and moved to Albany, New York, where she was the first Black student to graduate from what is now Albany High School. In 2014, the library's curator showed the albums to Martha Jones, who was, at that time, professor of African American history at the University of Michigan. Wanting to know more, Jones arranged for her students to research the photographs and find clues—inscriptions, for example, or captions—to help create a more complete picture of Chapman.

It's through Chapman's albums that one of the functions of cartes de visite comes into sharper focus. These small, portable portraits, when

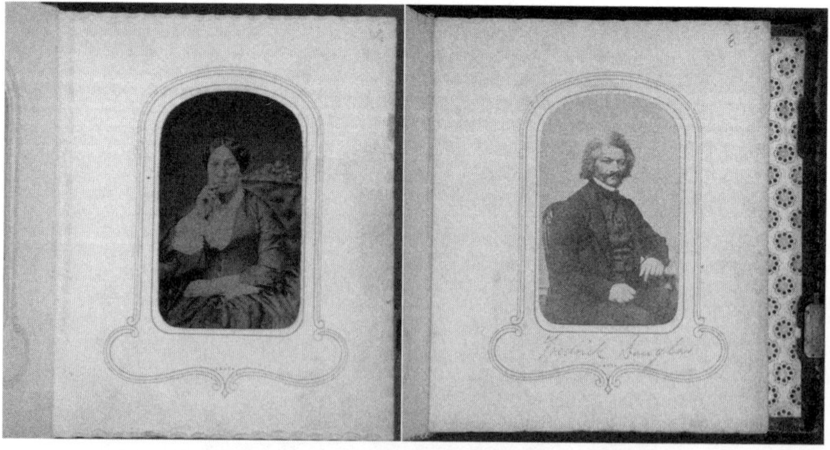

Two separate pages from Arabella Chapman's albums, showing a tintype portrait of Chapman and a carte de visite of Frederick Douglass.

viewed together in Chapman's leather-bound albums, provide a glimpse of who and what she cherished: her family, friends, neighbors, along with public figures. (In addition to Douglass, Chapman collected cartes de visite of John Brown and Abraham Lincoln.) The albums widen the aperture on her life. What makes them all the more extraordinary is that they are rare: according to Jones, in an interview with the Albany *Times Union*, it is unusual to discover a complete photo album showing an African American family from this era.

James Presley Ball, who took the photo of Douglass on the carte de visite shown earlier, was a prominent Black photographer, and one of four Black photographers who are known to have taken Douglass's portrait. Ball learned the daguerreotype process in Boston in the mid-1840s, from a Black photographer named John Bailey. Ball opened a small studio in Cincinnati; after that was unsuccessful, he worked for a stint as an itinerant photographer, then took a job in a hotel dining room in Richmond, Virginia, and saved up money for a furnished room to use as a photo studio. In the late 1840s, he returned to Cincinnati and opened a new studio, "Ball's Great Daguerrean Gallery of the West." The name was apt: Ball's studio had four rooms across three stories, gold leaf wallpaper, a gallery wall featuring Ball's photographs, panels painted with the Muses, and a piano. "There is scarcely a distinguished stranger that comes to Cincinnati but, if time permits, seeks the pleasure of Mr. Ball's artistic acquaintance," wrote *Gleason's Pictorial Drawing Room Companion* in April 1854.

Ball's customers were both Black and white, and included high-profile visitors to the city such as the performer Jenny Lind and Douglass. Ball was also an abolitionist, and in 1855, he worked with artists to produce a moving, painted panorama 600 yards long and 4 yards high of scenes from the slave trade, along with an accompanying pamphlet. The panorama was exhibited at Ball's gallery. There are few extant photographs of Ball himself, but one that does exist, with an inscription in Ball's handwriting, is, of course, a carte de visite.

The idea of shaping or controlling a narrative—whether in words or in pictures—seems to be a contemporary one, but it's not. With cartes de

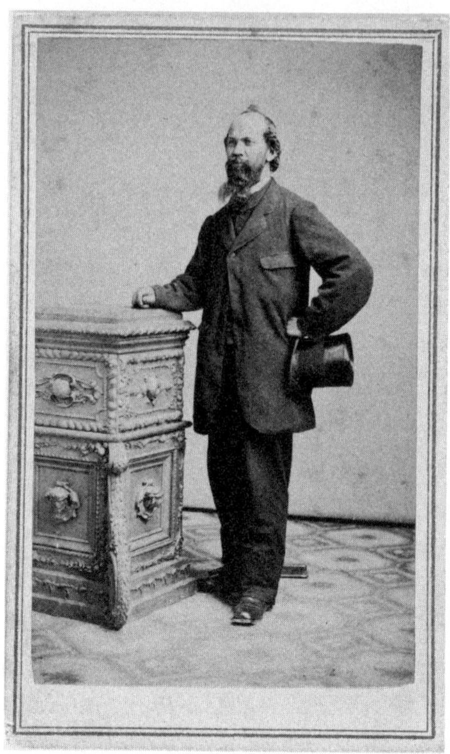

A carte de visite of photographer James Presley Ball, 1870.

visite, more people than ever before could engage with photography by purchasing a set of small, sharable portraits to show to the world. In the late 1860s, a new, larger format of the card photograph, the cabinet card, was introduced, but never reached the crazed heights of cartomania. In the 1880s, dry plates and small cameras exploded all previously established conventions of photography, and amateurs began snapshooting in droves, further democratizing a medium that was still only forty years old.

If you look at a carte de visite in isolation, it might be tempting to focus only on the superficial: an awkward pose, say, or perhaps an abnormally large papier-mâché rock. But this perception is deceiving. Within the cartes de visite lie notions about celebrity culture, self-imaging, authenticity, ownership, and representation that are deeply resonant today.

PART III

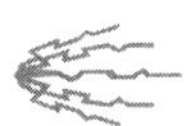

INTO FOCUS

Chapter 10

Running, Jumping, Galloping

Before photographic motion studies—instantaneous pictures of a moving subject—photographers were beholden to time. They carefully calculated their exposures as they fixed their lenses on their unmoving subject. But with motion studies experiments, photographers captured one fraction of movement—of a horse's galloping legs, say, or an athlete running. And after these experiments, it was only a matter of time before those moments were assembled and reconstructed, frame by frame, as the groundwork for a revolutionary new medium, the moving picture.

If you were to spend a day in Madison Square, New York, in 1885, watching the traffic on Fifth Avenue, you would see some 5500 horse-drawn vehicles passing by. Horses were an essential part of nineteenth-century life. They were used in agriculture and for transport. In cities like New York, they pulled carriages and delivery wagons and fire trucks, filling the streets with noise and manure. Unlike the modern equivalent, cars (which are judged by their horsepower), horses are in constant motion in every direction, even when they stand still. They expand their great nostrils as they inhale and exhale; their cheek muscles quiver; they stomp their hooves, sweat, bray, flick their tails, and twitch their ears. And when they trot or canter or gallop, it's difficult to discern what their legs are actually doing. If you haven't seen a horse galloping, go to YouTube and watch a horse race. Their legs move impossibly fast. But exactly how they move was a source of fascination to many in the nineteenth century, including artists and horse trainers, and in particular to millionaire railway magnate and horse fanatic Leland Stanford.

When Stanford started breeding racehorses, he studied their anatomy so thoroughly, he later boasted, "I could take the steed apart and put it together with the accuracy that distinguishes a skilled watchmaker who deals with the mechanism of a timepiece." But studying anatomy could not determine the much-debated question of "unsupported transit": whether, during a fast trot or a gallop, a horse had all four hooves off the ground. At some point afterwards, a story arose that Stanford had initiated a wager of $25,000 over unsupported transit—he believed that a horse had all four hooves off the ground—but that appears to be more photographic lore than fact. But he did want to know the answer, so in the early 1870s, Stanford approached a photographer named Eadweard Muybridge.

This wasn't the name he had been given at birth, in Kingston upon Thames, London, in 1830. His name then was Edward James Muggeridge, but he changed it several times throughout his life, including during a stint in Central America when he called himself Eduardo Santiago Muybridge. He added the extra, some might say unnecessary, vowels to his name from the Anglo-Saxon spelling. By the time Stanford contacted him, Muybridge was an established landscape photographer in San Francisco, signing his work as "Helios," the Greek god of the sun. Muybridge also followed in the footsteps of Carleton E. Watkins and produced beautiful mammoth format views, 20 by 24 inches, of Yosemite. (He also added in clouds to his landscapes.) But Stanford didn't want landscapes; he wanted Muybridge to photograph a horse in motion.

Muybridge's early attempts, from around 1872, are not extant, but they were evidently not successful enough for Muybridge to sell, or to merit the kind of fanfare that his later photographs received. After these attempts, Muybridge did attract a lot of attention, but not for his photography. In October 1874 Muybridge discovered that his wife Flora Stone, a photo retoucher half his age, and a caddish man-about-town named Harry Larkyns were having an affair, and that Flora and Muybridge's six-month-old son, Florado Helios, might be Larkyns's child. (Muybridge also discovered, on the back of a photograph of Florado, the inscription "Little Harry" in Flora's handwriting.) Muybridge left the city armed with a pistol. He tracked Larkyns down at a mine near Calistoga in the Napa Valley and shot him in the chest. Larkyns died almost instantly.

At Muybridge's trial for murder in February 1875, he pleaded not guilty on the basis of justifiable homicide and insanity. Muybridge took the stand not to testify about the murder but about the effects of a stagecoach crash. In 1860, he had been traveling overland to the east coast and had reached as far as northeastern Texas when the horses bolted. Muybridge was thrown from the stagecoach, landing, he recounted in court, "against my head." He woke up nine days later in Fort Smith, Arkansas, with double vision and having lost his senses of taste and smell. He eventually made it to New York, where he saw a physician and spent two months recuperating. Muybridge then went to England, where he sought the help of another physician. He didn't return to California until 1867. At his trial, various witnesses testified to Muybridge becoming more "eccentric" after his accident.*

At the end of the trial, the judge directed the jury to discount a "not guilty" verdict on the grounds of justifiable homicide. But the jury ignored the direction. They acquitted Muybridge on the basis that "under similar circumstances they would have done as Muybridge did," reported the *San Francisco Chronicle*. With the verdict Muybridge collapsed with emotion and "moaned and wept convulsively." Once he recovered, he descended the steps of the courthouse to cheers.†

Stanford was perhaps not overly perturbed by the trial: Muybridge's defense attorney was a friend of Stanford's. After the trial, Muybridge trav-

---

* In his paper from 2002, neuroscientist Dr. Arthur Shimamura noted that the area of the brain where Muybridge sustained damage typically causes changes in behavior such as emotional instability, heightened risk-taking, and obsessive-compulsive behavior, among others.

† Stone tried to divorce Muybridge but couldn't. He did not provide any financial support. Stone died in July 1875; the following year Muybridge put Florado in an orphanage.

eled to Central America. Back in San Francisco, in 1877, he tried to photograph a horse and sulky (a two-wheeled cart) in motion, but it yielded an image so inadequate Muybridge had it rendered as a painting—but he did paste in the photographed face of the sulky driver. The depiction strained credibility to the point that it was widely considered to be a fraud. By the following spring, he had rethought his approach.

On June 15, 1878, members of the press gathered by Stanford's racetrack in Palo Alto. The track was part of Stanford's stock farm, which eventually grew to over 8000 acres. (Today, it's the site of Stanford University.) After his earlier, disappointing attempts, and watched by members of the press he had invited, Muybridge had something to prove.

Alongside the track, which had been whitened with powered lime, Muybridge set up twelve cameras, twenty-one inches apart, in an open-front shed. Opposite the cameras he had erected an angled fifteen-foot-high white fence with black vertical lines, numbered and also spaced twenty-one inches apart. Each camera had an electromagnetic sliding shutter, specially designed by Stanford's railway engineers, connected to wires that stretched across the track.

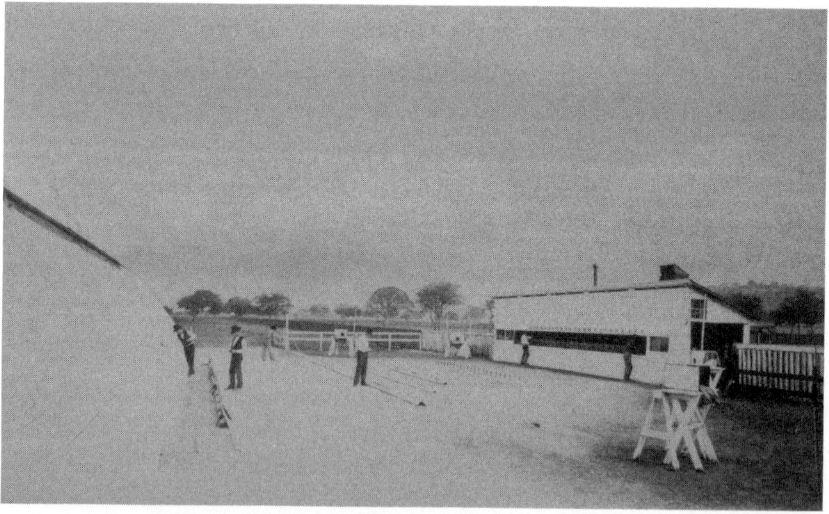

A view of Muybridge's setup from his 1879 experiments at the Palo Alto racetrack, showing a bank of twenty-four cameras.

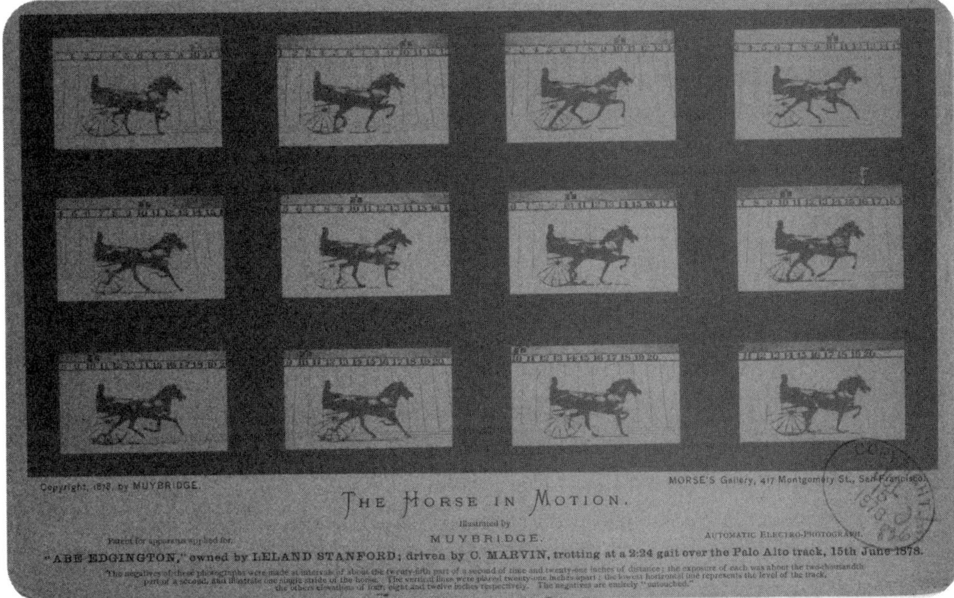

Muybridge sold prints from his 1878 motion experiments, such as this one, show-ing Abe Edgington, the fast trotter.

With the stage set, Stanford's head trainer barreled down the track in a sulky harnessed to a fast trotter named Abe Edgington. The power of the trotter's four fast-moving legs was such that he was traveling at about forty feet per second. As the sulky's wheels hit the wires, they triggered the shut-ters, and Muybridge made history.

Muybridge had set up a darkroom in the shed, and the images that he coaxed from the chemicals were astonishing: small in format and closer to silhouettes, they showed, in clear detail, the positions of the horse's legs in motion. A local California paper reported: "The negatives were very small, but perfect in outline and detail, even to showing the shape of each spoke in the wheel, and the whip in the driver's hand. These pictures, which could not be otherwise true, prove that the horse in trotting assumes positions never dreamed of before."

He repeated the experiment with a thoroughbred called Sallie Gard-ner, who barreled down the track "like a whirlwind," recalled one wit-ness, and galloped through black threads, now at chest height, which opened the twelve shutters for the briefest of moments. The photos clearly showed that in a gallop as well as in a fast trot, all four hooves do leave the ground.

It was a canny move to invite the press. The published accounts of the experiment further validated Muybridge's photographs and helped make him famous. The articles also expressed a widespread astonishment at how a horse moved. "They show the legs to be in almost all conceivable positions," reported one paper, adding that "while they rob the horse of that gracefulness generally credited to him when going at full speed," they would still be "invaluable to persons concerned in the care and training of trotters."

Others focused less on the utility of the photographs and more on the awkward poses. The *San Francisco Chronicle* described the legs as taking "the most incongruous attitudes that ever could be conceived by a disjointed imagination," before then comparing the positions to those of a "dilapidated marionette." The *Daily Alta California* described the leg positions as "ludicrous" and the *San Francisco Examiner* as "so devoid of all naturalness, so complex and ungraceful." Muybridge sent some of his photos to *The Illustrated London News*, which noted that "nearly every attitude of the galloping horse is grotesquely awkward and ungainly." (In a later article, it called the static leg position of a trotting horse "unutterably hideous.")

That viewers were startled had much to do with the preconceived ideas of how a horse moved. In European art, painters followed certain conventions when it came to horses, and none of them resembled a dilapidated marionette. For example, when depicting a gallop, artists had traditionally positioned the front legs extended forward and the back legs backward. As an archetypal example, think of the work of eighteenth-century painter George Stubbs, whose smooth-limbed, sprightly horses jaunted elegantly around the English countryside. Muybridge's horses looked nothing like that.

French artist Ernest Meissonier, a renowned painter of military scenes (and horses within them), reportedly cried out in disbelief when he saw Muybridge's photographs. Meissonier went to great pains to depict the horses in his paintings with accuracy. He had even installed a small rolling seat on rails next to a racetrack, so he could travel in parallel with the horses and observe their motions. (As for Stubbs, he showed his fealty to equine accuracy by killing and then dissecting horses as studies for his

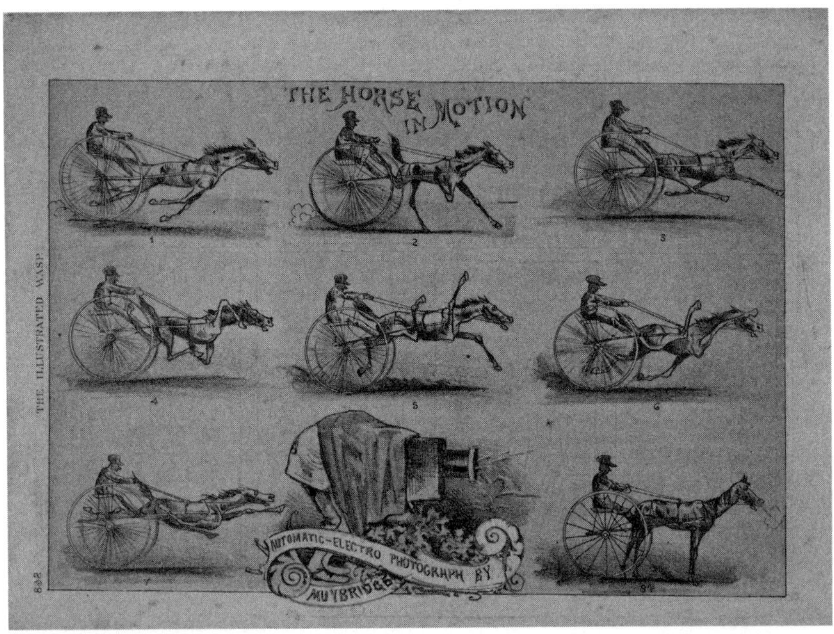

Muybridge's motion photographs destroyed the assumptions of equine elegance. This parody, from the San Francisco Illustrated Wasp, *July 27, 1878, reveals with humor the astonishment, and horror, of how a horse looked in motion.*

art—about twelve in total.) Once Meissonier recovered from his shock, he made a watercolor copy of his work *1807, Friedland*, which featured several galloping horses, but adjusted to his newfound knowledge. Other revered artworks came in for criticism. French painter Rosa Bonheur spent eighteen months at a market disguised as a man sketching horses for her work *The Horse Fair*. Post-Muybridge, one commentator said her horses were as fanciful as "unicorns."

Muybridge relished his newfound fame and reveled in the artistic crisis he had caused. In the early 1880s, he traveled to Paris and London and gave a series of grandstanding presentations on his work. In Paris, his audiences included artists, scientists, and—of course—Nadar, who would not have been able to resist the spectacle. In London, at the Royal Institution's imposing premises in Mayfair, Muybridge stated that "in no single instance" could he find a correct depiction of a horse. At the Society of

Arts, just off the Strand, Muybridge highlighted Bonheur's "glaring faults" in her depictions of horses. By now, Muybridge had more material to draw from. In 1879, he had taken more motion photographs, this time with a bank of twenty-four cameras to photograph horses and other animals. He had also created a new viewing device, which he presented in Paris and London, that he called the zoopraxiscope.

Muybridge did not possess a gift for names. The word "zoopraxiscope" was a mash-up of English and Ancient Greek (*praxis* means "action"), although knowing the etymology doesn't make *zoopraxiscope* any less of a mouthful. The zoopraxiscope was based on an optical toy called a zoetrope, one of several nineteenth-century devices that created the illusion of motion using persistence of vision. (This category includes the flip-book, which was invented in the 1860s and was also known as a *kineograph*, Latin for "moving picture.")

Muybridge had an artist copy his images, in sequence, around the edge of a sixteen-inch glass disk. On the zoopraxiscope, he positioned the glass disk next to a slotted disk, both of which were connected to a mechanism that made them spin in opposite directions. They were set in motion by cranking a handle, and the painted images rushed behind the window slots, appearing to the human eye as though they were moving. These sequences were projected onto a wall or screen, like a magic lantern presentation. A *moving* magic lantern presentation. Upon first seeing the zoopraxiscope in action in San Francisco, a reporter was awed: "Nothing was wanting but the clatter of the hoofs from the turf and an occasional breath of steam from the nostrils, to make the spectator believe that he had before him genuine flesh-and-blood steeds."

In London and Paris, the zoopraxiscope wonderment continued. In *The Illustrated London News*, the "unutterably hideous" positions of the horse's legs were transformed: "the ugly animals suddenly because mobile and beautiful, and walked, cantered, ambled, galloped, and leaped over hurdles in the field of vision in a perfectly natural and lifelike manner." Audiences gathered in front of the screen didn't see a static picture of a horse, as it had appeared in paintings and drawings for centuries; they saw a horse in motion—a horse as a moving picture. It was, wrote *The Photographic News*, "the essence of life and reality."

But those who prioritized the feeling of movement in art, rather than its literal depiction, remained unimpressed. "An artist paints what he sees and feels or what another observer sees and feels—not what rule and compass declare to exist," opined the *British Journal of Photography* in 1881, separating out "artistic truth" from "mathematical truth." No less a figure than Auguste Rodin agreed. "It is the artist who is truthful and it is photography which lies, for in reality time does not stop," he said, when discussing his sculpture *Saint John the Baptist*, whom he had depicted walking. This tension—between the sensation of movement and the facts of movement, between what was felt and what was mechanically observed—is awkwardly on display in a painting by Thomas Eakins from the late 1870s. For the horse's legs, he followed Muybridge's photographs, so they appear static. But the carriage wheels are painted with blurred brushstrokes, an artistic convention that suggests movement. It's a visual disconnect, and a literal depiction of the divide between those who emulated Muybridge's photographs and those who did not.

The 1880s were another momentous decade for Muybridge. He and Stanford had a falling out, after Stanford published a book about their locomotion studies called *The Horse in Motion* without crediting Muybridge. The author was J. D. B. Stillman, a friend of Stanford's, but "executed and published under the auspices of Leland Stanford." In the preface, Stanford relegated Muybridge to the position of an employee. Muybridge sued and lost. He found a new sponsor at the University of Pennsylvania, and in 1884 and 1885 he worked on a new series of motion studies that was broader in scope and more ambitious in execution.

This time he used three banks of cameras positioned around an open-fronted shed about 120 feet long. This was the stage for his subjects—not horses, but humans—to run, walk, leap, jump, climb, and more. Along the back wall of the shed, Muybridge positioned white threads horizontally and vertically to create a grid. With the three banks of cameras, he could capture up to thirty-six images from three different perspectives. Instead of a trip wire, the shutters were connected to an electromagnetic device, and they could be set off either in succession or simultaneously. The dry plate revolution had arrived, so he also didn't need to mess around with chemicals while under time pressure, as he had at Palo Alto.

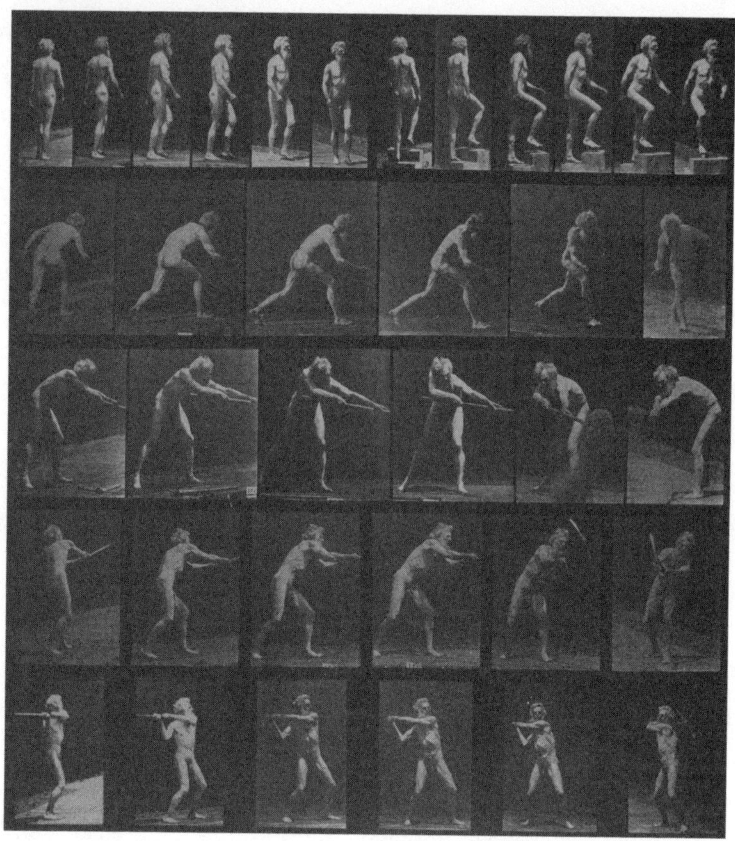

Muybridge in motion, as part of his University of Pennsylvania series, circa 1884.

Muybridge's goal was, he said, to show how "the movements appertaining to every-day life are performed." Some of his subjects were fully clothed, some draped in transparent fabric, and some were naked. And while some photos showed standard poses and activities, others do not— for example, the sequence of a naked woman kicking a hat. Muybridge himself also posed for several naked motion sequences, including one where he is seen swinging a hatchet.*

---

*   In his notes, Muybridge described himself as an "ex-athlete"; other men in the series were similarly described by their professions. The women were described by age, weight, and marital status.

In the summer of 1885, in the final year of his work with the University of Pennsylvania, Muybridge moved his equipment to the Philadelphia Zoo. Here he photographed a menagerie of animals in motion: oryx, elk, buffalo, elephants, a camel, lions, a baboon, a capybara, and a kangaroo, along with pigeons, vultures, hawks, cockatoos, ostriches, eagles, hawks, and swans. Most of these animals were pictured in expected situations, except for one particular sequence titled "Chickens Scared by a Torpedo."

At the end of this mammoth project, Muybridge had created around 100,000 negatives. In 1887 he published a book, *Animal Locomotion: An Electro-photographic Investigation of Consecutive Phases of Animal Movements*, which contained 781 plates, each showing a sequence of movement. The entire volume cost $600, but Muybridge also sold sections of the book for a lower price. To advertise it, Muybridge issued a prospectus; subscribers from the art world included John Everett Millais, James Whistler, John Singer Sargent, and Meissonier.

While studying Muybridge's work, historian Marta Braun has identified that, in about 40 percent of his plates, the sequences of motion are not actually continuous. He substituted, or missed out and renumbered, certain sets of images, which naturally calls into question their scientific utility. As Braun argues, Muybridge, "under the guise of offering us scientific truth,

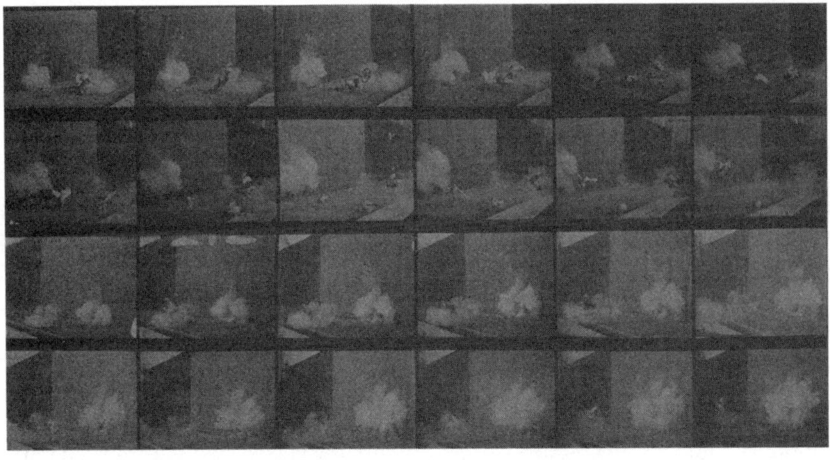

Muybridge's imaginatively odd series, "Chickens Scared by a Torpedo."

has, like any artist, made a selection and arranged his selection into his own personal truth."

But there was one scientist who drew inspiration from Muybridge to start his own revelatory photo investigations of movement: the French physiologist Étienne-Jules Marey.

Up until he saw Muybridge's work published in the French journal *La Nature* in 1878, Marey had largely resisted using photography to record motion. Marey was fascinated by movement, once writing, "From the invisible atom to the celestial body lost in space, everything is subject to motion." He devised instruments to record invisible movements, such as muscle contractions, or the wrist's pulse beats, as well as methods to study more visible forms of movement—like those of a horse in motion.

In the early 1870s, Marey attached a pneumatic rubber ball to the underside of each of a horse's hooves. Each ball was linked, via rubber tubes, to what Marey called a "registering apparatus," consisting of levers and a cylinder. When the horse's hoof struck the ground, it pushed air up the tube, which moved the corresponding lever to make a mark on the cylinder. From these annotations, Marey produced charts that tracked the horse's gait. His findings validated the idea of unsupported transit. Marey published these results in a paper in 1872, around the time that Muybridge was starting his first experiments. He also had them translated into a series of notations to use in a spinning zoetrope, the persistence-of-vision device that was the basis for Muybridge's zoopraxiscope.

But Marey didn't have photographs as visual evidence. He had graphs and charts. All his devices required some direct physical contact with the subject, such as harnesses he created to study the flight of birds. (He also constructed a mechanical bird and built artificial insect wings.) After seeing Muybridge's work in *La Nature*, he wrote to Muybridge to request instantaneous photographs of birds in flight. When Muybridge produced the photographs during his trip to Paris in 1881, Marey was disappointed. Muybridge had photographed a group of birds in flight, not a single bird flying through a succession of movements as Marey had wanted. To capture this trajectory, Marey realized, he would need to build a new kind of camera.

"I have a photographic gun that has nothing murderous about it and that takes a picture of a flying bird or a running animal in less than 1/500 of a second," Marey wrote to his mother, early in 1882. "I don't know if you can picture such speed, but it is something astonishing." The camera looked like a rifle: the barrel contained the lens, which was focused by either lengthening or shortening the tube. A wide cylindrical breech held a circular dry plate, a disk with twelve apertures, and a disk with a single slot, which served as a shutter. All three plates connected to a central axis, so that when Marey pulled the trigger, they rotated, sequentially imprinting, at rapid speed, the twelve views onto the plate. The exposure time was 1/720th of a second. Marey based the design on a photographic revolver created by the scientist Jules Janssen, which Janssen had built to capture the transit of Venus in 1874, with mixed results. (Janssen had based *his* camera on a Colt revolver, providing yet another dimension

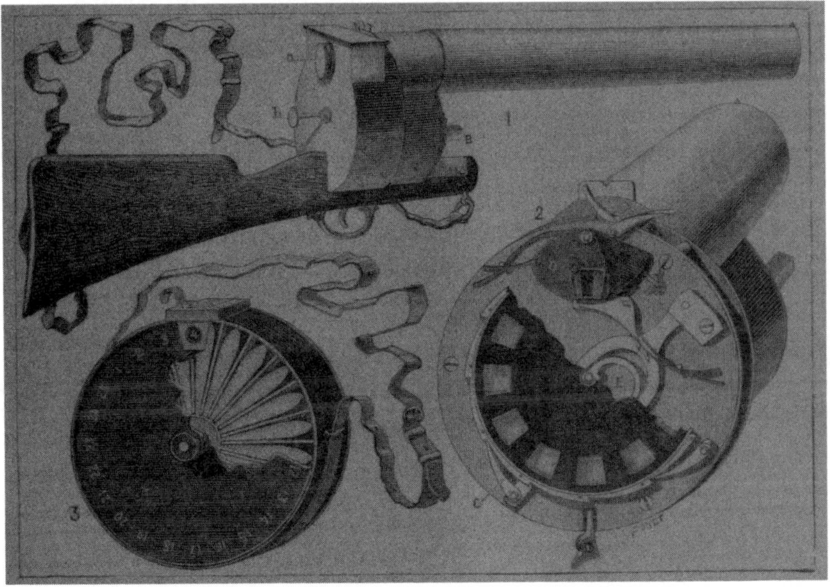

Marey's "photographic gun," as illustrated in La Nature, *1882.*

in the relationship between photography and guns—also embodied by terms like *snapshot, shooting, flashgun*—as one that is exceedingly, and sometimes uncomfortably, close.)

"The photographer raises the gun to his shoulder as if in the act of shooting a bird," reported *The Photographic News*, which also called Marey's photographic gun "so exquisitely ingenious." But Marey wasn't entirely satisfied. The photographs were small. To study side-by-side views of the wings in motion, Marey had to cut out each image from the disk and piece them together. When Marey tried to increase the plates' rotation speed in order to capture more images, the camera vibrated so badly that the images blurred.

"The instruments that are not there, he creates," Nadar wrote of Marey. "Those which are there, he perfects." Marey took the idea of the rotating, slotted disk shutter from the photographic gun and paired it with a regular camera and a static plate. He added a large handle to set the slotted shutter in motion. He called this new type of imagery *chronophotography, chronos* meaning "time." As the shutter ticked through its rotations, it captured the subject's intermittent movements, like a strobe light on a dancer in a nightclub. But instead of a sequence of photos showing slices of movement, Marey's chronophotographs showed the succession of movement on one plate. Marey began his experiments with a subject dressed in white, against a black background; he appears as clones duplicated across infinite space. One remarkable photo shows a man swinging on a rope, a pendulum in human form, a picture of halted oscillation.

Now that he had his basic technique, Marey continued to expand his research and his tools. He built a new camera with a larger shutter that ran on wheels on a track in front of a new background, a black open-front shed. He experimented with his subject's attire, by clothing one body part in black so it disappeared from the final picture. He also tried dressing his subjects entirely in black velvet, including their heads, but adding white cords along their limbs and white buttons on their joints. The photographs show a stick figure in motion, a sequence of lines and hinges stalking across the frame.

Throughout the 1880s, as he continued his experiments and improved his tools, Marey photographed the motions of gymnasts and army cadets, the

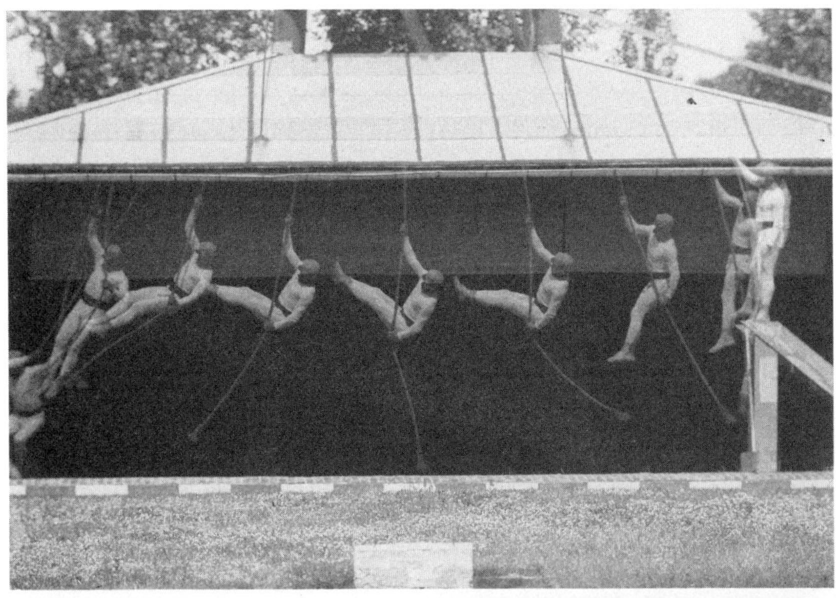

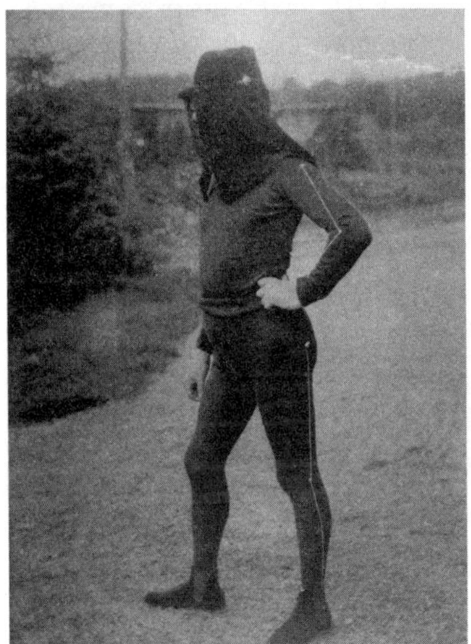

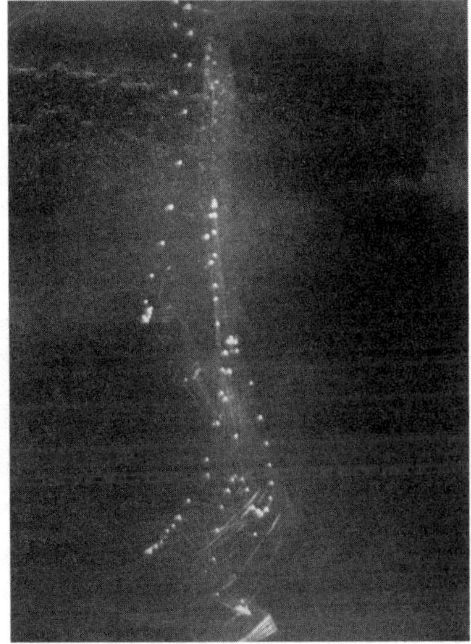

Marey captured motion in ways that had never been seen before. When he dressed his subject in black with white cords and buttons (bottom left), only the essentials of movement appeared on the plate (bottom right).

trajectory of batons and balls falling through the air, and the restless movements of birds, horses, and, from the zoo, two elephants and a water buffalo.

Once George Eastman's new roll film became available, Marey seized on its potential. He removed the fixed plate from his camera and added two mechanized rotating spools to feed the roll film across the lens. An electromagnetic clamp held the film to the lens momentarily, just to register the image, before the spool moved the film along. In the fall of 1888, at the Académie des Sciences in Paris, Marey presented a photographic sequence of a bird in flight, and a hand opening and closing. He didn't have the ability to project the images yet; they appeared as long strips of film, made with an exposure of 1/500th of a second. But the potential to project movement—to make the bird flap its wings, the hand open and stretch its fingers—must have been tangible.

Marey worked on creating a projector, but progress was frustratingly slow. He was also looking for a device to help him with scientific analysis, not reconstitute movement, which is the function of a film projector. As it so often happened in photographic history, several people were examining the same problem, around the same time, in slightly different ways. Muybridge, for example, had discussed with Thomas Edison combining the zoopraxiscope with the phonograph in 1888, to produce, for an audience, a combination of "visible actions and audible words." That same year, a French inventor named Louis Le Prince constructed a camera that captured short sequences of motion.

In 1889, at the Paris Exposition, Émile Reynaud projected animated strips of film with a magic lantern, pulled by small, evenly spaced perforations along the film's edges, or what we now call sprocket holes. While in Paris for the exposition, Edison visited Marey's laboratory and saw his strips of film with people and animals in motion. (After his visit, Edison reportedly said to an employee, "I knew instantly that Marey had the right idea.") Edison's experiments resulted in the Kinetoscope, where moving pictures were viewed through a hole in a box, one person at a time. It would be the Lumière brothers who would project moving pictures to an audience in 1895 with their cinematograph.*

---

* The night that Paul Martin photographed London's Empire Theatre, it was screening a cinematograph; the word is just visible on the marquee, gleaming in the darkness.

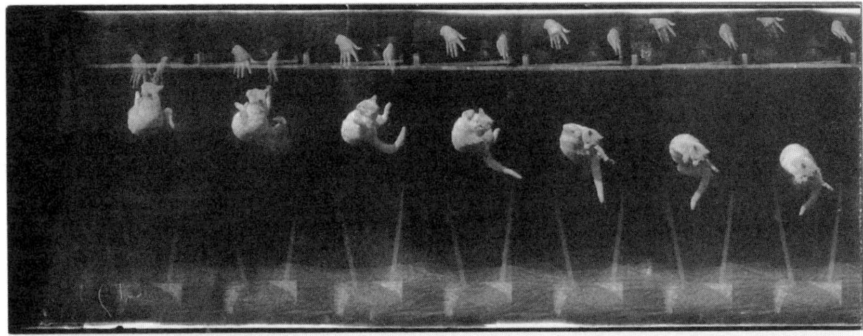

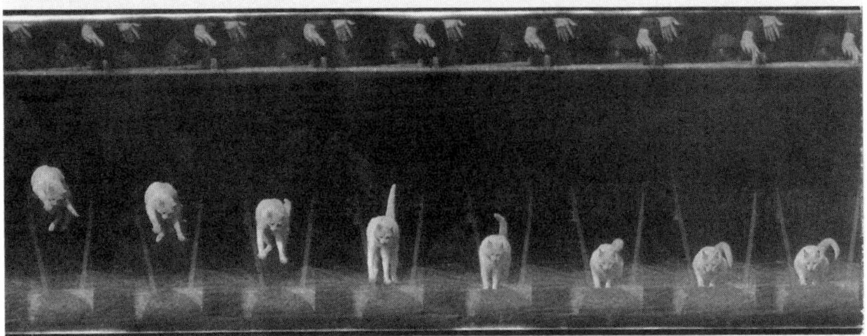

*One of Marey's more famous sequences, showing a cat twisting in the air and
landing on its feet.*

Marey remained steadfast in his interest in science, and dispassionate
about the evolution of moving pictures. In 1900 he wrote, "Animated projections, interesting as they are, are of little advantage to science, for they
only show what we see better with our own eyes."

Muybridge's early photographs had changed how artists depicted
horses; Marey's chronophotographs influenced a whole new artistic style.
Marcel Duchamp was inspired by Marey's work to create his 1912 masterpiece, *Nude Descending a Staircase (No. 2)*. Italian Futurists, obsessed with
time and modernity, also drew from Marey's work. As Aaron Scharf points
out in his book *Art and Photography*, naturalist painters embraced Muybridge's fixed moments in time, while Marey's images served those artists who focused on "the movements themselves rather than the objects in
movement, the fundamental rhythms and patterns of the universe." When

removed from the realm of science, Marey's chronophotographs become elegant, abstract artworks showing bodies in space.

Beyond their influence on artists, both men, separately but in tandem, contributed to the creation of an entirely new art form, cinema. Their work helped to pull photographs to life, moving pictures from isolated moments in the past to a smoothly running continuous present.

Chapter 11

Cells and Snowflakes

From photography's earliest days, practitioners sought to capture the wondrous visions that only a microscope could reveal. In the late 1830s, William Henry Fox Talbot photographed the cross-section of a plant stem. In the early 1840s, physician Alfred Donné and physicist Léon Foucault produced daguerreotype photomicrographs of bone tissue, teeth, and bodily fluids. But whether it was for the natural world or the medical, the intricacies of photomicrography challenged both photographers and microscopists. The results raised important questions about the accuracy, objectivity, and, most fundamental of all, the value of photomicrography.

Prior to photography, anyone who wanted a permanent visual record of what they saw through a microscope had to rely on illustration. The difficulties in drawing the singular view through a microscope were significant. A manual on microscopes from 1868 suggested a few different techniques. One involved fixing a beady eye to the microscope tube while employing the other eye to "govern the movements of the pencil." Another called for a camera lucida, the drawing aid that Talbot had used so unsuccessfully at Lake Como, after which he had the idea for photography. A third involved the use of a tinted glass reflector. All were prone to the errors of the human hand, and the subjectivity of the human eye.

But photomicrography had its own specific requirements, too. Steadiness was crucial. It helped to position the microscope horizontally, and for the operator to keep totally still during the exposure, as any movement vibrated through the apparatus and blurred the image. For the same reason, one journal suggested that anyone who lived near busy streets should work at night, or from a basement, to reduce the risk of any reverberations in the apparatus.

Light also needed careful consideration. The connection between the microscope's eyepiece and the camera needed to be covered with a light-proof material, but the specimen itself needed to be well illuminated and clearly focused. (It was sometimes necessary to rearrange the focus at the last minute, in case the equipment had expanded under the hot light.) Aside from the setup, the operator also needed to manage the chemicals and timing for the wet collodion process, at least until the dry plate process arrived in the late 1870s.

But the appeal of photomicrography is evident from the early English manual, *The Wonders of the Microscope, Photographically Revealed*, published in 1857. The author made his position on the value of photomicrography clear: "It must be evident, we think, to any one acquainted with the ordinary mode of copying microscopic preparations, that however great the attention and labour bestowed upon illustrations thus produced, they are necessarily always more or less defective and unnatural."

The publication of this manual coincided with a growing popularity of at-home microscope use. In that same year, 1857, *Household Words* declared, "We appear to be on the eve of a microscope mania." In the article, the author suggested ordering specimens through the mail; at the time of writing, he was waiting to receive the mouth of a leech. To vast numbers of amateurs, this small but mighty instrument offered education, entertainment, and an opportunity to be awed by the divine, or as one early microscope manual put it, "additional reverence and admiration for the beneficent and Almighty Creator." The entertainment continued at microscope societies, such as London's Quekett Microscopical Club, where, in January 1867, some 400 guests met, mingled, and viewed the toe of a mouse, the head of a beetle, and "a most extensive colony of mites

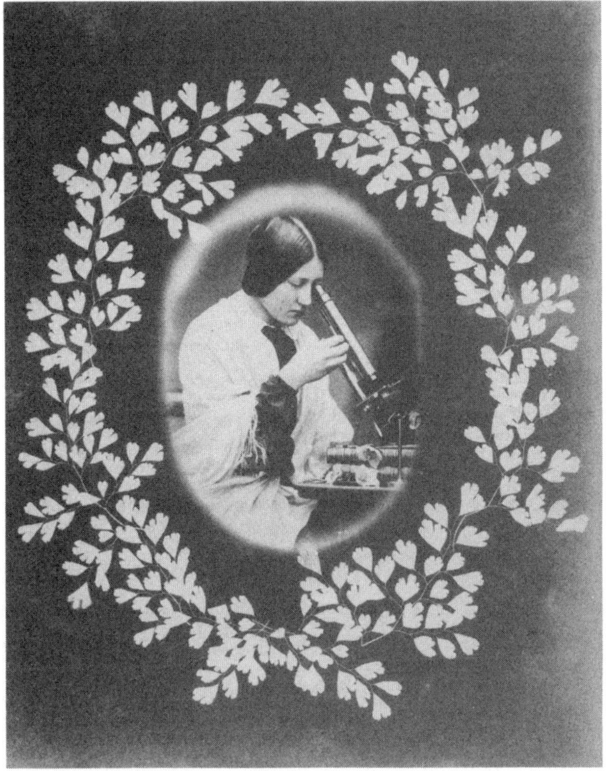

Scientist Thereza Dillwyn Llewelyn at a microscope, 1850s.

in a full quarter of a pound of cheese."* One Quekett soirée also featured magic lantern displays of enlarged photomicrographs by Richard Leach Maddox, the physician (and enthusiastic microscopist) who later experimented with gelatin for dry plates.

For those who wanted to put their home microscope to practical use, journals cited the instrument's utility in detecting food adulteration. In a warning against milk from London cows, for example, *Eliza Cook's Journal* noted, "The milk they yield shows constantly on examination by the

---

* The Quekett Microscopical Club is still running and holds meetings and workshops regularly.

microscope, that it contains particles of pus, or diseased matter." In 1862, *The English Woman's Journal* praised the microscope for "bringing to light the flour in our mustard, and the lime-white in our flour," before a call to arms: "Let it only be as common for a lady to have a microscope in her parlor as to have a pair of scales in her kitchen, and honesty would perforce become once more the policy of our traders." In the United States, the book *Food Materials and Their Adulterations* suggested that a trained eye and a microscope could detect additives in tea, mustard, pepper, and cinnamon.

Trickier, though, was marrying a microscope and camera with something that was as ephemeral as a quickly evaporating snowflake.

The English winter of early 1855 was astonishingly cold. The Thames froze, halting river traffic and forcing thousands out of work. Snow and frost delayed trains. Horses struggled along slippery roads. Omnibus drivers took to wearing veils to shield their eyes from the snow and the "terrific fury" of the wind. The city's death toll rose, which newspapers linked to the freezing conditions as well as the omnipresent "zymotic" diseases, such as cholera and dysentery, that persistently ravaged nineteenth-century populations.

Those more fortunate flocked to the city's parks by the thousands to ice-skate on frozen ponds. In late February, a brass band played to revelers on the Serpentine lake, as fireworks exploded in the sky above Hyde Park. Whether their experiences were happy or hazardous, Londoners endured weeks of freezing temperatures and extraordinary snowfall. And for James Glaisher, who later ascended so high in a balloon that he promptly passed out, it was a thrilling opportunity to observe and record the structure of snowflakes.

Glaisher set up a plate of glass outside an open window and, from inside, used a lens of "almost microscopic power" to look into the snow crystals. He methodically observed their accumulation and structure, noting temperatures and time of day. Yet even Glaisher, a man who, after regaining consciousness at some 29,000 feet above the earth, kept calm and carried on with his meteorological recordings, was swept away by the extraordinary vision the microscope provided: "It was difficult to conceive how such compound and solid structures could find existence in the almost imperceptible speck before me."

His subsequent papers included elegant illustrations of snow crystals. He asserted that the drawings were "faithful copies," even though what he was copying was slowly but surely evaporating on the glass. He gave his rough drawings to his wife Cecilia Glaisher, who worked them into beautifully detailed illustrations, sometimes with color. (He was also aided, for some specimens, by a "Mrs. King.") To this day, the Royal Microscopical Society has, as its emblem, a snow crystal by the Glaishers. It's easy to see why: the illustrations are outlandishly, absurdly beautiful. Recognizing this, Glaisher even sent some illustrations to *The Art-Journal*, to suggest snow crystals as an inspiration for ornamental designers. To our eyes, the shapes are instantly recognizable, a symbol of winter that appears everywhere from holiday cards to iPhone weather forecasts.

These beautiful illustrations were by no means the first to depict snow crystals. Some two hundred years earlier, Robert Hooke's *Micrographia*, published in 1665, included drawings of snow crystals that he observed with a microscope, along with other unseen wonders, such as the now familiar shape of a magnified flea, and the less familiar image of frozen urine. But accurately drawing a snow crystal, whether in the seventeenth century or the nineteenth, was a race against time; photographing a snow crystal, on the other hand, offered a quicker and more objective means to record its structure—as long as you could do so in the moments before it vanished forever.

If you have ever seen a photo of a snow crystal, it was probably one of Wilson Bentley's. They're famous, and rightly so: they're gorgeous. They show beautiful, elegant, multifaceted forms, in a range of shapes from lacy stars to triangles to hexagons. They represent the archetypal vision of wintry loveliness that you think of when someone says "snowflake."

Bentley's photos are also prodigious in number. From when he first started in the winter of 1885, the year he turned twenty, to 1931, when he died from pneumonia after walking through a blizzard, Bentley produced more than 5000 snow crystal negatives.

Unlike Glaisher, who had peered out a window into the cold, Bentley

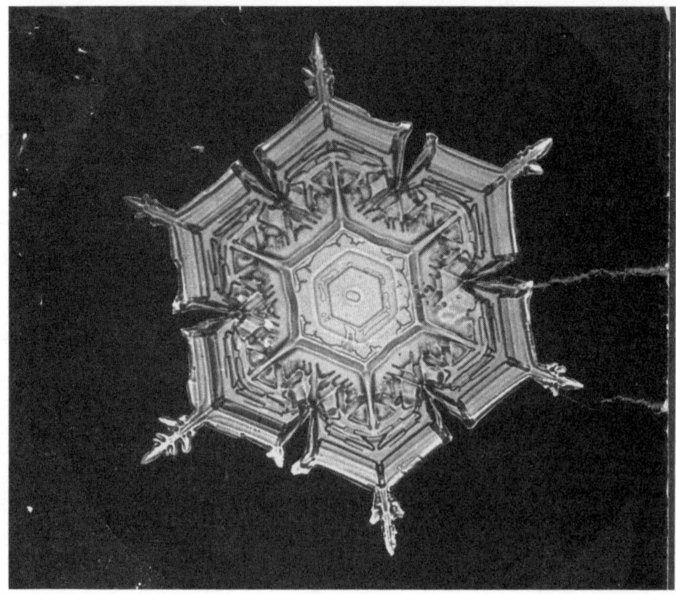

Bentley photographed some 5000 snow crystals over his lifetime.

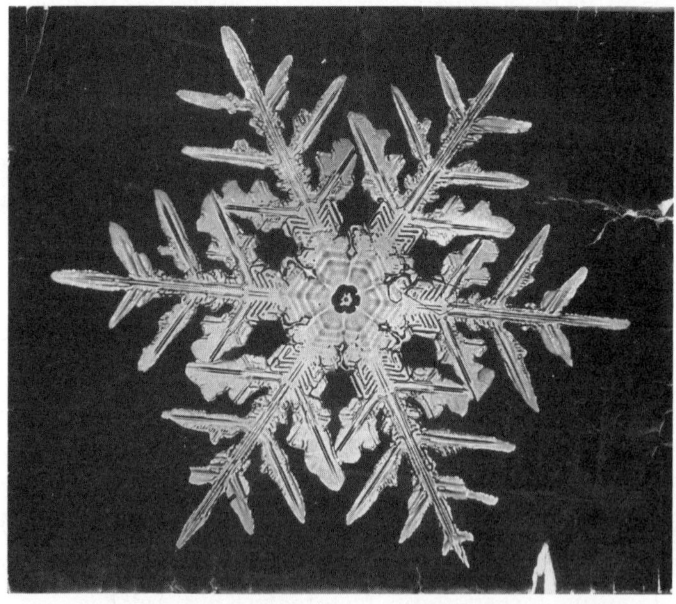

Improbable as it may seem, Bentley reported that one collector of his snow crystal photomicrographs was a Professor Snow, in Wisconsin.

photographed snow crystals in an unheated shed on his farm in Vermont. The process was intricate. He used a black tray to catch and select his snow crystals, and feathers to isolate the specimen, and to gently press it flat on the microscope slide. (The transfer from tray to slide was done using a splint from the bushy end of a broom.) His camera faced the shed's only window, to transmit light through the slide and specimen. For a shutter he used a black card over the microscope's objective, which he removed for the exposure time of, he later estimated, anywhere between eight and one hundred seconds. One essential item to this process, he once explained, was a pair of thick mittens.

In the late 1890s, Bentley started to sell his photomicrographs, some of which appeared in the *Monthly Weather Review*, a journal for meteorologists, and to write articles about snow crystals. Bentley is largely credited with the notion that no two snow crystals are alike, although Hooke, Glaisher, and others had also observed their many forms.

Despite the notion that photomicrography offered objectivity, Bentley's photomicrographs had a subjective element. His original negatives showed a transparent object photographed against a light background, but he wanted to portray snow crystals "in all their unrivalled beauty, with sharp outlines, white upon a dark ground." Bentley adopted a method known as blocking out, whereby on a copy negative, with a penknife, he scraped off the background around the edges of the snow crystal. When this negative was made into a print, the background would appear black and the snow crystal would be well defined. To Bentley, this was how a snow crystal looked when he first examined it on the black wooden tray.

But to Richard Neuhauss, editor of the German photo journal *Photographische Rundschau*, Bentley's approach was unacceptable. His photomicrographs, Neuhauss wrote sternly, were "almost completely worthless" because Bentley did not adhere to "one of the most elementary principles" of photomicrography: no retouching. (In these statements, Neuhauss reads like a German version of Thomas Sutton.) Neuhauss had made his own photomicrographs of snow crystals in the 1890s, which looked quite different from Bentley's, and were far from anyone's vision of wintry loveliness. Neuhauss also claimed that he was the first to photograph snow crystals,

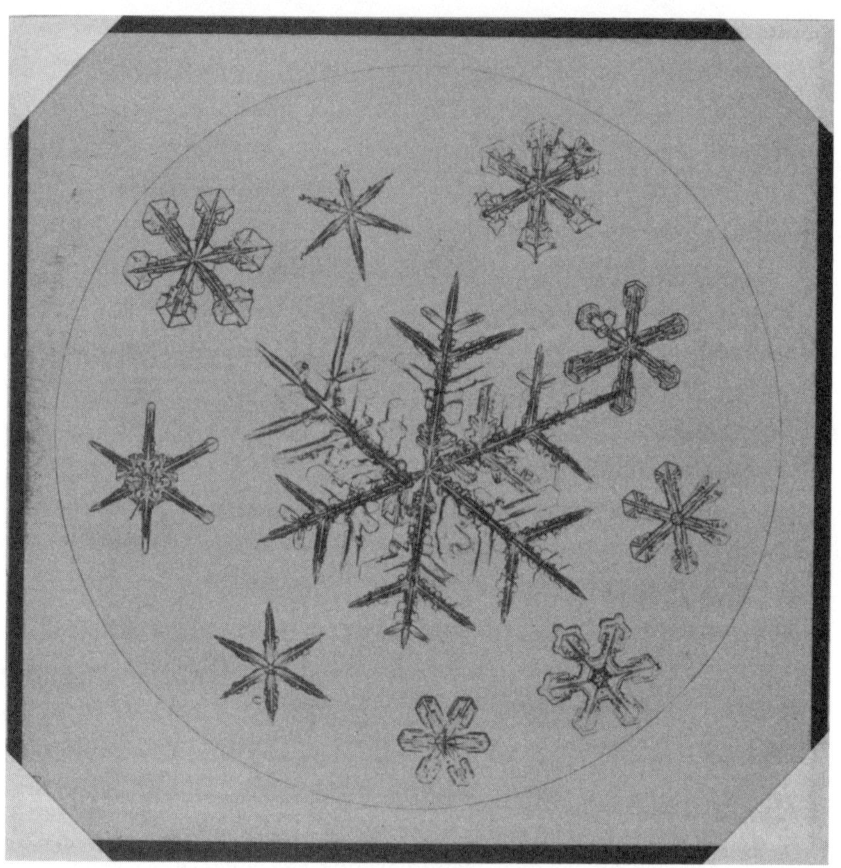

Neuhauss's snow crystal photomicrograph, 1892.

a claim that was sharply rebuked by Professor Gustaf Nordenskiöld of the Stockholm Academy of Science. (The early years of snow crystal photomicrography were surprisingly contentious.) Neuhauss's photos, Nordenskiöld wrote, are "few and rather poor, and the directions given by him for the photographing of snow crystals entirely misleading."

Bentley also preferred to photograph only the most symmetrical and unbroken snow crystals. He acknowledged his "old habit of seeking the beautiful and interesting, rather than the characteristic types" in a 1902 report for the *Monthly Weather Review*. That winter, he had photographed over 200 snow crystals.

Buried within the tempestuous debates over snow crystals were important questions about the value of photomicrography, which rested on notions of objectivity. "Photography certainly cannot lie, but the photographer may be a liar or a fool. For this reason the photomicrographer must not only be unprejudiced and honest, he must also be a microscopist and know his object," stated one manual. The author of an 1892 guide praised photomicrography's "great value" as providing results "free from bias." He continued: "It is a universally accepted canon that retouching, farther than the 'spotting out' of unavoidable pinhole in negatives, is inadmissible." Or, to paraphrase the author more succinctly, a photomicrograph should tell the truth, even if it cannot tell the whole truth.

Bentley responded to Neuhauss's accusations in the *Monthly Weather Review* by stating that he rejected the "ultrascientific viewpoint that insists on having a photograph presented exactly as it emerges from the chemical

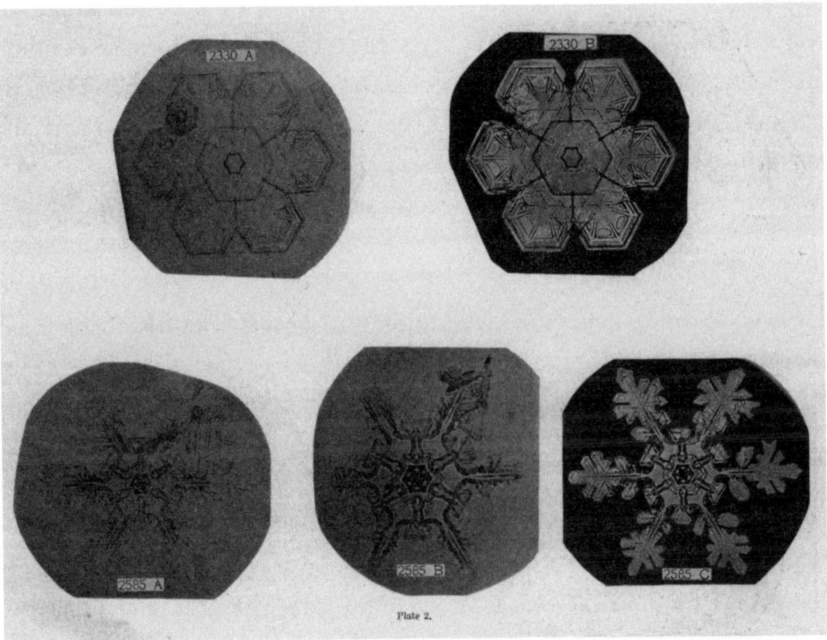

Bentley's response to criticisms of his method included these side-by-side photographs of snow crystals before and after retouching.

bath," before adding, "a true scientist wishes above all things to have his photographs 'as true to Nature' as possible; and if retouching will help in this respect, then it is fully justified." He included side-by-side before-and-after pictures so the reader could compare the difference.

In 1954, Japanese physicist Ukichiro Nakaya published his book *Snow Crystals: Natural and Artificial*. In the 1930s, Nakaya produced the world's first artificial snow crystals, which he created at Hokkaido University in a specially built laboratory (average temperature −35 Celsius), and took many photomicrographs of snow. In reviewing the history of snow crystal photomicrography, Nakaya noted that for scientific purposes, Bentley's retouching, and his failure to include magnification details, were "regrettable." He also noted that Bentley's photos had inspired people to see the beauty of snow crystals.

On this last point, there is no doubt. Today, you can buy mugs and tea towels and ornaments adorned with Bentley's snow crystals, just as Glaisher had suggested decades ago. It's precisely because Bentley's snow crystal photomicrographs are so beautiful that they have remained so popular. Part of the reason that they are so beautiful is because Bentley made sure they looked beautiful, with their rich dark backgrounds and their shining, well-defined shapes. The other reason they are so appealing is that they represent a hidden beauty, one that could be found on our doorsteps or sprinkled onto our windowpanes, if only we had a way to see.

In 1862, Joseph Janvier Woodward was transferred from his position as an assistant surgeon in the US Army to the Office of the Surgeon General, to work on *The Medical and Surgical History of the War of the Rebellion* (*MSHWR*). This vast, six-volume title, by multiple authors, was published between 1870 and 1888 and provided a detailed account of Civil War injuries and diseases based on statistics, reports, and photographs. (The six volumes were illustrated with a wide range of printing processes, including engravings and illustrations, so if you like a variety of gruesome imagery, this is the book for you.) Woodward's section was on camp diseases such as dysentery, and he saw photomicrography as "indispensable to the proper representation of microscopic objects." It was an area that was unfortunately rich with examples: of the Civil War's estimated 750,000 deaths, around two-thirds were from disease.

By 1865, Woodward oversaw the Medical and Microscopical division of the Army Medical Museum. (Stepping away from the microscope for a moment, Woodward was also in the unfortunate and unique position of attending to the autopsies of two assassinated presidents, Abraham Lincoln and James Garfield.) Woodward wanted to include photomicrographs in his section of *MSHWR*, but he was a surgeon, not a photographer. After consulting with Lewis M. Rutherfurd, a successful astrophotographer, he converted a room on the fourth floor of the Army Medical Museum into his photomicrography studio. The entire room was the camera, with darkened windows and a heliostat (an apparatus with a mirror to direct the sun's rays) positioned outside. Light traveled from the heliostat, through a color filter, onto the specimen, into the microscope, and projected the image onto the plate. Woodward clearly knew his limitations: he also employed what he called a "practical photographer" who could manage the wet collodion chemistry while he focused on his microscope. To that end, Woodward was an early adopter in the United States of aniline dyes to stain the tissue, which provided better contrast than natural dyes.

Like Nadar before him, Woodward grew tired of relying on the sun. He tried artificial light sources, including magnesium and an electric light powered by a 50-unit battery placed in a closet. (The battery fumes were piped into the chimney, unlike Nadar, who had to suffer while underground and surrounded by skeletons.) The electric light proved so effective that Woodward was able to knock out between twelve and thirty photomicrographs in a four-hour stint after the sun went down.

Woodward shared his photomicrographs with microscopists in Europe, and they were praised for decades. Those presented at a meeting of the Quekett Microscopical Club in 1869 were described as "extremely perfect and beautiful." As late as 1896, influential microscopist Simon Henry Gage wrote, in his book *The Microscope and Microscopical Methods*, "No photomicrographs of histological objects have ever exceeded those made by Woodward."

On a purely visual level, the photomicrographs that Woodward produced are gorgeous abstract patterns, some recognizable (a mosquito) and

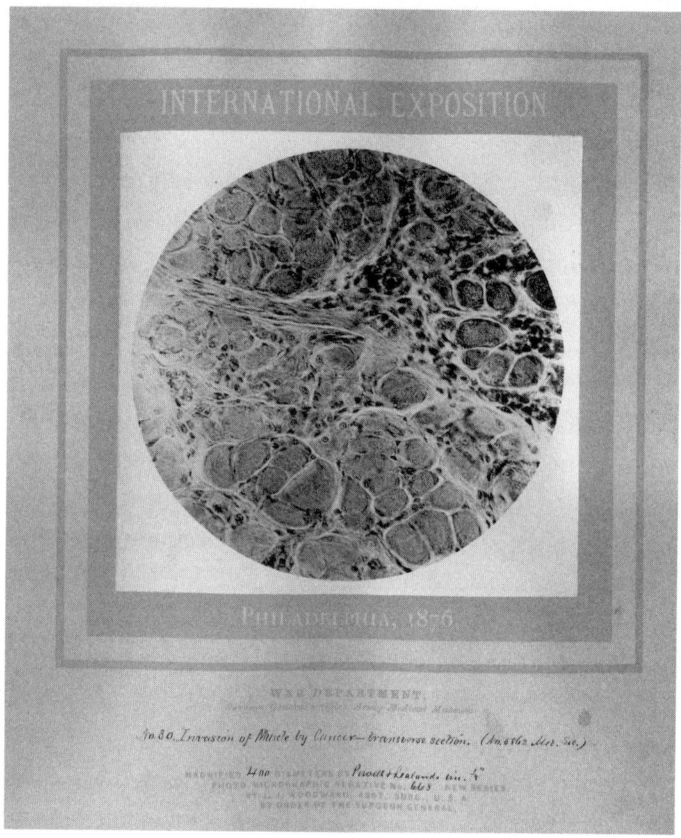

Woodward's photomicrographs were praised for decades.

some not (a cross-section of a colon). He produced insect photomicrographs simply to demonstrate the capacity of his microscope; they're captivating. But Woodward was not about aesthetics. Photomicrographs were for communication, diagnosis, and documentation. He would be intrigued, and possibly perturbed, to know that today, some of his photomicrographs are part of the collection at the San Francisco Museum of Modern Art.

Photomicrography was the earliest form of medical photography, a field that experienced some uncomfortable growing pains in the nineteenth century. Initially, doctors who wanted to photograph patients had to send them studio photographers, who used their own aesthetic, rather than clinical, conventions. The result saw patients posed next to props, or with a

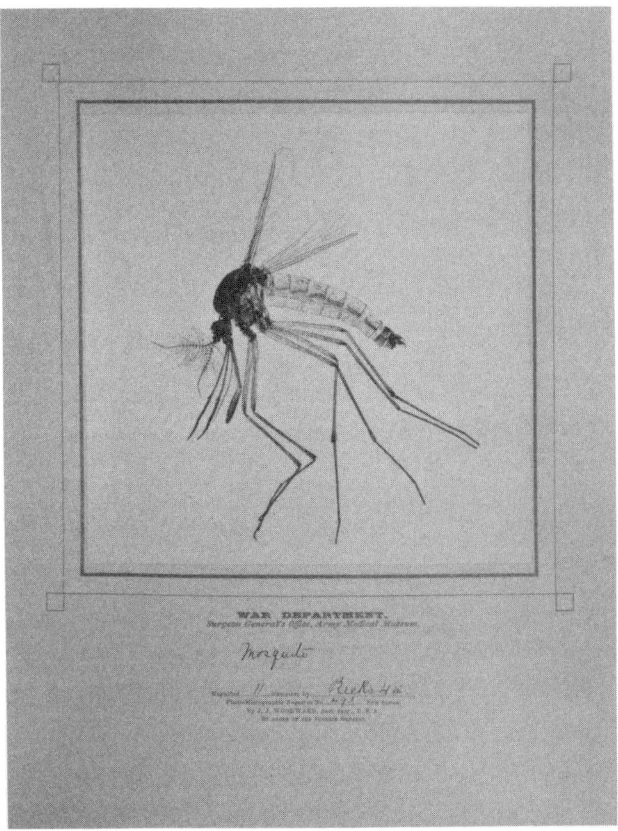

One of Woodward's test subjects.

background, appearing in every respect like regular studio portraits apart from, say, a distractingly large gunshot wound to the abdomen.

As with photomicrography, not everyone agreed that medical photographs were superior to illustrations. Depending on the subject, monochrome medical photographs could lack clarity; photographs of organs and tissues for morbid anatomy were especially difficult to decipher. The preference for illustration remains in some areas of medicine: consider, for example, the medical bible *Gray's Anatomy*, first published in 1858. Its forty-second edition, published in 2020, still includes illustrations.

Nor could early medical photographs highlight the area of interest or suppress that which was irrelevant. The photos showed whatever was in

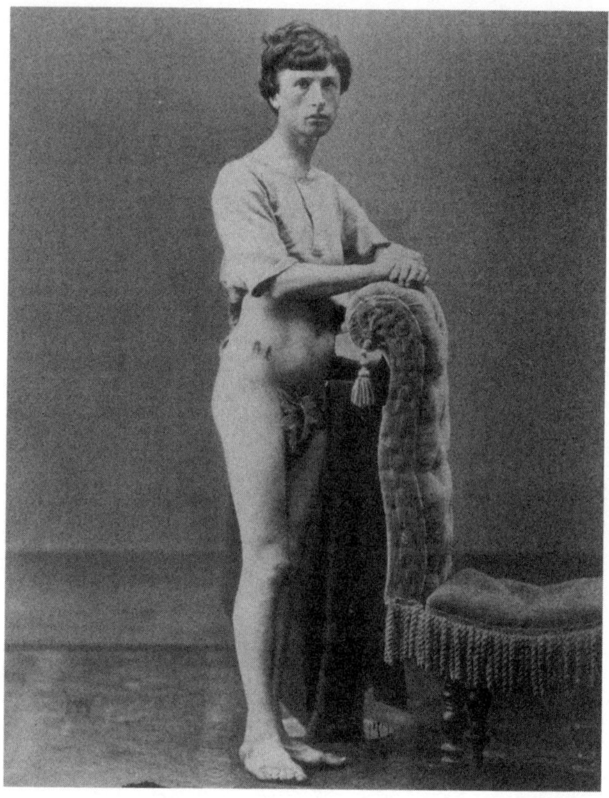

An early medical photograph of a Civil War soldier with an abdomen wound from a musket ball. Note the incongruous setting, the wound, and the fig leaf over his groin, which was drawn onto the print for modesty.

the camera's field of vision, including a patient's face and clothing, which gave away both their identity and clues as to their social class.

For doctors, photographs offered a permanent record, a way to diagnose, and to communicate—sometimes all three at once. One note that reportedly appeared on medical portraits of unfamiliar conditions was the scrawled question, "What is it?"

But photography had the lure of objectivity and speed, at least compared with the very manual process of illustration. One of the more successful applications of medical photography was in dermatology, an area of medicine with a clear visible component. In the mid-1860s, London dermatologist A. J. Balmanno Squire published a highly praised series of photographs that had been hand-tinted to more clearly depict various skin conditions and diseases. The *British Medical Journal* reviewed the book in glowing terms: "The photographs are singularly perfect representations of typical cases; and the whole work may be recommended as one of great beauty and usefulness." Dermatology and photography remained close: some twenty years later, the gun-toting, magnesium-flaring dermatologist Henry Piffard devised his pistol cartridge to photograph his patients.

Photography also played a role in diagnoses that we consider to be less visible, such as mental illness. In the nineteenth century, a person's facial features were believed to exhibit clues as to their personality, a practice known as physiognomy. These ideas were also applied to mental illness. (They were also used by Francis Galton, who originated eugenics, and who produced composite portrait photos to distill what he described as the "central physiognomical type of any race or group.")

In the late 1840s and 1850s, Dr. Hugh Welch Diamond, the medical superintendent of the women's ward of the Surrey County Lunatic Asylum, a public psychiatric hospital, photographed his patients as a form of diagnosis and treatment. In Diamond's view, photography's "unerring accuracy" could capture the features that revealed each patient's "internal derangement."

Diamond's patients were poor, and frequently came to the hospital from a local workhouse. In the portraits, his patients are typically seated, sometimes wrapped in a shawl or with loose hair; he also dressed one patient as Ophelia, with a crown of flowers. Whether any of the women gave their permission to have their photo taken is exceedingly unlikely. Nor is it likely they gave permission to have their portraits displayed at the Society of Arts photo exhibition in 1852, with Diamond's title "The Types of Insanity," nor at the subsequent exhibitions where they appeared.

In Diamond's view, the portraits were a form of treatment for his patients. In a paper that he presented to the Royal Society of Medicine in 1856, "On the Application of Photography to the Physiognomy and Mental Phenomena of Insanity," Diamond stated that in several cases, "photography unquestionably led to the cure." One example he gave was a patient whose mental delusions decreased after she looked at her portrait. This was a huge shift in treatment methodology. One of the long-standing practices at asy-

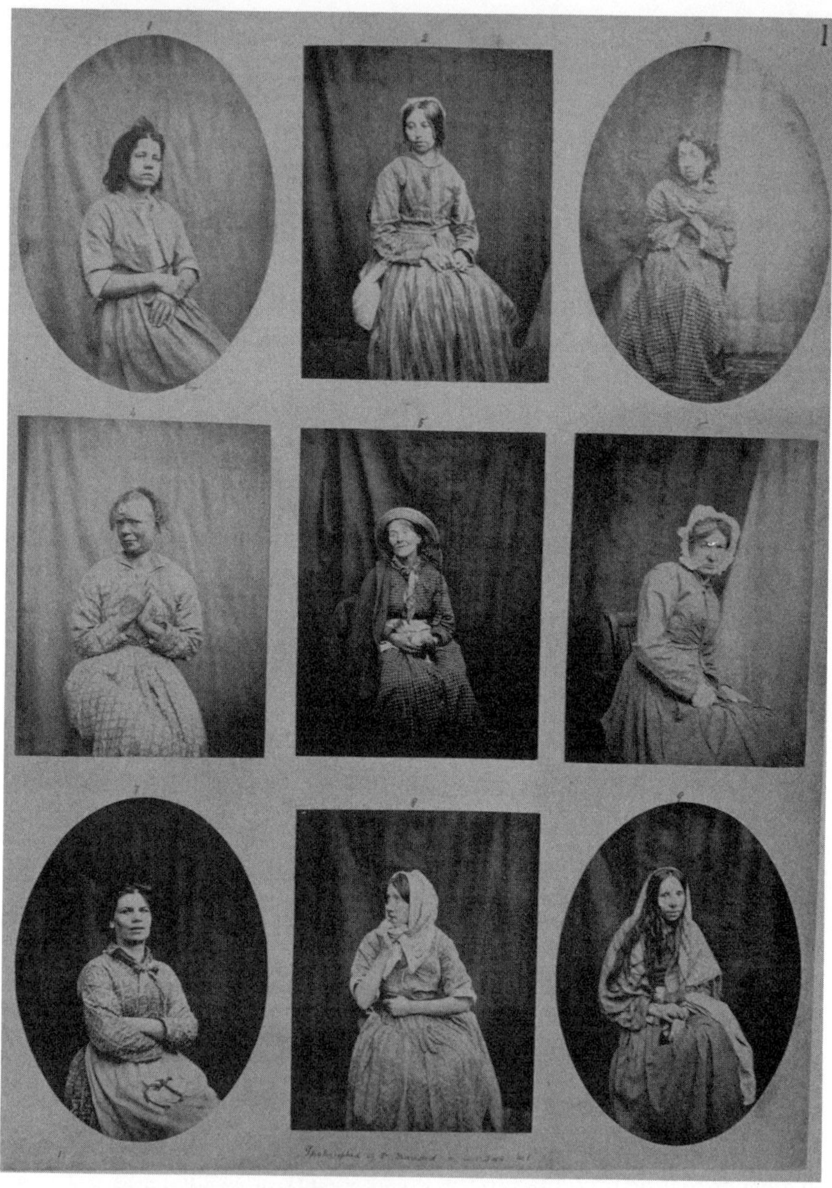

*Some of Hugh Welch Diamond's portraits of his patients at the
Surrey County Lunatic Asylum.*

lums had been to physically restrain patients, which Diamond was against. He also believed his portraits were a more accurate portrayal of mental illness than what he described as the standard "painful caricaturing."

Today there's a sense of unease looking at Diamond's portraits. Part of that stems from empathy for the women: who they were, how they lived, what they experienced. Part of it also stems from knowing that these women were on display—and were literally on display, at photo exhibitions. More broadly, as several historians have noted, the portraits indicate an unequal relationship: "Diamond's camera, and the cameras of other asylum superintendents who followed his approach, [were] equally a form of control and restraint, as was physiognomy itself," writes historian Sharrona Pearl.

At the end of the 1850s, Diamond took a new position at a private psychiatric hospital whose patients could afford to pay for their treatment. He didn't take their portraits. Not that photographing psychiatric patients ended with Diamond: in 1881, a doctor at an asylum in Yorkshire wrote a piece for the *British Journal of Photography* that grouped his patients by how difficult they were to photograph.

Photographs such as these, and those of diseases that afflicted skin like syphilis, invariably included the patients' faces, allowing viewers to assess and judge who they saw. Processes to protect patient anonymity in medical photography began in the 1890s. Doctors started to obscure the patient's eyes with black boxes. Backgrounds became plainer, suggesting a clinical setting rather than a studio. But the photographers themselves might still be amateurs: three London hospitals formed photographic clubs where photographs of patients were taken by medical students.

While photography helped develop and communicate medical knowledge, doctors also helped develop photography. Diamond had taught photography to Frederick Scott Archer, the inventor of the wet collodion process. (Diamond was also the jury member who sued Thomas Sutton for libel.) Richard Leach Maddox, physician and photomicrographer, tinkered with gelatin. Étienne-Jules Marey was a physiologist who created chronophotography. This may perhaps be expected: these were educated men with scientific knowledge, and the time and means to experiment with chemicals in darkrooms.

What is more surprising is the overlap between photography and dentistry, and not only through the oft-occurring joke that likened a visit to the photographer to a trip to the dentist. As photo historians Heinz Henisch and Bridget Henisch have pointed out, many photographers were also dentists. One Missouri practitioner advertised himself as a "dentist and photo artist." In another advertisement, a photographer offered a "first-class picture" at his studio, adding "while there, just step into his office where he is prepared to extract teeth and attend to all branches of dentistry." The subject of an obituary from 1892 had "for thirty years practiced dentistry and photography in the suburbs of Boston." Before he opened the first commercial daguerreotype studio in the United States, Alexander Wolcott was a manufacturer of dental supplies.

Patients and customers had a few things in common, too: both were required to sit still in uncomfortable chairs. They also faced the prospect of a disappointing outcome and a large amount of pain, physical or emotional (or both). One report from 1904 stated that some of the busiest professions on a Sunday were dentists and photographers. The article theorized this was because it was the only day off for many to get dental work or to get dressed up for a portrait.

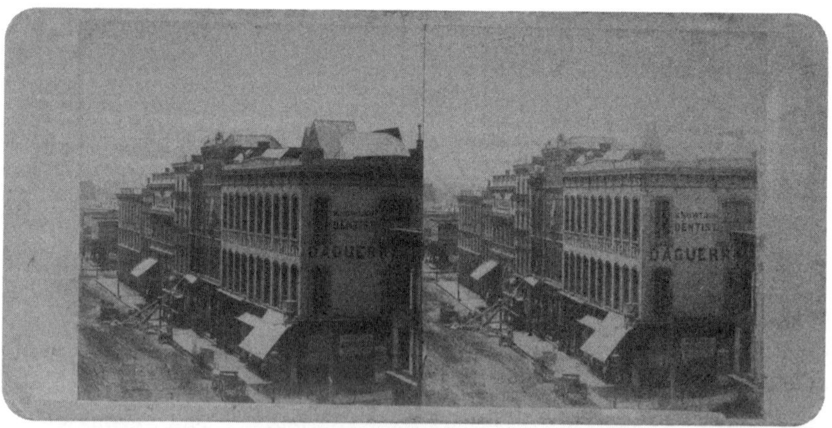

A building on Montgomery Street in San Francisco, with signs advertising a daguerreotype studio and a dentist.

The practices of dentistry and photography overlapped in other ways, too. An 1859 advertisement for a London gold and silver refining business requested "Photographic Refuse, Dentist Cuttings" from photographers and dentists. To bulk out a customer's sunken appearance, some New York photographers used dentists' cotton pads in their sitter's cheeks.

Medicine and photography started with photomicrography, but as these fields evolved and expanded, so too did one convention that is still with us today: the before-and-after portrait. It was used by surgeons and psychiatric doctors (including Diamond), and in early studies of anorexia nervosa. Dentists used it too, although more alarmingly to check that the post-procedure patient still largely resembled the patient pre-procedure. Today we see it in ads for teeth-straightening procedures and dermatological treatments like Botox. It is as instantly understood as that other nineteenth-century photographic revelation that is now a medical necessity, and one that, when it arrived, redefined medical vision: the X-ray.

Chapter 12

Bones, Skin, Eyes, and Brains

Without a shadow of a doubt, the most astonishing, revelatory, and notorious form of photography in 1896 was the X-ray.

On November 8, 1895, at the University of Würzburg, Wilhelm Röntgen, a German physicist, was experimenting with cathode rays in a Crookes tube (a partial-vacuum glass tube) when he noticed something unusual. To stop any light from emanating from the tube, he had covered it in black cardboard, and his laboratory was dark, with the curtains drawn. In the darkness, as he discharged an electric current through the tube, Röntgen noticed a glow.

It was coming from something that lay on a workbench not far from the Crookes tube—a piece of paper, on which one of Röntgen's students had written the letter "A" in a fluorescent substance called barium platinocyanide. Röntgen knew the effects of cathode rays on barium platinocyanide; what stupefied him was the fact that it was glowing even when the tube was covered in cardboard, blocking all visible light. The glow disappeared when he shut off the current, and reappeared when he turned it on. He increased the distance between the tube and the paper as far as two meters (about six feet), and still it glowed. Röntgen, who was, by this point, astonished and utterly mystified, continued to investigate late into the night and for weeks afterwards. And as he experimented, he noticed

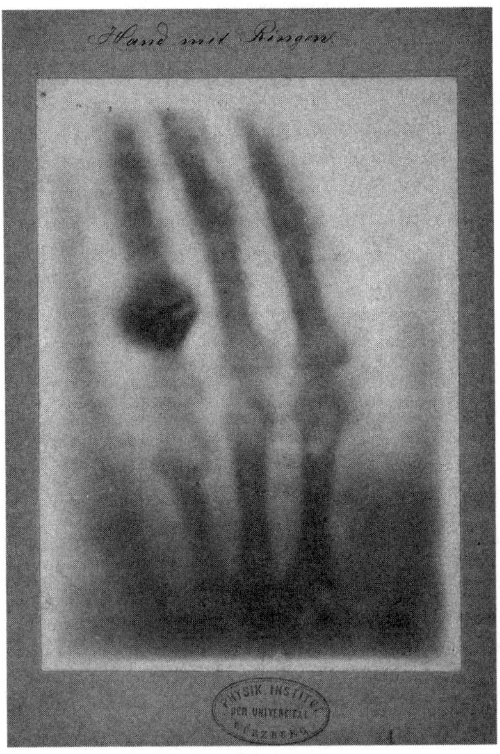

The first of many: Röntgen's X-ray of his wife's hand, from late 1895.

something else: when he held up various items between the tube and the screen to see how they were affected, he could see on the screen the bones of his own hand.

Röntgen experimented nonstop until Christmas, trying to figure out what, exactly, he had discovered. He also incorporated a photographic plate into his experiments to see if it was affected by these mysterious invisible rays. In the first X-ray images that the world had ever seen, Röntgen captured the contents of a closed wooden box, a rifle (where the X-rays revealed a flaw in the metal), and, most astonishing of all, the skeleton of a hand. For this final image, taken on December 22, 1895, Röntgen's wife Anna Bertha placed her hand on the plate, under the rays, for fifteen minutes. The result was a shadowy and opaque view of the delicate, tapering bones inside her living human hand, punctuated, on her fourth finger, by the hovering dark mass of a ring. Röntgen, still mystified about the exact nature of his discovery, called them X-rays.

Just after Christmas, Röntgen sent his paper, "On a New Kind of Ray," to the Würzburg Physical Medical Society. Within days, he sent copies to colleagues in Vienna, France, and Britain, along with some X-ray images; within weeks, newspapers and journals around the world had picked up the story.

Röntgen's discovery was utterly seismic. He received the first Nobel Prize in Physics, along with many other awards and honors. (Averse to the publicity he received, he gave only two lectures on X-rays, one of which was a request from the German emperor himself.) While the exact nature of the rays was unknown, the fact that they produced images of the insides of objects and living beings was an earth-shattering revelation. In the rush to understand and utilize this newfound knowledge, around fifty books and pamphlets, and 1000 articles, were published on the X-ray in 1896 alone. As *The London Quarterly Review* wrote in 1896, "Never has a scientific discovery so completely and irresistibly taken the world by storm."

The title of that article was "The Photography of the Invisible," and for good reason. From the very beginning, X-rays were perceived as a miraculous new type of photography, one that worked without a camera or visible light. This irked Röntgen; he was more interested in the rays than the photography. But to the general public, the fantastical nature of X-rays stemmed from what they revealed on a photographic plate. "Unlike most epoch-making results from laboratories," wrote a reporter for *McClure's Magazine*, "this discovery is one which, to a very unusual degree, is within the grasp of the popular and non-technical imagination." The *McClure's* article was titled "The New Marvel in Photography"; many other articles and books about X-rays invoked the phrases "the new photography," or "shadow photography," or "shadowgraphy," or some combination of the above, as in a May 1896 article titled "The New Shadow Photography."

As had been the case throughout photographic history, practitioners were swift to test out the boundaries of this "new photography." In February 1896, Austrian photo chemists Josef Maria Eder and Eduard Valenta published their book *Versuche über Photographie mittelst der Röntgen'schen*

Strahlen, or *Research on Photography with Röntgen Rays.* They had started their work on January 7, 1896, just ten days after Röntgen had submitted his paper. The book contained fifteen X-ray images, published as photogravures, which are as varied in subject as they are striking. They included six fish, three human hands, two frogs, a chameleon, a rat, and a snake, whose thin, delicate bones and distinctive curves were both dramatic and complementary to this new form of seeing. (As part of their experiments, Eder and Valenta also x-rayed an unopened Egyptian mummy that they borrowed from Vienna's Imperial-Royal Museum of Art History; the resulting image proved the hypothesis that it contained an ibis mummy.)

It must have been utterly astounding to see these first X-rays. Some of them are astonishing even now. Take, as an example, the X-ray view of a

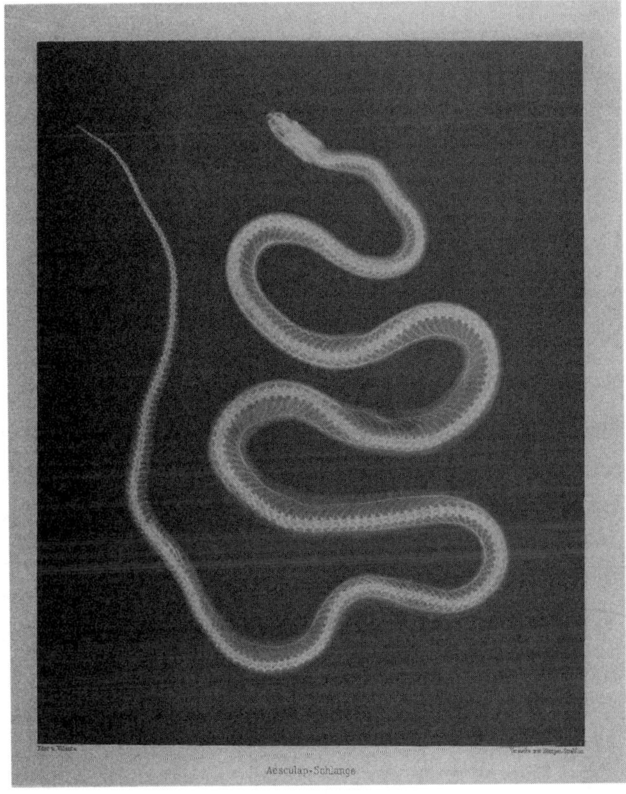

The startling, sinuous beauty of a snake X-ray, 1896.

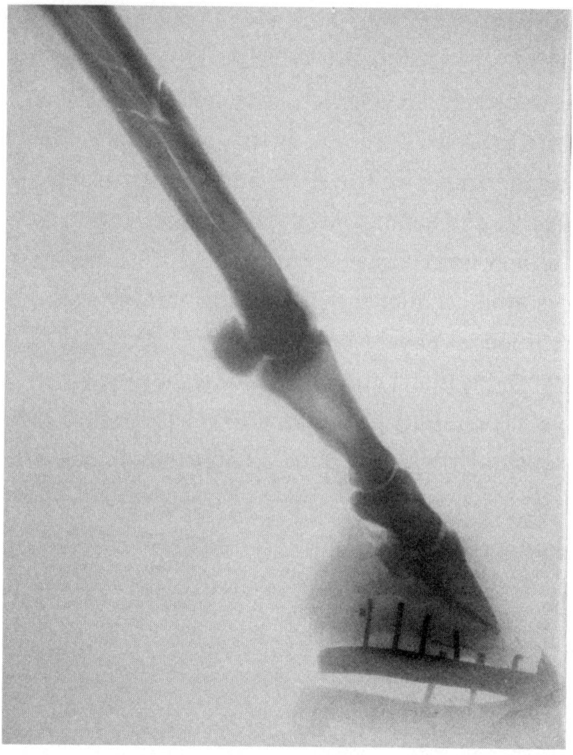

A report in 1896 on the use of X-rays by veterinarians noted that horses were dif-
ficult to X-ray because of their size and strength. This hoof was x-rayed by Arthur
Radiguet, circa 1898.

snake in profile with its jaws hinged open, perpetually ready to strike. Or
the gloriously absurd X-ray of crayfish on a dinner plate. Or the strangely
elegant X-ray image of a horse's hoof, extended out like a ballerina en pointe.

Aside from these wonderfully weird examples, X-ray imaging had a very
obvious practical use. For the first time, doctors could accurately pinpoint
foreign objects hidden inside the body prior to surgery. Dentists could now
see the roots of teeth while they were still in the gums. Customs officers
could check for smuggled goods without opening a single package. It wasn't
long before X-rays were used in less expected but ingenious ways: in 1897,
X-ray imaging authenticated a sixteenth-century painting by Albrecht
Dürer. It's no wonder that X-rays were sometimes called "the new sight."

Röntgen had not taken out a patent, instead choosing to publish his findings openly. This meant that anyone could produce their own X-rays so long as they had access to the necessary equipment. For those who didn't, and despite not knowing the nature of the rays, manufacturers forged ahead with producing do-it-yourself X-ray equipment, including those who typically produced cameras and photographic plates. In New York, E. & H. T. Anthony & Company (the factory-like purveyor of so many cartes de visite) advertised, in May 1896, "a full line of apparatus necessary for the production of X-ray pictures." In London, Marion & Company advertised its "Röntgen Ray Apparatus" next to its line of hand cameras and printing paper. Before hospitals established their own X-ray facilities, patients visited photographers to get their X-rays. One photographer in Hastings, England, ran a newspaper advertisement for portraits, family groups, and "the New Rontgen X Ray Photographs, Taken at this studio, or at one's own Residence."

Some operators established extremely effective X-ray techniques. Elizabeth Fleischmann, for example, opened California's first X-ray laboratory around 1896; her reputation grew after she x-rayed soldiers returning from the Spanish-American War. Her X-rays were so effective that in 1900 she was the subject of an article in the *San Francisco Chronicle* titled "The Woman Who Takes the Best Radiographs." It highlighted her skill in obtaining the right angle for an X-ray as well as her "native intelligence and a perseverance almost phenomenal." The article also credited her as making "the finest radiograph ever produced," showing the position of a bullet inside a soldier's skull.

Given the availability of do-it-yourself X-ray equipment, and the enormous appeal of seeing beyond human vision, X-rays had applications in the most varied, and sometimes strange, of ways. Staff in Queen Victoria's kitchen used an X-ray screen to check food for hidden dangers like stray fish bones, lest they get lodged in the royal throat. Queen Marie Amélie of Portugal procured an X-ray of a corseted woman to demonstrate the "extraordinary deformities" caused by a corset. (In both Germany and France, advocates for dress reform also used X-rays to show the physical effects of corsets.) In Britain, bootmakers used X-rays to show the ill effects of badly fitting shoes, and the Dorothy Dodd Shoe Company in the United States advertised for their "anatomically true" shoes, which they constructed according to foot

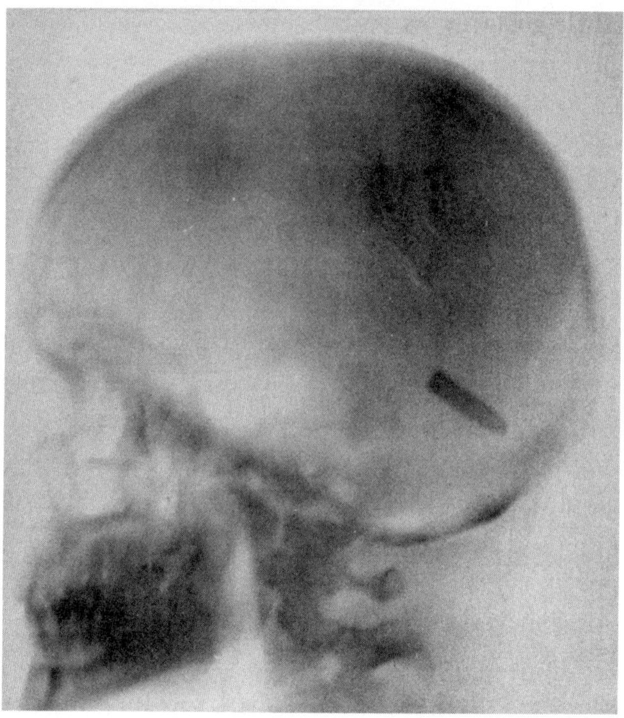

An X-ray by Elizabeth Fleischmann. One admirer of her work was US Surgeon-General George Sternberg. When he met her for the first time, according to the San Francisco Chronicle, *he said, "Do you mean to say that this woman—this little lady—is the one who made those admirable radiographs?"*

X-rays. This was a precursor to the "ped-o-scop" or "foot-o-scope," a shoe store fixture that instantly showed, with X-rays, whether shoes fit properly. (As late as 1960, the US Public Health Service was warning states to prohibit or "strictly regulate" the use of such devices in shoe stores because of risks from radiation.)

It's possible to detect a certain giddiness in some of the early uses of X-rays. During a lecture on X-rays, the English physicist John Henry Poynting displayed, to the delight of his audience, an X-ray of a bottle of beer to demonstrate its transparency. An English farmer had his chickens x-rayed to see which produced the most eggs. A woman in Sunderland lost her ring inside some cake batter, and only realized her mishap when it came out of the oven. Instead of breaking up the cake to find the ring, she took it to a photographer for an X-ray.

In February 1896, *The Photographic News* declared a full-blown "X-ray mania" from which there was no escape. That same author noted that he was already sick of seeing the bony skeleton hand portraits, but this was just the beginning; according to one report, the most widely published image of 1896 was the hand X-ray. The X-ray was a new form of entertainment, and was seized upon with glee. Visitors at London's Royal Aquarium could see a swimming performance, a tattooed lady, and "YOUR OWN BONES by the Rontgen 'X' Rays." An exhibition at Earl's Court included an X-ray kiosk "to photograph any portions of the skeleton!" In November 1896, the annual

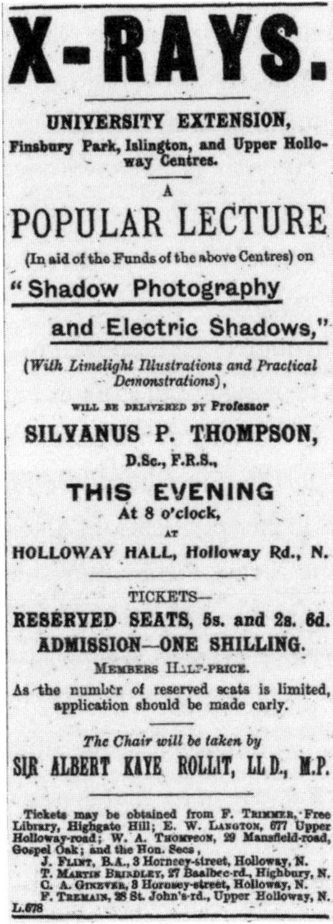

An advertisement from May 1896 for a lecture and demonstrations on X-rays.

exhibition of the Royal Photographic Society offered daily X-ray demonstrations; in the same month, *The Photographic News* reported that it was now possible to have your bones x-rayed on a major London street for a penny. By the end of 1896, hand X-rays were making an appearance on Christmas cards.

Across the Atlantic, in October 1896, shoppers at Bloomingdale's department store in New York City could witness an X-ray demonstration by a member of staff from Columbia University. That same year, Thomas Edison and his employees were hard at work on more X-ray innovations. Clarence Dally, who worked with Edison, experimented with an X-ray–powered light bulb, which failed, and something called the fluoroscope, which did not. The fluoroscope was a screen coated with calcium tungstate, another fluorescent substance, which displayed an X-ray image of anything that was placed between the screen and the X-ray equipment.

Edison exhibited the fluoroscope at the National Electrical Exposition in May, to an audience of nearly 2000 visitors. One by one they shuffled into a darkened room, illuminated by a single red light, to put their hand behind the fluoroscope screen as Edison switched the current on and off. "The fluoroscope exhibition was a great success from every point of view," reported *The New York Times*. "Everyone who wished saw the bones of his or her hand, wrist, and forearm."

But the power to see through objects, and especially people, was also met with deep unease. Upon hearing about Edison's new device, *The Pall Mall Gazette* stated that it was a "revolting indecency," and that looking through anyone's bones without permission "should be regarded as an aggravated form of indecent assault." Other publications also expressed concern about the intrusive nature of the X-rays. "They may show that a man has something inside him which was never intended for publication," reported *Photography: The Journal of the Amateur, the Professional, and the Trade*, in December 1896. "Sooner or later there will have to be a Rontgen Ray act to limit the use of this strange power, like the sale of poisons." In the same journal, one waggish writer declared that X-rays were "an absolute bore; everybody talks of them, nobody knows anything about them; everybody is theorizing," before proposing an "anti-X ray society." (This was in March 1896, mere months after Röntgen's discovery.) A tongue-in-cheek poem followed, titled "X-actly So":

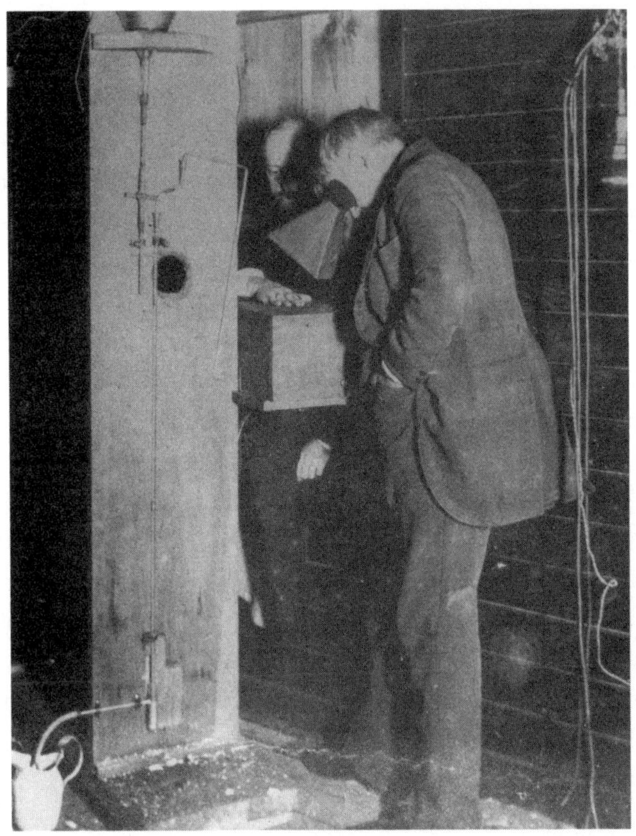

Thomas Edison looking at a hand through a fluoroscope, 1896.

The Rontgen rays, the Rontgen rays—
What is this craze?
The town's ablaze
With this new phase
Of X rays ways.
I'm full of daze
Shock and amaze;
For nowadays
I hear they'll gaze
Thro' clock and gown—and even stays!
These naughty, naughty Rontgen rays!

The "naughtiness" of Röntgen rays was a reference to the perceived indecency of X-rays. One article advised women to wear satin because, it claimed erroneously, satin blocked the X-rays.

The mysterious, freakish, all-seeing nature of X-rays naturally inspired writers. In 1896, *Longman's Magazine* published a short story titled "Röntgen's Curse," in which a scientist gave himself X-ray vision with eye drops of his own devising. After gaining X-ray vision (and dosing his poor dog with the drops too), he endured the horror of his new sight. "It was not merely that I saw my family in the form of skeletons sitting round me. The horror lay in the life of the skeletons. They were not like the dry bones in a museum of anatomy or in the valley of death. They looked fresh and clammy, and the skulls wagged and mouthed at me in a manner that made my skin creep with disgust to see them eating or pretending to eat, lifting the bony fingers to the gumless jaws, which they moved in the act of chewing." Most famously, H. G. Wells's novel *The Invisible Man*, published the year after X-rays were discovered, described a fictional type of ray that effected a pernicious invisibility.

Even outside of fiction, X-rays were imbued with powers they did not possess. By swallowing a "Röntgen Pill" that was "charged" with X-rays, a patient could illuminate their insides, or so the theory went. One scientist suggested that any regular photograph of a person could be made into an X-ray image to reveal their skeleton, with the right X-ray treatment. Others experimented with the notion that X-rays could restore vision to the blind. With the benefit of hindsight, it is tempting to gently scoff at such proposals. But consider the logic: if it was suddenly possible to see *through* an object, something that no human on earth had previously been able to do, all because of some hitherto unknown invisible rays, then perhaps those same rays had other, as yet undiscovered, qualities.

Of course, X-rays did have one major undiscovered quality, which was that excess exposure could kill you. Because it didn't hurt to get an X-ray, the general assumption was that they were not harmful. Yet accounts of X-ray "burns," as they were described, were reported as early as April 1896 in England and the United States. The following year, *The New York Times* ran a front-page article, "Burned by the X Rays," which detailed the "pitiable condition" of a woman who, at the recommendation of her dentist, went to a photographer for a dental X-ray. After attempting a shorter exposure time, the photographer took the X-ray image with a twenty-minute exposure. Afterwards she suffered from blistered skin, damaged hearing, and hair loss.

The intensity of the X-ray equipment varied wildly, and exposure times were a guessing game. One X-ray practitioner later recalled that the process was to "place the patient under the machine and hope for the best." In 1900, a woman in Hastings received an X-ray for a fracture, for which the photographer made two exposures, together lasting ninety minutes. Some weeks later, she died, and left a letter in which she claimed her demise was from "cruel over-exposure of the X-rays." Despite this message from beyond the grave, the coroner's report said she died from "shock and exhaustion" and "the effect of x-rays on a weakened system," and that neither her doctor nor the photographer was to blame. (The Hastings photographer responsible for her X-rays was the very same who had so enthusiastically advertised "the New Rontgen X Ray Photographs.")

Many injuries were to the hands of X-ray operators, which is not surprising when you consider they used their hands to check their X-ray machines. The X-ray demonstrators at Bloomingdale's and at Earl's Court both suffered skin damage. Edison's assistant Clarence Dally, who had worked with the fluoroscope and X-ray light bulb, reported pain after X-ray exposures in 1896, the same year as the Electrical Exposition in New York. He died from cancer, caused by X-ray exposure, in 1904, after he had lost hair on his hands, face, and scalp, suffered through multiple skin grafts, and had both his arms amputated. Elizabeth Fleischmann's symptoms also began in her hands. Five years after the *San Francisco Chronicle* described her as "the woman who takes the best radiographs," and after having her right arm amputated, Elizabeth Fleischmann died of radiation exposure. She wasn't yet forty years old.

The extraordinary qualities of X-rays—invisible, yet all-seeing—encouraged another form of photography that also relied on an unseen force. It is here that we depart from the world of physics to enter the world of psychics.

Early in 1896, as the world was gripped with X-ray fever, *The Amateur Photographer* published an article titled "Thought, the Invisible, and the Unknown." Thought photography, as it was known, was exactly what it sounded like: the idea that a thought could be directed from the brain onto a photographic plate, and the resulting image would depict that very thought. "Let one but grasp the notion that when the brain of a man thinks a great thought, something happens—that is to say, there is a physical side to it, as though a flash or current were sent forth—and then ask oneself, may it not be possible to find some means of making such emanation manifest?" queried *The Amateur Photographer.*

Thought photography was one strand of mystical photography that came into prominence around the time of the X-ray. The other was based around the idea of a vital force, sometimes called a "universal fluid" or "effluvia"—a psychic energy or force that radiates from living beings, and that could also be captured in a photograph, under the right circumstances. (It had all sorts of awkward names like effluviography and psychography; we'll just call it fluid photography.) The idea that living beings emanated psychic radiations that could be photographed predated the X-ray. In the 1850s, a German scientist named Karl von Reichenbach had experimented with capturing images of what he called the "Odic force," a life force that radiated from all living beings. His ideas were based on the pre–photography era work of Franz Anton Mesmer, creator of mesmerism. So the idea of photographing an invisible force was not new, but it was most certainly galvanized by the discovery of X-rays.

Hippolyte Baraduc, a French physician at La Salpêtrière, a psychiatric hospital in Paris, was deeply invested in the ideas of both fluid and thought photography. "The invisible fluid manifests itself by its own intimate, luminous, intrinsic force," he wrote in 1896. "In a word, one must know how to induce the psycho-odo-fluid current which the plate records as it passes by." Baraduc conducted some of his "psycho-odo-fluid" experiments at La Salpêtrière, which seems, in a word, unethical.

That 1896 quote was from Baraduc's book *L'âme humaine* (*The Human Soul*), which was published the same year that readers were flooded with books, articles, and pamphlets about X-rays, and which included some seventy photographs of what he called "psychicones." The 1913 English trans-

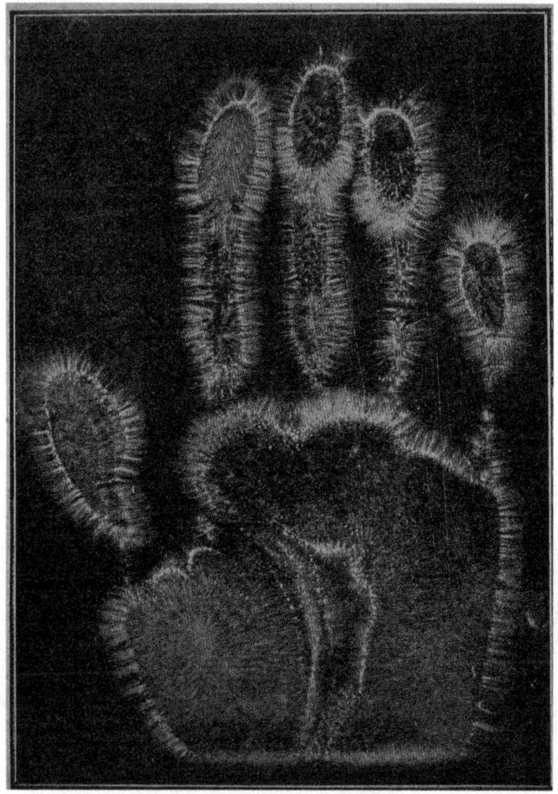

One of Hippolyte Baraduc's "fluidic" images.

lation of his book defined psychicones as "fluidico-vital images" created by "the spirit modulating the vital animistic force," which largely fails to clarify either their definition or creation. Baraduc's images typically involved touching the sensitized plate in some way, either with his hands or by pressing it against his forehead or above his heart. The resulting image showed whorls or spikes or mist, indistinct enough to be interpreted as anything (or nothing).

Baraduc also took his work home with him. He photographed his son in distress over a dead pheasant, and claimed that the haze in the resulting photo was his "vital force" arising from his "state of soul bewailing the death of the bird." Baraduc's beliefs were so strong that he photographed his wife twenty minutes after she died and claimed that the cloudy mass on the photographic plate was her soul. His experiments into his "fluidico-vital images" also took a cruel turn when he attached a photographic plate to a pigeon to record the impressions of its "wild terror," impressions that miraculously ceased once the bird expired. (A "wicked and unnecessary" experiment, reported *The Photogram*.)

Baraduc also dabbled in thought photography, although the main proponent in this field was Louis Darget, a soldier who spent three decades experimenting with thought and fluid photography. Darget's techniques varied: sometimes he focused his thoughts while touching the photographic plate in the developing bath liquid, or in front of his forehead. Sometimes he just touched the side of the plate, or just the developing liquid. In one experiment, a medium projected a current of her own "radiation" through Darget's arm and onto the plate.

According to Darget, strong emotions also created a strong vital force; after an argument, Darget once pressed a photographic plate against his forehead to record his rage. Darget and Baraduc also worked together to create a natty little device that held a photographic plate in place against the subject's forehead, using a box and a strap, to capture thoughts directly onto the plate. They called it the "portable radiographer."

Experimenters in psychic photography borrowed heavily from the logic and language of X-rays. One summary of Darget's theories directly invoked Röntgen's discovery: "The brain throws out rays which penetrate through the skull, even as the X rays do, and photography registers these rays." The word "radiation" appeared repeatedly in theories: "glowing radiation," "psychic radiations," "thought radiography"; Darget even used the term "v-rays," for the "vital force."

Röntgen's discovery also encouraged the experiments of W. Ingles Rogers in England. In February 1896, he described his work in an article for a photographic journal with the overly confident headline "Can Thought Be Photographed? The Problem Solved." For his experiment, Rogers took a rectangular box, painted black on the inside, and drilled two eyeholes at one end. Inside the box, he put a regular glass plate over the eyeholes and, at the other end, his photographic plate. Using a pump, he sucked the air out of the box to create what he called a "vacuum camera." He stared into the eyeholes at the photographic plate while thinking, at first, of a triangle, and then the far more emotionally charged thoughts of his deceased child. He had no existing photographs of the infant, he noted, but he could conjure up the scene of death in his mind. After he sat staring into the wooden box, the enlarged negative showed, according to Rogers, the face of his child. Rogers's article in *The Amateur Photographer* was immediately followed by a report on a recent lecture on X-rays called "The New Photography." Such was the proximity between these two worlds of invisible rays.

The experiments with thought and fluid photography certainly created *an* impression on the plate; but *what* that impression showed was entirely a matter of interpretation. In the case of Rogers, the image of his deceased child was, as a commentator in a subsequent article pointed out, "hardly fair to criticize severely. There are on the enlargement certain marks, but I, personally, cannot construe them into anything particular. They may or may not represent a child or almost anything else."

Darget sometimes wrote the circumstances of the experiment on his prints, which, you could argue, influenced their interpretation. An image shows a dark baton shape alongside the caption "La première Bouteille" and voilà, we see a wine bottle. Darget pressed a plate above his wife's forehead while she slept, and saw, in this "photographie du Rêve" (dream photograph), the shape of an eagle, a shape that could also be interpreted as some swirly lines. In other photos, small marks around fingers or hands were said to resemble magnet filings, demonstrating "the polarity inherent in psychic radiation."

To those with a working knowledge of photography, the images on the plates—the abstract whorls; streaky, glowing fingertips; ill-defined patches of light—were basic darkroom errors. Or, in the delightfully cutting words of French physicist Adrien Guébhard, "familiar accidents from photography's banal kitchen." For his part, Guébhard had investigated the fluid photographs taken by one of Baraduc's colleagues at La Salpêtrière, Jules Bernard Luys, and *The Photographic News* covered these findings with glee. "And now comes the brutal scientist, in the person of M. Adrien Guebhard, with his calm judgment, his thermometers, his weights and scales and test-tubes, and proves that after all the whole thing was due to *heat*."

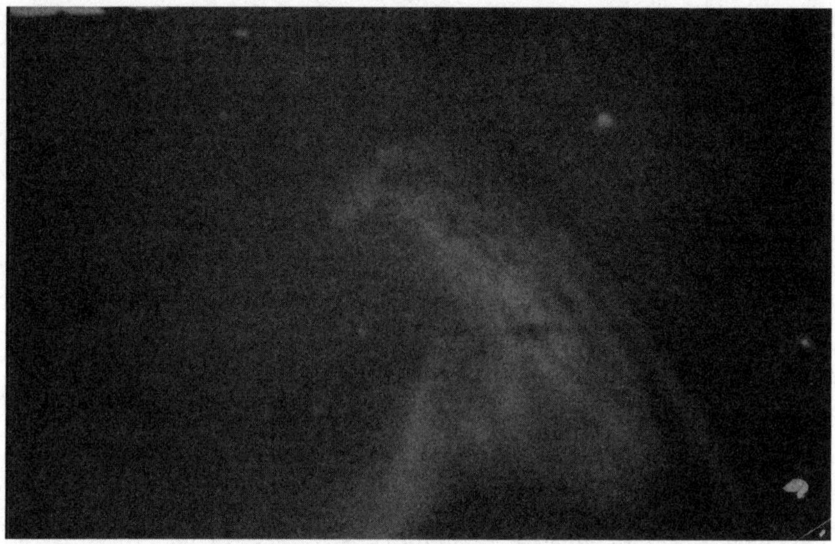

Louis Darget's "eagle" image, 1896.

In another inquiry, a French physician decided to test Luys's fluid photographs by carrying out his own experiments using a human hand severed from a dead body. To investigate the effect of body temperature, he reheated the dead hand and found it had the same effect on the photographic plate as a living (non-severed) hand.

Into this murky realm incautiously stepped Stanford University president David Starr Jordan, who wrote an article titled "The Sympsychograph: A Study in Impressionist Physics," which was published in September 1896 in *Appleton's Popular Science Monthly.* In it, he described an experiment with members of the newly formed Astral Camera Club, who were chosen for having the "greatest animal magnetism and greatest power of mental concentration." They gathered in a darkened room equipped with a photographic plate and a special lens, to which each member was connected via an insulated tube. Together they focused their thoughts on the "essence" of cat, in the hope of producing an image of a cat on the plate from just their minds, an image that Jordan called a "sympsychograph." The article included a photo of a cat as it might appear through a prism, multiple layers of eyes and ears and whiskers.

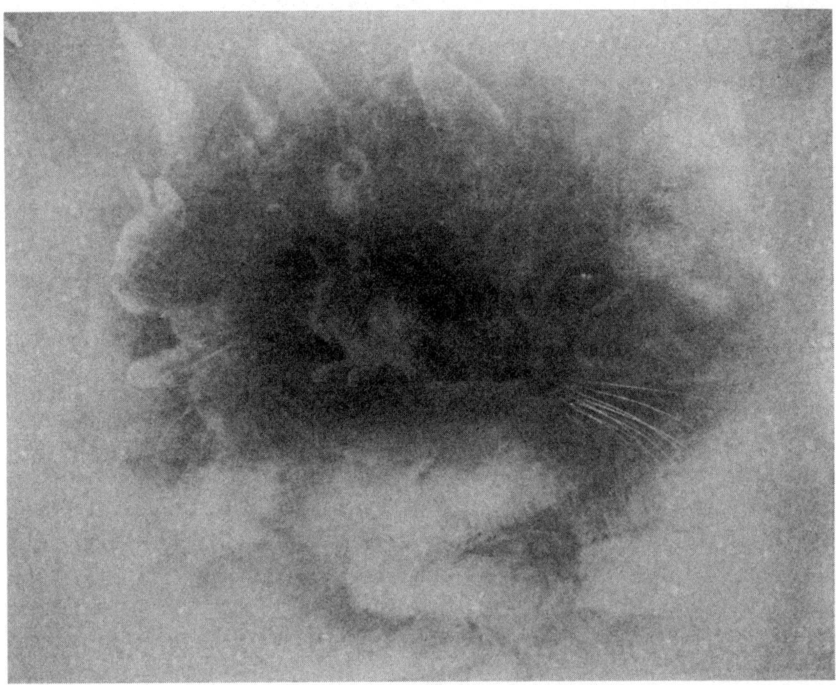

A "sympyschograph" of a cat, as imagined by David Starr Jordan and believed to be credible by many others.

Except the article was a parody. The Astral Camera Club did not exist, and the experiment never happened. The photograph was manipulated. One huge clue—other than the sheer ridiculousness of it—was the line, "The next experiment will be by similar means to photograph the cat's idea of man." Jordan believed no one would take it seriously and was horrified when they did. Shortly after it ran, the same magazine published a letter from Jordan where he professed that he was both "surprised and humiliated" that anyone had believed it. The editors of that magazine also issued an apology, with the title "An Almost Too Successful Joke."

In the wake of the discovery of X-rays, seemingly opposite forces swirled together in photography—physics and psychics, science and satire, facts and abstractions. What all these experiments had in common was the notion that the unseen could be harnessed by human hands and made visible on a photographic plate, to reveal a new understanding of the world.

Conclusion

A New Century

*T*he British Journal Photographic Almanac and Photographers Daily *Companion* was an annual roundup of that year's photographic advances. The 1898 issue, which ran to 1407 advertisement-filled pages, included a look at what it called "animated photography" or what we'd call the movies, radiography, and color photography ("of actual progress there is nothing to record"). The plethora of advertisements show that year's photographic preoccupations: hand cameras, Eastman Kodak film, the Lumières' cinematograph machine, and X-ray plates.

That same decade had revealed underwater vistas, candid camera fiends, and the continued risky attempts to use flash powder. Photographers peered up to the moon and down into the microscope to photograph worlds that were distant or invisible. And all of this was only possible because of what had come before. "In a retrospective glance at photography it is too common a practice to ridicule the trials and difficulties of early days and to laugh at the results," one photographic commenter wrote in 1886. He then marveled, "When we consider the immense difficulties in the way in the early times, the entire absence of books or literature of any kind on the subject, the impossibility of obtaining either pure chemicals or suitable apparatus, one can only wonder that photography ever made any progress at all."

Early photographic innovators had the overwhelming desire to improve and expand on something that was already astonishing. Sometimes they succeeded, and sometimes their attempts were ill-judged; in one instance, they created an entirely new art form, cinema. Photography has played an essential role in a broadening of perspectives through some of its earliest and most devoted practitioners, and those who sought social reform and justice. Together, these bold leaps in photography stretched the medium in previously unimaginable ways, and the results echoed into the twentieth century and beyond.

One summer night at the dawn of that new century, a French painter named Gabriel Loppé positioned his camera in the direction of the Eiffel Tower. A thunderstorm was raging above Paris. Loppé started the exposure; some twenty minutes in, lightning struck, illuminating the room as its deafening sound filled Loppé's ears. About two minutes later, Loppé ended the exposure, and so recorded forever, on a photographic plate, forks of lightning striking the Eiffel Tower. This was not the first photo of lightning, but it is surely the most dramatic. It also encapsulates just how far photography had come from its initial introduction in Paris, decades before, to this moment, in early June of 1902.

A photograph is a small miracle of chemistry, optics, and light. The experiments detailed in this book—whether they were inspired by something as romantic as a rainy night, or something as prosaic as better profit margins—all began with an idea. On that stormy summer night, Loppé did what practitioners had done for decades, and would continue to do: use photography to pursue knowledge, to create something beautiful, to record a moment in time, in a momentary, blinding flash of brilliance.

ACKNOWLEDGMENTS

Thank you to Laurie Abkemeier, for her guidance and patience. Thank you to my wonderful editor, Matt Weiland, and to a whole team at Norton, in particular Yumiko Gonzalez Rios, Huneeya Siddiqui, Jodi Beder, Steve Attardo, and Phil Pascuzzo.

I'm indebted to all the historians whose books have been a source of fascination and delight as I researched this one. I'm also indebted to the librarians who sourced books and rare materials, and whose work in sharing knowledge is so crucial. Special thanks to the Larchmont Public Library, the British Library, Julian Carr, the Thomas J. Watson Library at the Metropolitan Museum of Art, the New York Academy of Medicine library, and to the New York Public Library, whose Art and Architecture Room and online access to nineteenth-century journals were so essential to my research.

Thank you to all the archivists who assisted me, in particular Alex Bainbridge at the Pennsylvania State Archival Collections; Sandrine Bodin at the Bibliothèque de L'Observatoire de Banyuls-sur-mer; Maria McEachern from the John G. Wolbach Library (which has subsequently closed); Anne Anderton at the John Rylands Research Institute and Library; Bill Cantine at the DeSoto National Wildlife Refuge; and Arne Langleite at the Norwegian Museum of Science and Technology. I'm grateful to those individuals who kindly gave permission to use photos from their private collections, and to those museums whose open access policies make photo history widely available for all.

Photography overlaps with many specialized fields. Thank you to all those who lent their time and expertise to answering my questions: Dr. Michael Jung, Helen Rozwadowski, Gerry Beegan, Garry Alan Milius, Paul E. Smith, Dr. Tom Smith, and the Schmidt Ocean Institute. Thank you also to Janine Freeston, Nick Brandreth, and Dr. Geoffrey Belknap for discussing early technical history. Any errors that have crept in are mine. I'm also grateful to the editors at Atlas Obscura who worked with me on early iterations of pieces about Louis Boutan, Julius Neubronner, and Anna Atkins.

I wrote this holed up in a small office with dark blue walls, and which very much resembled a cave. To all those friends who offered encouragement, whenever I emerged, squinting in the light, thank you. I'm especially grateful for the wisdom and insight of Caroline Hickey, Charles Arrowsmith, Betsy Capes, and Meredith Carlisle. Thank you to my mother Sue and my sister Gemma. My deepest gratitude is for my husband Matt, who had always championed this book, and to Harry, whose curiosity is as endless as it is inspiring.

NOTES

A note on sources: To access many photography journals, I frequently made use of Gale Nineteenth Century Collections Online. To access periodicals and newspapers online, I predominantly used the British Newspaper Archive online, the California Digital Newspaper Collection, TimesMachine, and Newspapers.com.

Epigraphs

vii **"We know that knowledge":** A. A. Campbell Swinton, "The New Shadow Photography," *Journal of the Camera Club*, vol. 10, no. 119, Apr. 1896, 66.

vii **"Any man who believes":** Walter D. Welford, "The Influence of the Hand Camera," *Photography: The Journal of the Amateur, the Professional, and the Trade*, vol. 5, no. 234, May 4, 1893, 278.

Introduction: Sights Unseen

1 **Imagine, for a moment:** Descriptions of the daguerreotype studio and process are based on the following materials: *Practical Hints on the Daguerreotype*, 2nd ed. (London: Sherwood, Gilbert & Piper, 1846); Samuel Dwight Humphrey, *A System of Photography* (Albany: C. Van Benthuysen, 1849); Helmut Gernsheim in collaboration with Alison Gernsheim, *The History of Photography: From the Camera Obscura to the Beginning of the Modern Era* (London: Thames & Hudson, 1955), 130–54; and from advertisements for photo studios and their opening hours, for example, in *Hull Advertiser and Exchange Gazette*, Nov. 3, 1843; *Dundee, Perth and Cupar Advertiser*, Oct. 2, 1849; *Wolverhampton Chronicle and Staffordshire Advertiser*, Mar. 7, 1849.

1 **the word *prunes*:** Saying the word *prunes* was a "recipe for a pretty expression," according to Andrew Winter, "The Pencil of Nature," *People's Journal*, vol. 2 (London: People's Journal Office, 1847), 228, quoted in Gernsheim and Gernsheim, *History of Photography*, 144. (Winter also advised wearing satin.)

2   **anything to stop the merest tremble:** Naomi Rosenblum, *A World History of Photography*, 3rd ed. (New York: Abbeville Press, 1997), 43.

2   **a mere twenty seconds:** Exposure times varied enormously depending on the month and time of day (Gernsheim and Gernsheim, *History of Photography*, 138). There were no hard and fast rules for exposures—guidance from an 1846 manual stated that exposure times for a portrait could be anywhere from twenty seconds to a minute and a half: *Practical Hints*, 20.

2   **"The daguerreotype does not flatter":** "From the French," *Photographic and Fine Art Journal*, vol. 7, no. 4, Apr. 1854, 101.

2   **"resemble fried fish":** Francis Way, "Theory of Portraiture," trans. Ambrose Andrews, *Photographic Art Journal*, vol. 5, no. 2, 1853, 107.

3   **"It is hardly saying too much":** Letter from Herschel to Talbot, May 9, 1839, quoted in Beaumont Newhall, *Latent Image: The Discovery of Photography* (Albuquerque: University of New Mexico Press, 1983), 84.

3   **"It baffles belief":** Letter from Forbes to his sister, May 20, 1839, in Newhall, *Latent Image*.

3   **"it would be impossible":** Walsh's Paris letter of March 5, 1839, published in *New-York American*, May 22, 1839, from the online resource *The Daguerreotype: An Archive of Source Texts, Graphics, and Ephemera*.

3   **"Briefly to explain it":** "Foreign Correspondence," *Athenaeum*, no. 587, Jan.–Dec. 1839, 69.

3   **"drawings from nature":** "The Daguerreotype," *Morning Chronicle*, Sept. 14, 1839.

3   **"This machine, or rather apparatus":** "The Daguerreotype," *Sun*, Sept. 13, 1839.

4   **"In Daguerre's camera":** Arago's address quoted in Newhall, *Latent Image*, 6.

5   **negotiate a purchase of the rights:** Newhall, *Latent Image*, 7, 87–89.

5   **Daguerre launched the Diorama:** For Daguerre's early life, see Roger Watson and Helen Rappaport, *Capturing the Light: The Birth of Photography, a True Story of Genius and Rivalry* (New York: St. Martin's Press, 2013), 32–33 and 48–51.

5   **communicated by letter, in code:** Newhall, *Latent Image*, 40–41, and Watson and Rappaport, *Capturing the Light*, 80.

5   **agreed on their terms:** The French government offered an annual pension to Daguerre and to Niépce's son, for life. In return, they received the rights to publish the daguerreotype. Newhall, *Latent Image*, 88.

5   **The event was so popular:** Newhall, *Latent Image*, 90, and Aaron Scharf, *Art and Photography* (London: Penguin Books, 1968), 36.

5   **published a manual . . . public demonstrations:** Newhall, *Latent Image*, 98, 104.

5   **Talbot was an exceedingly well-educated:** Watson and Rappaport, *Capturing the Light*, 46, 59–60.

6   **he attempted a sketch:** Talbot wrote that his sketches were "melancholy to behold." Quoted in Watson and Rappaport, *Capturing the Light*, 92.

6   **Talbot experimented:** Watson and Rappaport, *Capturing the Light*, 108; Roger Watson, "Talbot, William Henry Fox (1800–1977)," in *Encyclopedia of Nineteenth-Century Photography*, ed. John Hannavy (New York, London: Routledge, 2008); Newhall, *Latent Image*, 11.

7   **He later claimed:** Watson and Rappaport, *Capturing the Light*, 114.

7   **"I shall *ever* wish":** Letter dated February 1839, quoted in Watson and Rappaport, *Capturing the Light*, 138.

7   **photography also has a prehistory:** On Thomas Wedgwood, whose experiments were published in 1802, see Newhall, *Latent Image*, 14, and Brian Coe, *The Birth of Photography: The Story of the Formative Years 1800–1900* (New York: Taplinger Publishing Co, 1977), 10–13. On Elizabeth Fulhame, see Keith J. Laidler, *To Light Such a Candle: Chapters in the History of Science and Technology* (Oxford: Oxford University Press, 1998), 68, and Watson and Rappaport, *Capturing the Light*, 101.

7   **To make a daguerreotype:** Gernsheim and Gernsheim, *History of Photography*, 71; Newhall, *Latent Image*, 4–5. On the innovations to the daguerreotype process, including the use of bromine and toning with gold chloride, see Marcy J. Dinius, "Daguerreotype," in *Encyclopedia of Nineteenth-Century Photography*.

8   **"observation and experiment":** *Practical Hints*, 20.

8   **"mirror with a memory":** Oliver Wendell Holmes coined the phrase in an article on photography, titled "The Stereoscope and the Stereograph," that appeared in *The Atlantic* in June 1859.

8   **Experiments with the photographic chemicals:** Pauline F. Heathcote, "The First Photographic Portrait Studios in the British Isles: Professional Foundations," in *Technology and Art: The Birth and Early Years of Photography*, ed. Michael Pritchard (Bath: Royal Photographic Society Historical Group, 1990), 84; Newhall, *Latent Image*, 109.

8   **a daguerreotypist in London:** Gernsheim and Gernsheim, *History of Photography*, 138–39.

8   **To make a calotype:** Gernsheim and Gernsheim, *History of Photography*, 82; Lee Ann Daffner, "Calotype and Talbotype," in *Encyclopedia of Nineteenth-Century Photography*. Talbot initially used potassium bromide, before switching to hypo: Newhall, *Latent Image*, 117.

9   **Daguerre took out a patent:** Newhall, *Latent Image*, 106–7.

10   **Talbot, perhaps in reaction . . . loudly criticized:** Newhall, *Latent Image*, 137, and Watson and Rappaport, *Capturing the Light*, 191.

10   **"secures to himself":** "The Photographic Patents," *Humphrey's Journal* (reprinted from *London Art Journal*), vol. 6, no. 10, Sept. 1, 1854, 155.

10   **"practice of intimidation":** "The Stereoscope," *Photographic Art Journal* (reprinted from *Illustrated London News*), vol. 3, no. 5, May 1852, 291–92.

10   **"The work is not always pleasant":** John Thompson and Adolphe Smith, *Street Life in London* (London: Sampson Low, Marston, Searle & Rivington, 1877), 25.

11   **in breach of his patent and sued:** The case of *Talbot v. Laroche* began in December 1854, in which Talbot sued a photographer, Martin Laroche, for patent infringement for using the collodion process. Talbot lost, and the outcome meant that British photographers could use the collodion process without fear of being sued. Talbot did not appeal, nor did he renew his calotype patent when it expired. For details, see Newhall, *Latent Image*, 128–38, in which Newhall writes, "Talbot's defeat was hailed by the profession" (137).

11   **The wet collodion process:** Newhall, *Latent Image*, 126–27. See also Bryan Clark Green, "Wet Collodion Negative," in *Encyclopedia of Nineteenth-Century Photography*.

11   **One photographer recalled:** "Photography in the Field," John Nicol, *British Journal of Photography*, vol. 34, no. 1403, Mar. 25, 1887, 184, quoted in Gernsheim and Gernsheim, *The History of Photography*, 277.

11   **Although it was possible to enlarge a print:** "Before the advent of fast bromide paper in the 1880s, enlarging was troublesome, and apart from leading portrait photographers who had frequent demand, it was confined to specialists who took in outside work": Gernsheim and Gernsheim, *History of Photography*, 315. See also John Hannavy, "Enlarging and Reducing," in *Encyclopedia of Nineteenth-Century Photography*.

12   **"The camera is as constant":** "Camera Amateurs," *Photographic Times and American Photographer* (reprinted from *San Francisco Chronicle*), Sept. 4, 1885, 514.

12   **albumen paper:** J. M. Reilly, "Manufacture and Use of Albumen Paper," *Journal of Photographic Science*, vol. 26, no.4, 1978. Reilly writes that albumen paper was "the most widely used photographic printing material in the nineteenth century" (156).

13   **"photographer's cheesecake":** Recipe quoted in Heinz K. Henisch and Bridget A. Henisch, *The Photographic Experience, 1839–1914: Images and Attitudes* (University Park:

Pennsylvania State University Press, 1994), 57. The following recipe is from the *British Journal Photographic Almanac*, 1862, 68: "The Photographer's Cheesecake. To convert the yolks of eggs used for albumenizing to useful purposes: Dissolve a quarter of a pound of butter in a basin placed on the hob, stir in a quarter of a pound of pounded sugar, and beat well together; then add the yolk of three eggs that have previously been well-beaten; beat up together and thoroughly; throw in half a grated nutmeg and a pinch of salt; stir, and *lastly* add the juice of two fine-flavoured lemons, and then rind of one lemon that has been peeled very thin; beat all up together thoroughly, and pour into a dish lined with puff-paste, and bake for about twenty minutes. This is a most delicious dish."

13   **quantities of egg use:** James M. Reilly, *The Albumen & Salted Paper Book: The History and Practice of Photographic Printing, 1840–1895* (Rochester, NY: Light Impressions, 1980), 31–34.

13   **The separating of those eggs:** Details on the tanners' and bakers' use of egg yolks, and the smell, from Henisch and Henisch, *Photographic Experience*, 56–57.

14   **photographs printed onto their skin:** Bill Jay, *Cyanide and Spirits: An Inside-Out View of Early Photography* (Munich: Nazraeli Press, 1991), 41. See also *The Year-Book of Photography and Photographic News Almanac*, 1885, 127.

14   **"To the fickle":** "Notes," *Photographic News*, London, vol. 35, no. 1720, Aug. 21, 1891, 592, quoted in Jay, *Cyanide*, 41.

14   **Fingernails were also used:** Jay, *Cyanide*, 42; see also "Prints on Finger Nails," *American Amateur Photographer*, vol. 18, no. 3, Mar. 1906, 110, and "Fingernail Photographs," *Abel's Photographic Weekly*, vol. 9, no. 213, Jan. 27, 1912, 509.

14   **Jeremiah Gurney nearly died:** Bill Jay, "Death in the Darkroom: Poisonings of Nineteenth Century Photographers," *Phoebus 3: A Journal of Art History*, 1981, 89.

14   **"I have only to enumerate":** R. J. Flower, "Concerning Accidents Arising from the Employment and Manipulation of Chemical Products," *British Journal of Photography*, vol. 16, no. 466, Apr. 9, 1869, 169.

14   **Just one example, as noted by photo historian Bill Jay:** Recounted in Jay, *Cyanide*, 153.

15   **"the story of a photographer":** *Photographic Notes*, vol. 3, no. 65, Dec. 15, 1858, 290.

15   **"cyanide of potassium":** D. D. T. Davie, *The Photographer's Pocket Companion; Being a Practical Treatise on the Collodion Process* (New York: H. H. Snelling, 1857), 66–67.

15   **"the smarting pain":** "Dangers of Cyanide of Potassium," *British Journal of Photography*, no. 11, June 1, 1857, 108.

15   **Multiple photographers suffered:** M. Carey Lea, "Poisons in Photography," *Philadelphia Photographer*, vol. 1, no. 11, Nov. 1864, 163.

15   **Some photographers experienced:** Jay, "Death," 95–96; Jay, *Cyanide*, 179.

15   **"not only the antidotes":** "Photographic Poisons and Their Antidotes," *British Journal Photographic Almanac*, 1869, 38.

16   **photographic annuals:** "Elsen's Table of Poisons and Antidotes" appeared in both British and American photographic annuals from the 1880s through to the 1900s. See, for example, *The Year-Book of Photography and Photographic News Almanac*, 1883, and *The American Annual of Photography*, 1908.

16   **pyrogallic acid on a patch of eczema:** A. Sufferer, "Dark-Room Disease," *Photographic News*, vol. 26, no. 1235, May 5, 1882, 254.

16   **accidentally drank pyrogallic acid:** Jay, *Cyanide*, 175–76.

16   **Beyond the terrible smell of collodion:** George Kemp, "Practical Remarks on the Various Operations of Photography in Relation to Health," *British Journal of Photography*, vol. 9, no. 205, Jan. 1, 1864, 2; Todd Gustavson, *Camera: The History of Photography from Daguerreotype to Digital* (New York: Sterling Publishing, 2009), 28.

16   **Ammonia, if spilled:** "Dangerous Chemicals," *British Journal of Photography*, vol. 30, no. 1230, Nov. 30, 1883, 721.

16   **tickling the throat:** "Photographic Recipes," *Humphrey's Journal*, vol. 19, no. 18, Jan. 5, 1868, 284.

16   **"the air gets loaded":** W. J. Hickmott, "The Dark Room in Theory and Practice," *The American Annual of Photography*, 1892, 51.

Chapter 1: Under Darkness

21   **Toward the end of 1861:** Sources vary on when Nadar photographed the catacombs; it was either late in 1861 or early in 1862. See Sylvie Aubenas, "Electric Light," in *The Nadars, a Photographic Legend* (Les Galeries Virtuelles de la Bibliothèque Nationale de France, 2018), and Sylvie Aubenas, "Beyond the Portrait, beyond the Artist," in Maria Morris Hambourg, Françoise Heilbrun, and Philippe Néagu, *Nadar* (New York: Metropolitan Museum of Art, distributed by Harry N. Abrams, 1995), 95.

21   **vast, red, gaslit letters:** Adam Begley, *The Great Nadar: The Man Behind the Camera* (New York: Tim Duggan Books, 2017), 106.

21   **"a veritable carnival of red":** "Visits to Noteworthy Studios," *Photographic News*, vol. 18, no. 814, Apr. 10, 1874, 174.

21   **red graced the borders:** Ernest Lacan, "Foreign Correspondence," *British Journal of Photography*, vol. 9, no. 176, Oct. 15, 1862, 397.

21   **"a stage devil":** "Visits to Noteworthy Studios," 174.

21   **He had been brought:** In his autobiography, he writes of the "deficiencies of daylight." Nadar, *When I Was a Photographer*, trans. Eduardo Cadava and Liana Theodoratou (Cambridge, MA: MIT Press, 2015, originally published in 1900), 84, and Aubenas, "Electric Light."

22   **With an electric arc lamp . . . white linen reflector:** Chris Howes, *To Photograph Darkness: The History of Underground and Flash Photography* (Carbondale: Southern Illinois University Press, 1989), 7–8.

22   **"the alliance of the two":** Ernest Lacan, "Foreign Correspondence," *British Journal of Photography*, vol. 8, no. 136, Feb. 15, 1861, 76.

23   **request from the director of the catacombs:** Aubenas, "Electric Light."

23   **"vomiting forth an unbearable stench":** Quoted in Erin-Marie Legacey, *Making Space for the Dead: Catacombs, Cemeteries, and the Reimagining of Paris, 1780–1830* (Ithaca, NY: Cornell University Press, 2019), 17.

23   **By 1860, the catacombs:** Shao-Chien Tseng, "Nadar's Photography of Subterranean Paris: Mapping the Urban Body," *History of Photography*, vol. 38, no. 3, Aug. 2014, 234–35.

23   **In the 1860s, the British press:** See, for example, "The Catacombs of Paris," *Illustrated Times*, Nov. 18, 1865, 317, and "The Catacombs of Paris," *Pall Mall Gazette*, Aug. 19, 1868, 3.

23   **"the inevitable band of English tourists":** Nadar, *When I Was a Photographer*, 75–76.

23   **Nadar's equipment:** Aubenas, "Electric Light," and Howes, *To Photograph Darkness*, 9.

23   **For Nadar, the catacombs:** Nadar, *When I Was a Photographer*, 92–93, and Tseng, "Nadar's Photography," 242.

23   **The catacombs . . . "dim light":** "Les catacombes de Paris" website, last accessed May 26, 2024.

24   **Exposure times were long:** Howes, *To Photograph Darkness*, 10–11.

25   **"He is nothing":** "Visits to Noteworthy Studios," *Photographic News*, vol. 18, no. 814, Apr. 10, 1874, 174.

25   **Another photo reveals wires:** Nadar produced seventy-three plates in the catacombs and showed at least one at the London Exhibition of 1862: Tseng, "Nadar's Photography," 233, and Aubenas, "Beyond the Portrait, beyond the Artist," 99.

26   **An American visitor noted:** "A Ride through the Sewers of Paris," *The Methodist*, vol. 10, no. 12, Mar. 20, 1869, 89.

26   **the shoot was so nightmarish:** Nadar, *When I Was a Photographer*, 93–94.

27   **"Pity the Poor Fogged-Out Photographers!":** *Punch*, Jan. 21, 1871, in Henisch and Henisch, *The Photographic Experience*, 291–92, and Jay, *Cyanide*, 211.

27   **"the terror of the photographic studio":** "Concerning Magnesium," *British Journal of Photography*, vol. 11, no. 240, Dec. 9, 1864, 494, quoted in Howes, *To Photograph Darkness*, 30.

27   **"scrupulously clean":** Lake Price, *A Manual of Photographic Manipulation, the Treating of the Practice of the Art; and Its Various Applications to Nature* (London: John Churchill & Sons, 1868), 69–70.

27   **"the prevailing character":** W. B. Parker, "Glass Rooms, and Lighting the Sitter," *Photographic News*, vol. 7, no. 276, Dec. 18, 1863, 607.

27   **"I am a little bewildered":** H.G.B., "The Glass Room—Modes of Lighting," *Photographic News*, vol. 6, no. 211, Sept. 19, 1862, 453.

27   **You might have seen magnesium:** Sparklers need metallic fuel such as aluminum, iron titanium, zinc, or magnesium. Joe Schwarcz, "How Do Sparklers Work?," McGill Office for Science and Society, Sept. 17, 2021.

28   **"I can draw back my blinds":** "Dull Days," *Photographic News*, Apr. 2, 1875, 158.

28   **The downside to magnesium:** Gernsheim and Gernsheim, *The History of Photography*, 427.

28   **With later designs:** Howes, *To Photograph Darkness*, 101.

28   **a New York dermatologist:** Henry G. Piffard, "Cutaneous Photography Made Easy," *Journal of Cutaneous and Genito-Urinary Diseases*, vol. 6, 1888, 41–43, and Henry G. Piffard, "Photogenic Cartridge," Patent No. US383984A, 1888–06–05. The quote "good sized pistol" is from "The Society of Amateur Photographers of New York," *Photographic Times and American Photographer*, vol. 17, no. 318, Oct. 21, 1887, 531.

29   **"tolerable flash light":** "Sport with the Camera: Amateur Photographers and the Pastimes They Indulge In," *New York Times*, Nov. 25, 1889, 2.

29   **twelve fireflies:** Henisch and Henisch, *The Photographic Experience*, 15.

30   **"to discover what it was":** "A somewhat 'fishy' story reaches us," *British Journal of Photography*, vol. 29, no. 1144, Apr. 7, 1882, 194, quoted in Jay, *Cyanide*, 38.

30   **"Bengal light":** Howes, *To Photograph Darkness*, 4–6.

30   **Flash powder combined:** E. J. Wall, "Early Photography by Magnesium Light," *British Journal of Photography*, vol. 68, no. 3198, Aug. 19, 1921, 497; Jay, *Cyanide*, 156.

30   **"pyrotechnic compounds":** Charles L. Mitchell, "Flash Powder Explosions," *St. Louis and Canadian Photographer*, vol. 8, no. 3, Mar. 1890, 97.

30   **"dangerous in every sense of the word":** "The Magnesium Flash and Its Danger," *Anthony's Photographic Bulletin*, vol. 19, no. 2. Jan. 28, 1888, 33.

30   **potassium chlorate is deemed too reactive:** Dr. Tom Smith, email to author, June 25, 2024. See also Neil Withers, "Potassium Chlorate," *Chemistry World*, Nov. 22, 2011.

30   **picric acid . . . powerful explosive:** "Information on Picric Acid," Stanford Environmental Health and Safety, last accessed May 27, 2024; "Picric Acid Found in Warehouse near Uckfield," BBC News, Nov. 3, 2017; Khalida Volou, "Bomb Squad Rushes to School in Montgomery County to Dispose Chemical in Chemistry Lab," ABC 7 News, June 22, 2022; Nikki Preston, "Defense Force Called to Museums after Explosive Found in Collections," *New Zealand Herald*, July 20, 2021.

30   **Flash powder exploded:** On flash powder accidents, see, for example, "The Third Fatal Accident from Magnesium Explosion," *Photographic Times and American Photographer*, vol. 18, no. 373, Nov. 9, 1888, 530; "Spirit of the Times," *Photography: The Journal of the*

Amateur, the Professional, and the Trade, vol. 9, no. 434, Mar. 4 1897, 132; "Hand Blown Off by Flash Powder," *St. Louis and Canadian Photographer*, vol. 27, no. 3, Mar. 1903, 130; "An Explosion in a Studio," *Photography: The Journal of the Amateur, the Professional, and the Trade*, vol. 11, no. 548, May 11, 1899, 318.

31 **Denver Fire-brick and Chemical Supply House:** "Fatal Chemical Explosion," *St. Louis and Canadian Photographer*, vol. 8, no. 10, Oct. 1890, 403.

31 **Pulitzer Building:** Jay, *Cyanide*, 156, and "An Unexpected Boom: The Pulitzer Celebration Comes Near Blowing Up the City Hall," *New York Times*, Dec. 11, 1890, 3.

31 **The circumstances of some accidents:** "Flashlight Accident," *British Journal of Photography*, vol. 46, no. 2024, Feb. 17, 1899, 104.

31 **free "from the danger":** *International Annual of Anthony's Photographic Bulletin*, vol. 9, 1897, 23.

31 **it contained potassium chlorate:** Romyn Hitchcock, "Flash-Light Powders," *International Annual of Anthony's Photographic Bulletin*, vol. 10, 1898, 17.

32 **"the most powerful light":** *The American Annual of Photography*, 1889, 86.

32 **the chemical factory in Philadelphia:** Jay, *Cyanide*, 158; Howes, *To Photograph Darkness*, 106; "Fatal Explosion of Flash Powder," *Photographic Times and American Photographer*, vol. 19, no. 427, Nov. 22, 1889, 582; "Literary and Business Notes," *American Journal of Photography*, vol. 10, no. 11, Nov. 1889, 389.

32 **dispose of flash powder down a sink:** "Fatal Explosion," 582.

32 **"All forms of flash powder":** Charles L. Mitchell, "Flash Powder Explosions," *St. Louis and Canadian Photographer*, vol. 8, no. 3, Mar. 1890, 101.

32 **"If this is going to be":** "Views Caught with the Drop Shutter," *Anthony's Photographic Bulletin*, vol. 19, no. 21, Nov. 10, 1888, 672.

32 **"The abuse of flash powders":** Samuel Rodman, "Flashlight Photography," *Sun* (New York), Jan. 6, 1898, 6.

32 **the flash bulb became commercially available:** Gernsheim and Gernsheim, *The History of Photography*, 429.

32 **Cunard banned:** "9 Hurt in Pier Blast; Ford and Son Escape," *New York Times*, June 22, 1929, 1.

33 **Riis had arrived:** Alexander Alland Sr., *Jacob A. Riis: Photographer and Citizen* (New York: Aperture Inc., 1974), 17, 20–23.

33 **"gripped my heart". . . . "There it was":** Jacob Riis, *The Making of an American* (New York: Macmillan, 1901), 267.

33 **It's unclear exactly which flash formulae:** The flash powder that Riis used has been described as magnesium flash powder by Bonnie Yochelson and Daniel Czitrom, *Rediscovering Jacob Riis: Exposure Journalism and Photography in Turn-of-the-Century New York* (Chicago: University of Chicago Press, 2007), 147. Kate Flint suggests he might also have used magnesium and gun cotton, in "Flashes of Light," *Flash! Photography, Writing and Surprising Illumination* (Oxford: Oxford University Press, 2017), 24. On Riis's accidents with, and use of, flash powder, see Riis, *The Making of an American*, 271.

33 **As art historian Bonnie Yochelson:** Yochelson and Czitrom, *Rediscovering Jacob Riis*, 142–43.

33 **"When the report was submitted":** Riis, *The Making of an American*, 273.

34 **"It was hard to realize":** "The Society of Amateur Photographers of New York: Lantern Exhibition," *Photographic Times and American Photographer*, vol. 18, no. 333, Feb. 3, 1888, 58.

34 **Riis continued to publicize:** Yochelson and Czitrom, *Rediscovering Jacob Riis*, 87, 154–57.

34 **halftone process . . . majority of the images:** Yochelson and Czitrom, *Rediscovering Jacob Riis*, 87, 157–59.

35 **"He goes into the slums":** "Matters We Ought to Know," *New York Times*, Jan. 4, 1891, 19.

35 **social reform photography:** See also Rosenblum, *A World History*, 359–61, and Beaumont Newhall, *The History of Photography: From 1839 to the Present Day*, revised and enlarged ed. (London: Secker & Warburg, 1964), 139–42.

35 **"To watch the picture":** Riis, *The Making of an American*, 265.

35 **Henry van der Weyde:** Rogier van der Weyden, "Henry van der Weyde 1838–1924," *History of Photography*, vol. 23, no. 1, Spring 1999, 69; "Portraits by the Electric Light," *Anthony's Photographic Bulletin*, vol. 9, Feb. 1878, 40. Re "light that would 'embrace' . . . rather than 'strike,'" see Van der Weyde, "Photography by the Electric Light," *Journal of the Society of Arts*, Feb. 24, 1882, 370.

35 **"Not only is every high light":** "Portraits by the Electric Light," 40, quoted in Van der Weyden, "Henry van der Weyde," 70.

36 **Central electricity stations:** Wolfgang Schivelbusch, *Disenchanted Night: The Industrialization of Light in the Nineteenth Century* (Berkeley: University of California Press, 1988), 65.

36 **Eugène Ducretet . . . Étienne Trouvelot :** Jimena Canales, *A Tenth of a Second: A History* (Chicago: University of Chicago Press, 2009), 140.

36 **"sinuous lines":** Marie-Sophie Corcy, "Electricity and Magnetism," in *Brought to Light: Photography and the Invisible, 1840–1900*, ed. Corey Keller (New Haven and London: San Francisco Museum of Modern Art in association with Yale University Press, 2008).

36 **icon of a lightning bolt:** Kate Flint, "Victorian Flash," *Journal of Victorian Culture*, vol. 23, no. 4, Oct. 2018, 486.

36 **Trouvelot and others produced of electric sparks:** See Étienne Trouvelot, "Sur la form des décharges electriques sur les plaques photographiques," *Lumière Electrique*, no. 45, Nov. 10, 1888, 269–73.

36 **The person who most successfully captured:** The descriptions of Martin's work are from his book *Victorian Snapshots* (New York: Arno Press, 1939, reprint 1973), 24–25.

37 **"a good stock of patience":** Paul Martin, "Nocturnal Photography," *The American Annual of Photography*, 1898, 60.

37 **"a ghost of daylight":** Martin, *Victorian Snapshots*, 25.

37 **He also encountered:** Martin, *Victorian Snapshots*, 25–26.

37 **For maximum effect:** Roy Flukinger, Larry Schaaf, and Standish Meacham, *Paul Martin: Victorian Photographer* (Austin: University of Texas Press, 1977), 56.

Chapter 2: To the Moon

39 **Over six million people:** Ben Johnson, "The Great Exhibition 1851," Historic UK, last accessed Jan. 11, 2024, and Ed King, "The Crystal Palace and Great Exhibition of 1851," British Library, last accessed Jan. 11, 2024.

39 **31-ton broad-gauge railway engine to a fruit dish:** Christopher Hobhouse, *1851 and the Crystal Palace, Being an Account of the Great Exhibition and Its Contents; of Sir Joseph Paxton; and of the Erection, the Subsequent History and the Destruction of His Masterpiece* (New York: E. P. Dutton & Company, 1937), 72 and 103.

40 **Britain was, at this point, the richest country:** Susie L. Steinbach, *Understanding the Victorians: Politics, Culture and Society in Nineteenth-Century Britain*, 2nd ed. (London and New York: Routledge, 2017), 95. Steinbach writes that the exhibition was "intended as a display of Britain's supremacy on the world stage."

40 **This exhibit, the *Official Catalogue*:** *Official Descriptive and Illustrated Catalogue of the Great Exhibition of the Works of Industry of All Nations 1851*, by authority of the Royal Commission (London: William Clowes and Sons, 1851), 210.

40   **"Manufacturing Machines and Tools":** Hobhouse, *1851 and the Crystal Palace*, 74.

40   **a machine folded envelopes:** Hobhouse, *1851 and the Crystal Palace*, 91, and David Le Conte, "Warren De La Rue," *Antiquarian Astronomer*, no. 5, Feb. 2011, 16 and 32.

40   **In the same year as the Great Exhibition:** Le Conte, "Warren De La Rue," 15–16; E.B.K., "Obituary Warren De La Rue," *Monthly Notices of the Royal Astronomical Society*, vol. 50, Feb. 1890, 155.

40   **One of Whipple's early-1850s lunar daguerreotypes:** At the time of accessing the daguerreotype, it was held at the John G. Wolbach Library of Harvard University. The library closed on March 22, 2024.

40   **To nineteenth-century visitors:** Jennifer Tucker notes that the exhibition "offered unprecedented access to important examples of contemporary photographs, particularly those of scientific phenomena such as astronomical photographs and photomicrographs": Jennifer Tucker, *Nature Exposed: Photography as Eyewitness in Victorian Science* (Baltimore: Johns Hopkins University Press, 2005), 22.

40   **Except for a few women:** "A Brief History of the Royal Astronomical Society," Royal Astronomical Society, last accessed Mar. 26, 2024.

41   **strong impression on De La Rue:** E.B.K., "Obituary," 157, and Warren De La Rue, "Report on the Present State of Celestial Photography in England," in *Report of the Twenty-Ninth Meeting of the British Association for the Advancement of Science; Held at Aberdeen in Sept. 1859*, 131.

41   **"one of the most satisfactory attempts":** *Exhibition of the Works of Industry of All Nations, 1851, Reports by the Juries on the Subjects in the Thirty Classes into Which the Exhibition Was Divided, by Authority of the Royal Commission*, vol. 1 (London: Spicer Brothers, 1851), 602.

41   **"it will be possible to accomplish":** Dominique François Arago, "Report," in Alan Trachtenburg, ed., *Classic Essays on Photography* (New Haven: Leete's Island Books, 1980), 20.

41   **John William Draper . . . Lyceum of Natural History:** Beth Saunders, "Mapping the Moon," in *Apollo's Muse: The Moon in the Age of Photography*, ed. Jennifer Bantz (New York: Metropolitan Museum of Art, 2019), 21, and Don Trombino, "Dr. John William Draper," *Journal of the British Astronomical Association*, vol. 90, no. 6, 1980, 567.

41   **"like the tail of a comet":** Quoted in Omar W. Nasim, "James Nasmyth on the Moon: Or on Becoming a Lunar Being, without the Lunacy," in *Selene's Two Faces: From 17th Century Drawings to Spacecraft Imaging*, ed. Carmen Pérez González (Leiden: Koninklijke Brill, 2018), 150.

42   **During the exposure:** "Mr. De La Rue, on Lunar Photography," *Monthly Notices of the Royal Astronomical Society*, vol. 18, no. 1, Nov. 1857, 16.

42   **"willing to lose":** De La Rue, "Report," 131.

42   **In the intervening years:** Saunders, "Mapping the Moon," 22.

42   **150 moon "portraits":** "Liverpool Photographic Society," *Liverpool Photographic Journal*, no. 6, June 10, 1854, 71.

42   **"as large as the room would allow":** Quoted in Ann Thomas, "Capturing Light: Photographing the Universe," in *Beauty of Another Order: Photography in Science*, ed. Ann Thomas (New Haven: Yale University Press in association with the National Gallery of Canada, 1997), 202.

42   **This was not even the largest:** "Liverpool Photographic Society," *Liverpool Photographic Journal*, no. 10, Oct. 14, 1854, 128; also Saunders, "Mapping the Moon," 22.

42   **magic lantern:** "Photographs of the Moon," *Illustrated London News*, Sept. 30, 1854, 302.

42   *The Illustrated London News* **covered:** Saunders, "Mapping the Moon," 22.

42   **It showed a beautifully detailed:** "Meeting of the British Association of Science at Liverpool," *Illustrated London News*, Sept. 30, 1854, 300–301.

42 **Draper's daguerreotype was considered:** "The John Draper Lunar Daguerreotype: Digital Exhibition," NYU Libraries, last accessed Mar. 26, 2024; also, "1860 Sun Photo for Smithsonian," *Washington Post*, July 5, 1962, and Mark Holborn, *Sun and Moon: A Story of Astronomy, Photography and Cartography* (London: Phaidon Press, 2019), 190.

43 **"woodpeckers":** Roy Flukinger and Larry Schaaf, "Paul Martin: A Critical Biography," in Flukinger, Schaaf, and Meacham, *Paul Martin*, 8.

43 **Later technology:** Gerry Beegan, "The Mechanization of the Image: Facsimile, Photography, and Fragmentation in Nineteenth-Century Wood Engraving," *Journal of Design History*, vol. 8, no. 4, 1995, 257–74.

43 **there wasn't a need for engravers either:** Flukinger, Schaaf, and Meacham, *Paul Martin*, 11–15.

44 **several blocks bolted together:** Gerry Beegan, email to author, Mar. 4, 2024.

44 **Once De La Rue resumed:** Le Conte, *Warren De La Rue*, 18–19.

44 **clockwork mechanism:** De La Rue, "Report," 131.

44 **"almost imperceptible mist"... "very capricious in its properties":** De La Rue, "Report," 136–37.

44 **He estimated that his success rate:** De La Rue, "Report," 135.

44 **"perseverance, long practice":** De La Rue, "Report," 133.

44 **De La Rue's moon negatives were small:** De La Rue, "Report," 139–40.

44 **"gave great promise":** "Mr. De La Rue, on Lunar Photography," *Monthly Notices of the Royal Astronomical Society*, vol. 18, no. 1, Nov. 1857, 17.

45 **the moon as a solid object with mass:** Thomas, "Capturing Light," 205–6.

45 **"Seems, Madam! Nay, it IS!":** Advertisement from the London Stereoscopic Company, *Art-Journal Advertiser*, June 1856; also, John Plunkett, "'Feeling Seeing': Touch Vision and the Stereoscope," *History of Photography*, vol. 27, no. 4, 2013, 396.

45 **"You are not sensible of a picture":** "A Word on the Stereoscope," *Leisure Hour: A Family Journal of Instruction and Recreation*, no. 336, June 3, 1858, 346–48.

45 **"in use and demand":** "A Word on the Stereoscope," 348.

46 **"expressed her gratification":** "A Word on the Stereoscope," 348.

46 **a thousand different scenes:** "Paper Read by Sir David Brewster, at the First Monthly Meeting of the Photographic Society of Scotland, Held April 8th, 1856," *Photographic Notes*, vol. 1, no. 1–2, Jan. 1 and Jan. 25, 1856, viii.

46 **"a thousand hungry eyes":** Quoted in Peter E. Palmquist, *Carleton E. Watkins: Photographer of the American West* (Albuquerque: Published for Amon Carter Museum by University of New Mexico Press, 1983), 18.

46 **"globe-like representation":** Warren De La Rue, "Correspondence: Lunar Stereographs," *British Journal of Photography*, vol. 6, no. 97, July 1, 1859, 168.

46 **"The appearance of rotundity":** "There were exhibited at the Meeting a series of stereoscopic photographs of the moon, which had been executed in the first instance by Mr. De La Rue," *Monthly Notices of the Royal Astronomical Society*, vol. 19, no. 1, Nov. 1858, 40.

46 **"Nothing can surpass":** E.B.K., "Obituary," 159.

46 **"presented to us with a convex disc":** A.B.G., "On the Advantageous Employment of Stereoscopic Photographs for the Representation of Scenery," *Photographic Notes*, Sept. 1, 1858, 199.

46 **Sales were reportedly enormous:** "Who Photographed the Moon," *Photographic Journal* [reprinted from *The London Review*], vol. 8, no. 131, Mar. 16, 1863, 247.

46 **"bring this luminary so near":** A. Claudet, "Physiology of Binocular Vision," *Art-Journal*, Mar. 1, 1867, 75.

47 **"No drawing or engraving":** "Photographs of the Moon, Enlarged from Original Nega-

tives Taken by Warren De La Rue, F.R.S.," *British Journal of Photography*, vol. 10, no. 187, Apr. 1, 1863, 148.

47   **"brought to light details of dykes":** "Address delivered by the President, Dr. Lee, on presenting the Gold Medal of the Society to Mr. Warren De La Rue," *Monthly Notices of the Royal Astronomical Society*, vol. 22, no. 4, Feb. 1862, 136.

47   **some obituary writers:** "zeal was most intense": "The Late Thomas Sutton, B.A.," *British Journal of Photography*, vol. 22, no. 782, Apr. 30, 1875, 211; "keen and warm controversialist": W. Jerome Harrison, "The Literature of Photography," *Anthony's Photographic Bulletin*, vol. 18, 1887, 427; "trenchant vigour": "Obituary.—Thomas Sutton, B.A.," *Photographic News*, Apr. 2, 1875, 162. For another carefully worded obituary, see "The Late Thomas Sutton, B.A.," *British Journal of Photography*, vol. 22, no. 777, Mar. 26, 1875, 147.

48   **He was once sued for libel:** *Photographic Notes*, vol. 10, no. 232, Dec. 1, 1865, 318, and *Photographic Notes*, vol. 11, no. 237, Feb. 15, 1866, 48.

48   **"caricature":** Thomas Sutton, "Reminiscences of an Old Photographer," *British Journal of Photography*, Aug. 30, 1867, p. 413, quoted in Gernsheim and Gernsheim, *History of Photography*, 139.

48   **He also later suffered health problems:** Jay, "Death," 88.

48   **"merely illustrate the fact":** Thomas Sutton, "On Some of the Uses and Abuses of Photography," *Photographic Notes*, vol. 8, no. 163, Jan. 15, 1863, 18.

48   **"inferior to those":** Quoted in Gaston Tissandier, *A History and Handbook of Photography*, translated from the French (New York: Scovill Manufacturing Company, 1977), 264.

48   **astronaut Buzz Aldrin:** Accounts of Aldrin's photo, and concerns from scientists about the moon's surface, are from Buzz Aldrin with Ken Abraham, *Magnificent Desolation: The Long Journey Home from the Moon* (New York: Harmony Books, 2009), 31–39. Information about the NASA equipment and use of photographs is from James H. Sasser, "Photographic Summary of Apollo 11 Mission," 9–34, in *Apollo 11 Preliminary Science Report*, NASA, October 31, 1969, and *Apollo 11 Lunar Photography, Data Users' Note (NSSDC ID No. 69-059A-01)*, National Space Science Data Center, NASA, April 1970. To see Armstrong's descent, search YouTube for "Neil Armstrong first step on the moon."

49   **Nasmyth exhibited a fifteen-foot-high steam hammer:** *The Crystal Palace, and Its Contents: Being an Illustrated Cyclopaedia of the Great Exhibition of the Industry of All Nations, 1851* (London: W. M. Clark, 1852), 15.

49   **He built his first telescope in 1827:** Details on Nasmyth's career and astronomy from James Nasmyth, *James Nasmyth Engineer: An Autobiography*, ed. Samuel Smiles (New York: Harper & Brothers, 1883).

50   **"I endeavoured":** Nasmyth, *James Nasmyth*, 334–35.

50   **Nasmyth had seen his father:** Frances Robertson, "Science and Fiction: James Nasmyth's Photographic Images of the Moon," *Victorian Studies*, vol. 48, no. 4, Summer 2006, 596 and 608.

51   **"faithfully reproduce the lunar effects":** James Nasmyth and James Carpenter, *The Moon: Considered as a Planet, a World, and a Satellite*, 3rd ed. (London: John Murray, 1885), ix.

52   **"were hailed":** Nasmyth and Carpenter, *The Moon*, 382.

52   **Nasmyth's goal:** Robertson, "Science and Fiction," 609.

52   ***The Strines Journal*:** Email to author from Anne Anderton, Special Collections Curator, John Rylands Research Institute and Library, Mar. 6, 2024; "The Moon and Its Surface Inequalities," *Strines Journal: A Monthly Magazine of Literature, Science and Art*, vol. 4, 1855, 111; Nasim, "James Nasmyth on the Moon," 160.

53   **innovations in photomechanical printing:** Gernsheim and Gernsheim, *History of Photography*, 340–41; Robertson, "Science and Fiction," 610.

53   **Nasmyth wrote to Darwin's publisher:** Robertson, "Science and Fiction," 610.

53   **"thirty years of assiduous observation":** Nasmyth and Carpenter, *The Moon*, ix.

53   **"No more truthful or striking representations":** J. Norman Lockyer, "The Moon," *Nature*, vol. 9, no. 228, Mar. 12, 1874, 358, quoted in Saunders, "Mapping the Moon," 25.

53   **The views of lunar scenery are "wondrous":** "An Illustrated History of the Moon," *Popular Science Review*, vol. 13, 1874, 193–94.

53   **"The illustrations are magnificent":** Nasmyth, *Autobiography*, 400.

54   **smoke created with fingerprints:** Robertson, "Science and Fiction," 617.

Chapter 3: Up in the Air

57   **On the afternoon of Monday:** Details of Shadbolt's first trip are from Cecil V. Shadbolt, "A Photographic Trip to Cloud-Land," *British Journal of Photography*, vol. 29, no. 1155, June 23, 1882, 354.

57   **"there is no telling what":** Cecil V. Shadbolt, "Balloon Photography," *British Journal Photographic Almanac*, 1884, 152.

58   **"hanging motionless in space" . . . "horrified to see":** Shadbolt, "Photographic Trip."

58   **"instantaneous map-photograph":** "On the Table," *Photographic Journal*, vol. 7, no. 1, Oct. 7, 1882, 12.

58   **only with the wet collodion process:** On daguerreotypes and aerial photography, see Beaumont Newhall, *Airborne Camera: The World from the Air and Outer Space* (New York: Hastings House, 1969), 19.

58   **The "lighter-than-air" gas:** C. H. Gibbs-Smith, *A History of Flying* (New York: Frederick A. Praeger, 1954), 68.

58   **One aeronaut instructed his passengers:** John A. Tennant, ed., "Aerial Photography," *The Photographic Miniature: A Magazine of Photographic Information*, vol. 5, no. 52, July 1903, 165.

58   **Shadbolt recounted one attempted departure:** Cecil V. Shadbolt, "A Balloon Trip," *British Journal of Photography*, vol. 31, no. 1250, Apr. 18, 1884, 248.

59   **because it was so far from the sea:** Jennifer Tucker, "Voyages of Discovery on Oceans of Air: Scientific Observation and the Image of Science in an Age of 'Balloonacy,'" *Osiris*, vol. 11, Science in the Field (1996), 169; and James Glaisher et al., *Travels in the Air*, ed. James Glaisher (London: Richard Bentley, 1871), 30.

59   **Glaisher and aeronaut Henry T. Coxwell:** Glaisher et al., *Travels in the Air*, 67–71.

59   **Shadbolt was once dragged:** Shadbolt, "A Balloon Trip," 248.

59   **Such were the logistics of ballooning:** J. L. Hunt, "James Glaisher FRS (1809–1903) Astronomer, Meteorologist and Pioneer of Weather Forecasting: 'A Venturesome Victorian,'" *Quarterly Journal of the Royal Astronomical Society*, vol. 37, 1996, 329.

60   **"levitating into the silent immensity":** Nadar, *When I Was a Photographer*, 60.

60   **Nadar predicted:** Robert M. Doty, "Aloft with a Balloon and Camera," *The View from Above: 125 Years of Aerial Photography*, ed. Rupert Martin (London: The Photographers' Gallery, 1983), 9.

60   **According to Nadar:** Nadar, *When I Was a Photographer*, 66–67.

60   **Nadar hatched a plan:** Richard Holmes, *Falling Upwards: How We Took to the Air: An Unconventional History of Ballooning* (New York: Pantheon Books, 2014), 163.

60   **When inflated, Le Géant:** Holmes, *Falling Upwards*, 164–65; Newhall, *Airborne Camera*, 31; Gernsheim and Gernsheim, *History of Photography*, 507; and Martyn Barber, *A History of Aerial Photography and Archaeology* (Swindon: English Heritage, 2011), 67–68.

63 **Amazingly, no one died:** Details about Le Géant's voyage from Eugène d'Arnoult, *Voyage du "Géant": de Paris à Hanovre en ballon* (Paris: E. Dentu, 1863); "Dangerous Balloon Ascension," *New York Times*, Nov. 8, 1863, 3; Holmes, *Falling Upwards*, 167–69; Newhall, *Airborne Camera*, 31; Gibbs-Smith, *A History of Flying*, 157–58.

63 **Any photographs Nadar took:** Newhall, *Airborne Camera*, 32.

64 **underneath it got so hot:** Nadar, *When I Was a Photographer*, 63.

64 **The French newspaper *Le Petit Figaro*:** Newhall, *Airborne Camera*, 32.

64 **"Boston, as the eagle":** Oliver Wendell Holmes, "Doings of the Sunbeam," *Atlantic Monthly*, vol. 12, July 1863, quoted in Newhall, *Airborne Camera*, 24–25.

64 **After the start of the Civil War:** Newhall, *Airborne Camera*, 27, and Doty, "Aloft with a Balloon," 11.

64 **"If lawn tennis grounds":** Cecil V. Shadbolt, "Rambles in Cloudland," *The Yearbook of Photography and Photographic News Almanac*, 1884, 48.

64 **earliest surviving aerial photographs:** "Ballooning in Bowler Hats: Early Images from Victorian Skies," Historic England blog, Apr. 13, 2016.

64 **"the best pictures taken from a balloon":** "Annals of Photography for 1883," *The Yearbook of Photography and Photographic News Almanac*, 1883, 27.

65 **"beyond doubt, the most remarkable":** "A French Litterateur on Mr. Cecil V. Shadbolt's Balloon Pictures," *British Journal of Photography*, vol. 31, no. 1245, Mar. 14, 1884, 166.

65 **"excited a lively interest":** "Exhibition of Photographs at the Minneapolis Convention," *Photographic Times and American Photographer*, vol. 17, no. 358, July 27, 1888, 351.

65 **Shortly after takeoff:** "Terrible Balloon Accident at the Crystal Palace," *The Times*, June 30, 1892, 9, and "Death of Mr. C. V. Shadbolt," *British Journal of Photography*, vol. 39, no. 1680, July 15, 1892, 452.

66 **a photographer named Arthur Batut:** On Batut's camera, see Geoffroy De Beauffort and Michel Dusariez, *Aerial Photographs Taken from a Kite: Yesterday and Today* (n.p.: KAPWA Foundation Publishing, 1995), 121; Newhall, *Airborne Camera*, 42; Tennant, "Aerial Photography," 168.

66 **The line needed to have enough pull:** William A. Eddy, "The Naval and Military Use of Kite Photography in Time of War," *International Annual of Anthony's Photographic Bulletin*, vol. 11, 1899, 105.

66 **"He let the string slip":** *Photographic News*, vol. 45, no. 281, May 17, 1901, 317.

66 ***La photographie aérienne par cerf-volant*:** Translated in De Beauffort and Dusariez, *Aerial Photographs*, 136–37.

66 **William A. Eddy photographed a parade:** Newhall, *Airborne Camera*, 42–43.

67 **he offered to teach aerial photography:** "Eddy and His Kites," *New York Times*, July 2, 1898, 3.

67 **worked at the Blue Hill Observatory:** Bob White, "Diamonds in the Sky: The Contributions of William Abner Eddy to Kiting," Best Breezes blog (originally appeared in *Kitelift Magazine* in Nov. 1999), last accessed May 20, 2024.

67 **According to *The Jersey Journal*:** "Hist! Ice Cream Swiped! Kite as Detective," *Jersey Journal*, Mar. 31, 1908, 5; "Sent Kite after Thieves," *Times and Democrat* (Orangeburg, SC), Apr. 24, 1908, 6.

68 **"My grandfather, as he recounted later":** Christopher de Hamel, "My Grandfather, the High-Flying Photographer," *Daily Telegraph*, Jan. 21, 2017. The photo appears in the book by Douglas Archibald, *The Story of the Earth's Atmosphere* (London: George Newnes, 1897), 186.

68 **"Cody Aeroplane":** S. F. Cody, "The Kite That Lifts a Man," *Pearson's Magazine*, vol. 10, 1903, 353, and Clive Hart, *Kites: An Historical Survey* (New York: Frederick A. Praeger, 1967), 149.

69   **Although he was not the first:** Frank H. Winter, "Camera Rockets and Space Photography Concepts before World War II," *History of Rocketry and Astronautics: Proceedings of the Seventh and Eighth History Symposia of the International Academy of Astronautics, Baku, U.S.S.R. 1973, Amsterdam, The Netherlands 1974*, AAS History Series, vol. 8, 1989, 73–74.

69   **But Maul's early experiments:** Frank E. Rietz, "Alfred Maul—A Pioneer of Camera Rocket," *Acta Astronautica*, vol. 24, 1991, 364.

70   **"an impressive number":** Maul, quoted in Winter, "Camera Rockets," 80.

70   **a camera inside:** Winter, "Camera Rockets," 84, and Rietz, "Alfred Maul," 364.

71   **"The exposure, the opening":** "The Military Rocket Camera", *Scientific American*, Feb. 6, 1915, 125, quoted in Winter, "Camera Rockets," 84.

71   **120 soldiers:** Rietz, "Alfred Maul," 366.

71   **As he fine-tuned his apparatus:** Winter, "Camera Rockets," 80–81.

71   **The largest rocket he built:** Willy Ley, *Rockets, Missiles and Men in Space* (New York: Viking Press, 1968), 98–99.

71   **"The pigeon is the first bird":** "The Bird Photographer: German Druggist's Successful Experiment with Pigeon Camera," *New-York Tribune*, Dec. 13, 1908, 8.

72   **"photographic record of its journey":** Dr. Neubronner, "The Carrier Pigeon Photographer: A New Use for the Homing Pigeon," *Scientific American Supplement No. 1716*, Nov. 21, 1908, 332.

72   **cameras that he attached to the pigeons:** Neubronner, "The Carrier Pigeon Photographer"; "News and Notes," *Bulletin of Photography*, vol. 3, no. 66, Nov. 11, 1908, 66; "Pigeon Photographers," *Strand Magazine*, vol. 10, no. 1, July 1911, 3; Newhall, *Airborne Camera*, 48.

74   **Wartime aerial photography:** Steven Cable, "Aerial Photography and the First World War," National Archives blog, June 30, 2015.

74   **German planes were photographing:** Denis Cosgrove and William L. Fox, *Photography and Flight* (London: Reaktion Books, 2010), 35.

74   **"The military aerial photograph":** Major Edward J. Steichen, "American Aerial Photography at the Front," *The Camera*, July 1919, 360.

74   **"allows for a visual comparison":** Kieran Baxter, "What a Century of Climate Change Has Done to France's Biggest Glacier," *The Conversation*, Mar. 15, 2018.

Chapter 4: Into the Water

75   **On a bright September morning:** Louis Boutan, *La photographie sous-marine et les progrès de la photographie* (Paris: Schleicher Frères, 1900), 205.

75   **Previous attempts by various photographers:** W. D. Valentine, "On Submarine Photography," *Photographic News*, vol. 26, no. 1237, May 18, 1882, 277; John Ingham, "Submarine Photography," *British Journal Photographic Almanac*, 1885, 102; "Subaqueous Photography," *Photographic News*, vol. 39, no. 1902, Feb. 15, 1895, 104.

76   **"almost as an impossibility":** "Some Difficulties with Gelatin Plates," *British Journal of Photography*, vol. 29, no. 1170, Oct. 6, 1882, 568.

76   **Boutan was not a photographer, but he was a scientist:** Louis Boutan, "Submarine Photography," *Century Magazine*, vol. 56, 1898, 42; Lucien Pohl, "Hommage à Louis Boutan," *Institute Océanographique de l'Indochine*, Sept. 30, 1936, 7–9; James Dugan, *Man Explores the Sea* (Middlesex: Penguin Books, 1956), 168.

76   **Boutan completed his doctoral thesis:** Alejandro Martinez, "'A Souvenir of Undersea Landscapes': Underwater Photography and the Limits of Photographic Visibility, 1890–1910," trans. Catherine Jagoe, *História, Ciências, Saúde—Manguinhos*, Rio de Janeiro, vol. 21, no. 3, July–Sept 2014, 5.

76 **Situated on the Mediterranean:** Martinez, "'A Souvenir,'" 4–5.

76 **Arago was a place:** Boutan, *La photographie*, 116–25.

77 *Haliotis* **mollusk:** Martinez, "'A Souvenir,'" 5, and Robert H. Sherard, "The Experiments of Louis Boutan," *Pearson's Magazine*, Nov. 1900, 462.

77 **having learned to dive:** Rahmy Elkays, "Louis Boutan et la photographie sous-marine (1886–1900)," *Focales*, June 1, 2021, 2–3.

77 **"the beauty, the extreme beauty":** Sherard, "The Experiments," 462.

77 **a solicitor named William Thompson:** William Thompson, "On Taking Photos Underwater," *Journal of the Society of Arts*, vol. 4, no. 181, May 9, 1865, 425, and Gernsheim and Gernsheim, *History of Photography*, 511–12.

77 **There wasn't a blueprint:** Thompson, "On Taking Photos Underwater," 425–26.

78 **"a very good photograph of the rocks":** "Photography under Water," *Photographic and Fine Art Journal*, vol. 9, no. 8, Aug. 1856, 242.

78 **"This remarkable picture":** "Extracts from Foreign Publications: Meeting of the Liverpool Photographic Society," *Humphrey's Journal*, vol. 8, no. 5, July 1, 1856, 65.

78 **"incalculable benefit to science":** Thompson, "On Taking Photos Underwater," 426.

80 **Boutan's first camera:** For details of Boutan's first camera and his solution to water pressure, see Boutan, *La photographie*, 163–69; Sherard, "The Experiments," 462.

80 **its six plates, measuring:** Dimitri Rebikoff, "The History of Underwater Photography," *Photogrammetic Engineering*, Aug. 1967, 898.

80 **"enemies of the fish tribe":** "A Deep Sea Diver's Experience," *Edinburgh Evening News*, Jan. 2, 1897, 6.

80 **diving suit was about 170 pounds:** Dr. Michael Jung, email to author, Feb. 23, 2024.

80 **Boutan likened his rubberized canvas suit:** Boutan, *La photographie*, 143.

81 **Boutan developed a methodology:** Boutan, "Submarine Photography," 45.

82 **diving watches:** The Omega Marine was the first watch to be specifically tested for water pressure, in 1932. See Anthony Turner, James Nye, and Jonathan Betts, eds., *A General History of Horology* (Oxford: Oxford University Press, 2022), 490.

82 **The experiments with this first camera . . . "an extremely vague outline":** Boutan, *La photographie*, 171, 186; Boutan, "Submarine Photography," 43–45.

82 **the real problem was light:** For exposure times, see Boutan, *La photographie*, 171. For the underwater lamps, see *La photographie*, 227–36.

84 **During the 1890s:** *Catalogue of Scientific Papers: Fourth Series (1884–1900)*, compiled by the Royal Society of London, Cambridge University Press, vol. 13, 1914, 741.

84 **Boutan had tried a completely different tack:** Boutan, *La photographie*, 173–74, and Dugan, *Man Explores the Sea*, 169–70.

84 **Boutan's third and final device:** Boutan, *La photographie*, 175; Louis Boutan, "Submarine Photography and Its Results," *Photography: The Journal of the Amateur, the Profession and the Trade*, vol. 13, no. 634, Jan. 3, 1901, 12–13; and Martinez, "'A Souvenir,'" 7–9, 17.

85 **Instead of a fixed-focus lens:** Boutan, "Submarine Photography," 14, and Martinez, "'A Souvenir,'" 8–9.

85 **He pulled the cord:** Boutan described this experiment in "L'instantané dans la photographie sous-marine," *Comptes Rendus Hebdomadaires des Séances de l'Académie des Sciences*, no. 19, Nov. 7, 1898, 731, and Boutan, *La photographie*, 204–6.

86 **Boutan's response to a newspaper cartoon:** Dugan, *Man Explores the Sea*, 171.

87 **one of the least maneuverable devices:** Boutan, *La photographie*, 195–99.

88 **Boutan theorized:** Boutan, *La photographie*, 238–39.

89 **François Deloncle:** Boutan, *La photographie*, 239.

89 **Chaufour built two spherical:** Boutan, *La photographie*, 240–48, and Martinez, "'A Souvenir,'" 11.

89 **Boutan first tested:** Boutan, *La photographie*, 249–54.

90 **At least, that was the plan:** Details of this experiment from Boutan, *La photographie*, 257–65.

92 **lit by magnesium powder:** W. H. Longley, "Life on a Coral Reef," *National Geographic Magazine*, vol. 51, 1927, 73.

92 **Boutan published a book:** Details about Boutan's life after underwater photography are from the James Dugan Papers, 1933–1977, boxes 12 and 20, Pennsylvania State University Libraries; Pohl, "Hommage," 9–14; Dugan, "Man Explores the Sea," 74–77; Rachel Trethewey, *Pearls before Poppies: The True Story of the Red Cross Pearls* (Cheltenham: History Press, 2019).

93 **"Will not such photographs":** Boutan, "Submarine Photography," 559.

93 **In 2021, extraordinary footage:** Randi Rotjan, "244 Coral Samples, 14 Seamounts, and 2 Glass Octopuses," Schmidt Ocean Institute, July 19, 2021.

Chapter 5: Intrigue and Impact

97 **range of standard sizes:** There were some slight variations in daguerreotype size between Europe, England, and America. For more details, see Charles A. Long, *Practical Photography, on Glass and Paper, a Manual*, 2nd ed. (London: Bland & Long, 1856), 25, and Todd Gustavson, *Camera*, 352.

97 **reduced a daguerreotype using a microscope lens:** Hardwicke Knight, "Early Microphotographs," *History of Photography*, vol. 9, no. 4, Oct.–Dec. 1985, 311, and Frederic Luther, "The Earliest Experiments in Microphotography," *Isis*, vol. 41, no. 3/4, Dec. 1950, 278.

97 **John Benjamin Dancer:** Details about Dancer from Frederic Luther, *Microfilm: A History, 1839–1900* (Annapolis, MD: National Microfilm Association, 1959), 16, and "John Benjamin Dancer: Originator of Microphotography," *British Journal of Photography*, Feb. 16, 1973, vol. 120, no. 5874, Feb. 16, 1973, 138.

97 **In April 1853:** J. B. Dancer, *British Journal of Photography*, vol. 6, no. 94, May 15, 1859, 126, and Luther, "The Earliest Experiments," 278.

98 **Queen Victoria owned:** Gernsheim and Gernsheim, *History of Photography*, 318.

99 **"A family group":** David Brewster, "On the Photomicroscope," *British Journal of Photography*, vol. 11, no. 206, Jan. 15, 1864, 22; also quoted in Luther, *Microfilm: A History*, 28.

99 **"little or no practical utility":** Thomas Sutton, *A Dictionary of Photography* (London: Sampson Low, Son, and Co., 1858), 295; also quoted in Michael Hallett, "John Benjamin Dancer 1812–1887: A Perspective," *History of Photography*, vol. 10, no. 3, July–Sept. 1986, 250.

99 **During a visit to Paris:** Brewster, "On the Photomicroscope," 22; Frederic Luther, "Rene Dagron, Inventor of Microfilm," *History of Photography*, vol. 20, no. 4, 1996, 345.

99 **René Prudent Patrice Dagron:** Jean Scott, *Stanhopes: A Closer View* (Essex: Greenlight Publishing, 2002), 9–10, and Luther, "Rene Dagron," 347.

99 **Photos had been incorporated:** See, for example, L. West and P. Abbott, "Daguerreian Jewelry: Popular in Its Day," *Daguerreian Annual: Official Yearbook of the Daguerreian Society*, 1990, and Larry J. West and Patricia A. Abbott, *Tokens of Affection and Regard: Antique Photographic Jewelry* (New York: Carl Mautz Publishing, 2005).

100 **To produce microphotographs en masse:** Scott, *Stanhopes*, 21–22; Sol Legault, "Making the Invisible Visible: Exhibiting Erotic Stanhopes" (Master of Arts thesis, Toronto Metropolitan University, 2012), 13, 18; on the exposure times for the Taupenot process, see Gernsheim and Gernsheim, *History of Photography*, 324–25.

101   **cylinder of optical glass:** This is the design from Dagron's 1860 patent, which had evolved and become simpler since his 1859 patent, according to Jean Scott, author of *Stanhopes: A Closer View*, 11–12.

102   **"By placing the eye":** Brewster, "On the Photomicroscope," 22.

102   **patent was successfully challenged:** On the challenges to Dagron's patent, see Scott, *Stanhopes*, 18, and Luther, *Microfilm: A History*, 35–38.

103   **150 employees:** Ernest Lacan, "Foreign Correspondence," *British Journal of Photography*, vol. 8, no. 155, Dec. 2, 1861, 432.

103   **"rapid, simple, and ingenious":** A. Claude and P. Le Neve Foster, "Dublin Exhibition," *Photographic Journal*, vol. 10, no. 162, Oct. 16, 1865, 164.

104   **"Several Customs seizures":** "Customs Seizures," *South Australian Register*, Sept. 11, 1869, 4, quoted in Scott, *Stanhopes*, 27.

104   **"I happen to know":** "Importation of Immorality," *Adelaide Observer*, Sept. 11, 1869, 3.

104   **In the case of the watch keys:** "Police Courts," *South Australian Register*, Sept. 21, 1869, 4.

104   **By their very nature:** Legault, "Erotic Stanhopes," 28–29.

104   **One notable collection:** Legault, "Erotic Stanhopes," 29; Matt Bloom, "Stanhopes: Hidden Erotica of the 19th Century," *Kinsey Today*, vol. 18, no. 2, Spring 2014, 16–17; Garry Alan Milius, Special Collections Manager, Kinsey Institute, phone call with the author, Feb. 29, 2024.

104   **Dagron's company also produced erotic Stanhopes:** Scott, *Stanhopes*, 110.

105   **salvaged pocketknives:** Conservation efforts in the 1970s—before anyone was aware of the Stanhopes—destroyed some of the images and damaged the remaining four: Bill Cantine, Museums Specialist, DeSoto National Wildlife Refuge, emails to author, June 24 and Oct. 8, 2024. See also Jeanne M. Harold, "Have You Checked Your Pocketknife Lately?," *Cultural Resource Management*, no. 7, 1996, 15, and Hunter Oatman-Stanford, "Royalty, Espionage, and Erotica: Secrets of the World's Tiniest Photographs," *Collector's Weekly*, Dec. 2, 2014.

105   **"twelve obscene and indecent":** "Indecent Photographs," *Humphrey's Journal*, vol. 19, no. 15, Dec. 1, 1867, 234.

106   **Sellers were prosecuted:** Simon Popple, "Indecent Exposures: Photography, Vice and the Moral Dilemma in Victorian Britain," *Early Popular Visual Culture*, vol. 3, no. 2, Sept. 2005, 113–15.

106   **Their usual mode of operation:** Lisa Z. Sigel, *Governing Pleasures: Pornography and Social Change in England, 1815–1914* (New Brunswick, NJ: Rutgers University Press, 2002), 21–22.

106   **In an irresistible if admittedly juvenile detail:** "Indecent Photographs and Prints," *British Journal of Photography*, vol. 19, no. 631, June 7, 1872, 274.

106   **At the SSV's annual meeting:** "Talk in the Studio," *Photographic News*, vol. 14, no. 613, June 3, 1870, 264.

106   **Yet the material still proliferated:** "Indecent Photographs," *British Journal of Photography*, vol. 21, no. 727, Apr. 10, 1874, 178.

106   **it was destroyed during:** Popple writes that "evidence was almost universally destroyed as part of court proceedings." Popple, "Indecent Exposures," 127.

106   **"photographs of young ladies":** J.B., "Abuse of Photography," *Photographic Notes*, vol. 10, no. 226, Sept. 1, 1865, 23.

106   **In 1877, a woman:** "Charge of Selling Indecent Photographs," *British Journal of Photography*, vol. 24, no. 900, Aug. 3, 1988, 371.

107   **"of a very objectionable description":** "Photography in Court," *British Journal of Photography*, vol. 27, no. 1060, Aug. 27, 1880, 419.

107 **When the Prussian army laid siege:** Details of the Siege of Paris are from Scott, *Stan-hopes*, 41–42, and Luther, *Microfilm: A History*, 58–83 and 140.

107 **"Microscopic copies of despatches":** Brewster, "On the Photomicroscope," 22.

107 **possible to enlarge an image:** Gernsheim and Gernsheim, *History of Photography*, 315.

108 **"mammoth-plate" photograph:** "Mammoth plate" included sizes slightly larger or smaller, but 18 x 22 was the size of Watkins's plates. See Weston J. Naef and James N. Wood, *Era of Exploration: The Rise of Landscape Photography in the American West, 1860–1885* (Buffalo, NY: Albright-Knox Art Gallery, 1975), preface.

109 **either Emmons or Eugene:** Palmquist, *Carleton E. Watkins*, 3, and Nanette M. Sexton, "Carleton E. Watkins: Pioneer California Photographer (1829–1916): A Study in the Evolution of Photographic Style during the First Decade of Wet Plate Photography" (PhD thesis, Harvard University, 1982), 39.

109 **We also can't confirm:** For example, Tyler Green has Watkins leaving Oneonta in the spring of 1849, as does Maria Morris Hambourg. See Green, *Carleton Watkins: Making the West American* (Oakland: University of California Press, 2018), 24–25, and Hambourg's essay in Douglas R. Nickel, *Carleton Watkins: The Art of Perception* (San Francisco and New York: San Francisco Museum of Modern Art and Harry N. Abrams, 1999), 13. Peter Palmquist dates his departure to 1851 (Palmquist, *Carleton E. Watkins*, 3), and so does Christine Hult-Lewis in Weston Naef and Hult-Lewis, *Carleton Watkins: The Complete Mammoth Photographs* (Los Angeles: J. Paul Getty Museum, 2011), 1.

109 **Collis P. Huntington, a onetime tin peddler:** Sexton, "A Study," 39.

109 **In his book:** J. D. Borthwick, *Three Years in California, 1851–1854* (Edinburgh: William Blackwood and Sons, 1857); specifically, pages 7, 32, 35, 39.

109 **one historian has estimated:** Green, *Carleton Watkins*, 26.

109 **city of staggering growth:** See Lula May Garrett, "San Francisco in 1851 as Described by Eyewitnesses," *California Historical Society Quarterly*, vol. 22, no. 3, Sept. 1943, 253–80.

109 **In a biography published after he died:** Charles B. Turrill, "An Early California Photographer: C. E. Watkins," *News Notes of California Libraries*, vol. 13, no. 1, Jan. 1918, 30, and Richard Rudisill, "Watkins and the Historical Record," *California History*, vol. 57, no. 3, Fall 1978, 216.

110 **many land grant cases:** Between 1852 and 1856, there were 800 land grant cases in San Francisco. Historian Kevin Starr has noted that "when it was over, very few of the original grantees had come through the process with their holdings intact." Kevin Starr, *California: A History* (New York: Modern Library Chronicles, 2005), 103–4. See also Sexton, "A Study," 75–77.

110 **In his deposition:** Peter E. Palmquist, "Carleton E. Watkins's Oldest Surviving Landscape Photograph," *History of Photography*, vol. 5, no. 3, July 1981, 223–24.

110 **"wild sublimity":** J. M. Hutchings, "Second Tourist Party to Yosemite Valley," *Mariposa Gazette*, Aug. 9, 1855, 2.

110 **Mariposa Battalion, who had entered Yosemite:** Information about the Battalion and Yosemite's Native American inhabitants is from: Rebecca Solnit, *Savage Dreams: A Journey into the Hidden Wars of the American West* (San Francisco: Sierra Club Books, 1994), 270–85; Linda W. Greene, *Yosemite: The Park and Its Resources* (U.S. Dept. of the Interior, National Park Service, 1987); Margaret Sanborn, *Yosemite: Its Discovery, Its Wonders and Its People* (New York: Random House, 1981); Dayton Duncan, *Seed of the Future: Yosemite and the Evolution of the National Park Idea* (Yosemite: Yosemite Conservancy, 2013); Carmen George, "American Indians Share Their Yosemite Story," *Fresno Bee*, September 30, 2015; Daniel Duane, "Goodbye, Yosemite. Hello, What?," *New York Times*, Sept. 2, 2017.

110 **"but one alternative" . . . "cheap labor":** Redick McKee, Geo. W. Barrour, and O. M.

Wozencraft, "Address of the Indian Agents," *Daily Alta California*, vol. 2, no. 35, Jan. 14, 1851, 2, quoted in Solnit, *Savage Dreams*, 271.

110 **"wild animals"**: Hutchings, "Second Tourist Party", 2.

111 **"It is beyond the power"**: "The Great Yo-Semite Valley," *Hutchings' Illustrated California Magazine*, vol. 4, no. 5, Nov. 1859, 195.

111 **"imperial plate" size:** See Naef and Wood, *Era of Exploration*, preface.

111 **The mammoth camera:** Size and weight from Nickel, *Carleton Watkins: The Art of Perception*, 210; Sexton, "A Study," 131; and George Philip Lebourdais, "Notes on Process," in *Carleton Watkins: The Stanford Albums* (Stanford, CA: Stanford University Press, 2014), 16. As regards who built it, see Green, *Carleton Watkins*, 84; Sexton, "A Study," 112; and Palmquist, *Carleton E. Watkins*, 12.

112 **financial backer:** Green, *Carleton Watkins*, 85–86, and Naef and Hult-Lewis, *Carleton Watkins*, 45–46.

112 **"so steep, so rough":** Horace Greeley, *An Overland Journey from New York to San Francisco in the Summer of 1859* (New York: C. M. Saxton, Barker & Co., 1860), 303.

112 **"To an unaccustomed rider":** Helen Hunt Jackson, *Bits of Travel at Home* (Boston: Roberts Brothers, 1878), quoted in Sanborn, *Yosemite*, 82.

112 **Hutchings clambered:** "The Great Yo-Semite Valley," 200–220.

112 **Watkins also had to consider the wind:** Palmquist, *Carleton E. Watkins*, 17–18.

114 **"rare skill and indomitable energy":** Rev. H. J. Morton, "Yosemite Valley, *Philadelphia Photographer*, vol. 3, no. 36, Dec. 1866, 379.

114 **In the spring of 1862:** Nickel, *Art of Perception*, 10.

115 **"The views of lofty mountains":** "Fine Arts," *New York Times*, Dec. 12, 1862, 2.

115 **Israel Ward Raymond . . . John Conness:** Kate Nearpass Ogden, "The Park Idea," in *Yosemite* (London: Reaktion Books, 2015), 106–7.

115 **Some have suggested that Lincoln:** On whether Conness showed Lincoln the photos, see Green, *Carleton Watkins*, 168.

115 **"the point of departure":** Hans Huth, "Yosemite: The Story of an Idea," *Sierra Club Bulletin*, vol. 33, no. 3, Mar. 1948, 68.

115 **Olmsted had first seen:** Sexton, "A Study," 190.

115 **Olmsted wrote to Watkins:** Huth, "Yosemite," 70.

115 **According to Whitney:** Palmquist, *Carleton E. Watkins*, 25.

116 **"one of the most persevering":** "California Photographs," *Humphrey's Journal*, vol. 18, no. 2, May 15, 1866, 21.

116 **over 1300 mammoth views:** Green, *Carleton Watkins*, 6.

116 **Mount Shasta . . . the Pacific Northwest:** Palmquist, *Carleton E. Watkins*, 41 and 72.

116 **His last decades:** Palmquist, *Carleton E. Watkins*, 81–83.

117 **"abounds in photographers":** Kate Nearpass Ogden "Yosemite Valley as Image and Symbol: Paintings and Photographs from 1855 to 1880" (study for PhD, Columbia University, 1992), 98.

Chapter 6: Trick or Truth

118 **"Photography is truth embodied":** "The Society's Exhibition," *Photographic News*, vol. 11, no. 481, Nov. 22, 1867, 560.

119 **"This magnificent picture":** "Birmingham Photographic Society," *Photographic Notes*, June 1, 1857, 202.

119 **To produce the work:** Rejlander described the process in "On Photographic Composition; with a Description of *Two Ways of Life*," *Photographic Journal*, vol. 4, no. 65, Apr. 21, 1858, 193–94.

120   **more than thirty separate negatives:** Mia Fineman, *Faking It: Manipulated Photography before Photoshop* (New Haven: Yale University Press, 2012), 75.

120   **"I regard art":** O. G. Rejlander, "What Photography Can Do in Art," *The Year-Book of Photography and Photographic News Almanac*, 1867, 50, quoted in Stephanie Spencer, "Art and Photography: Two Studies by O. G. Rejlander," *History of Photography*, vol. 9, no. 1, Jan.–Mar. 1985, 52.

120   **"plasticity" of photography:** Rejlander, "On Photographic Composition," 190.

120   **Prince Albert acquired three copies:** Curator Lori Pauli estimated that Rejlander produced at least nine different versions of *Two Ways of Life*, all with similar compositions. Pauli, *Oscar G. Rejlander: Artist Photographer* (New Haven: Yale University Press, 2018), 32–33.

121   **"is in many points masterly":** "Fine Art Gossip," *Athenaeum*, no. 1539, Apr. 25, 1857, 538, quoted in Pauli, *Rejlander*, 33.

121   **The Photographic Society of Scotland:** Pauli, *Rejlander*, 33.

121   **"not by its merits":** "Fine Arts: Photography," *Edinburgh News and Literary Chronicle*, Dec. 26, 1857, 4.

121   **the "vice" portion, and its naked bodies:** "Notes on the Exhibition of the Photographic Society of Scotland," *British Journal of Photography*, vol. 6, no. 85, Jan. 1, 1859, 8, and Gernsheim and Gernsheim, *History of Photography*, 247.

122   **"Works of high art":** "The Photographic Exhibition," *Art-Journal*, Apr. 1, 1858, 121.

122   **"The man who is capable":** "The Exhibition," *Photographic Journal*, no. 66, May 21, 1858, 207.

122   **"It would be difficult":** "Correspondence: Exhibition of the Photographic Society," *Photographic News*, vol. 2, no. 27, Mar. 11, 1859, 8.

122   **"dull recorder":** H. P. Robinson, "The Limitations of Photography," *The Year-Book of Photography and Photographic News Almanac*, 1885, 33.

122   **"has been a failure":** Sutton, "On Some of the Uses and Abuses of Photography," 202.

122   **"a mistake and failure":** *Photographic Notes*, no. 166, Mar. 1, 1863, 55.

122   **"no better than caricatures":** *Photographic Notes*, no. 167, Mar. 15, 1863, 69.

122   **"in which degraded females were exhibited":** Sutton, "On Some of the Uses and Abuses," 203.

123   **the "patchwork" nature:** Wall's "patchwork" comments appeared under his byline in two different issues of *British Journal of Photography*, vol. 7, no. 120, June 15, 1860, 176, and vol. 7, no. 121, July 2, 1860, 190–91. Robinson's talk had appeared in print in April: Henry P. Robinson, "On Printing Photographic Pictures from Several Negatives," *British Journal of Photography*, vol. 4, no. 96, Apr. 16, 1860, 201.

123   **"The mass of chalky white sky":** Samuel Fry, "Instantaneous Photography and Composition Printing," *British Journal of Photography*, vol. 7, no. 131, Dec. 1, 1860, 849.

123   **The problem stemmed from the photo emulsions:** Fineman, *Faking It*, 9, and Gernsheim and Gernsheim, *History of Photography*, 264.

123   **red safe light:** Antoine Claudet lodged a patent for this at the end of 1841. Gernsheim and Gernsheim, *History of Photography*, 141.

124   **a French photographer named Gustave Le Gray:** Fineman, *Faking It*, 46–47; Nils Ramstedt, "An Album of Seascapes by Gustave Le Gray," *History of Photography*, vol. 4, no. 2, Apr. 1980, 131; see also Ken Jacobson, *"The Lovely Sea-View . . . Which All London Is Now Wondering At": A Study of the Marine Photographs Published by Gustave Le Gray, 1856–1858* (Petches Bridge: Ken and Jenny Jacobson, 2001).

124   **"cannot be used indiscriminately":** V. Blanchard, "On the Production and Use of Cloud Negatives," *Photographic News*, vol. 7, no. 261, Sept. 4, 1863, 424.

124   **"indiscriminately printing-in clouds":** C. Jabez Hughes, "Art Photography: Its Scope and Characteristics," *Photographic News*, vol. 122, Jan. 4, 1861, 2.

124 **they could purchase stock photographs:** Gernsheim and Gernsheim, *History of Photography*, 265.

124 **Carleton E. Watkins's landscapes:** Fineman, *Faking It*, 11–13.

124 **"Cloud Negatives":** "Perry's Cloud Negatives—100 Subjects," *British Journal of Photography*, vol. 25, no. 923, Jan. 11, 1878, 923, and "Cloud Negatives. No More Inartistic, Monotonous Backgrounds to Landscapes, Views, &c," *The Year-Book of Photography and Photographic News Almanac*, 1877, xlix.

124 ***A Few Plain Answers:*** Ross and Thomson, *A Few Plain Answers to Common Questions Regarding Photography* (Edinburgh: MacPherson & Syme, 1853), 5.

124 **As curator Sylvie Aubenas:** Sylvie Aubenas, *Gustave Le Gray 1820–1884* (Los Angeles: Getty Publications, 2002), 291.

127 **"Honor the photographer":** Wilhelm Busch, quoted in Henisch and Henisch, *The Photographic Experience*, 286.

127 **a photo "that made him look":** *Pittsburgh Morning Post*, Dec. 2, 1863, quoted in Henisch and Henisch, *The Photographic Experience*, 286.

127 **"very awkward customer":** "Windus v. Peat: An 'Awkward Customer,'" *British Journal of Photography*, vol. 23, no. 835, May 5, 1876, 215.

127 **"vague likeness":** N. P. Lerebours, *A Treatise on Photography, Containing the Latest Discoveries and Improvements Appertaining to the Daguerreotype* (London: Longman, Brown, Green, and Longmans, 1843) 5, quoted (in part) in Newhall, *The History of Photography*, 59.

127 **"It must want color":** "Photography," *Blackwood's Edinburgh Magazine*, vol. 51, no. 318, Apr. 1842, 518, quoted in Brian Coe, *Colour Photography: The First Hundred Years 1840–1940* (London: Ash & Grant, 1978), 10.

127 **add color with a variety of materials:** Henisch and Henisch, *The Photographic Experience*, 94–95, and Fineman, *Faking It*, 9.

127 **"tinted face and hands":** Quoted in Coe, *Colour Photography*, 12.

127 **fix the "disagreeable expression":** "Lessons on Colouring Photographs," *Photographic News*, vol. 2, no. 37, May 20, 1859, 126.

128 **In the wet collodion years:** John Plunkett, "Retouching," in *Encyclopedia of Nineteenth-Century Photography*.

128 **"there is no photographic establishment":** Elizabeth Eastlake, "Photography," *London Quarterly Review*, no. 101, June–Sept. 1857, 467.

128 **"Fading Committee":** Henisch and Henisch, *The Photographic Experience*, 91–92, and Grace Seiberling, *Amateurs, Photography, and the Mid-Victorian Imagination* (Chicago: University of Chicago Press, 1986), 36–37.

128 **other hazards:** John Eastham, "Some Causes of Fading Prints," *The Year-Book of Photography and Photographic News Almanac*, 1869, 72.

128 **Retouchers used desks:** Burrows and Colton, *Concise Instructions in the Art of Retouching* (London: Marion & Co., 1876), ix.

128 **"a tool which, in educated hands":** Special Correspondent, "Photographers' Association of America: Fifth Annual Convention," *British Journal of Photography*, vol. 31, no. 1268, Aug. 22, 1884, 539.

128 **"The essential value of photography":** "Photographic Process," *Humphrey's Journal*, vol. 7, no. 16, Dec. 15, 1855, 260.

128 **Some argued that limited retouching:** "Retouching: Its Use and Abuse," *Photographic News*, vol. 15, no. 682, Sept. 29, 1871, 460.

129 **"not flattery":** Ernest C. Morgan, "Retouching," *Anthony's Photographic Bulletin*, vol. 24, no. 19, Oct. 14, 1893, 617.

129 **"All life is gone":** "Retouching: Its Use and Abuse," 460.

129 **"desire to be made good-looking"**: "The Views of a Cynic upon Portraiture," *British Journal of Photography*, vol. 33, no. 1379, Oct. 8, 1886, 631.

130 **"Do not on any account"**: J. Hubert, *The Art of Retouching: With Chapters on Portraiture and Flash-Light Photography* (London: Hazell, Watson, & Viney Ltd, 1891), 48–49, quoted in Plunkett, "Retouching."

130 **"considerable pruning"**: W. Hartmann, "The Purpose and Limits of Retouching," *Photographic News*, vol. 15, no. 658, Apr. 14, 1871, 172–73.

130 **"becomes one of the most"**: "Retouching: Its Use and Abuse," 460.

130 **"The rendering of"**: "Retouching and Photographic Truth," *Photographic News*, vol. 16, no. 698, Jan. 19, 1872, 25.

130 **Upon his arrest:** Clément Chéroux, "Ghost Dialectics: Spirit Photography in Entertainment and Belief," in *The Perfect Medium: Photography and the Occult*, Clément Chéroux, Andreas Fischer, and Pierre Apraxine (New Haven: Yale University Press, 2005), 50.

130 **The full extent of Buguet's:** Chéroux, "Ghost Dialectics," 49.

131 **about 240 enlarged portraits:** John Warne Monroe, *Laboratories of Faith: Mesmerism, Spiritism, and Occultism in Modern France* (Ithaca, NY: Cornell University Press, 2018), 177.

131 **"men, women, boys, and girls of all ages":** "'Spirit Photographers' Sent to Prison," *British Journal of Photography*, vol. 22, no. 790, June 25, 1875, 312.

131 **During this exposure:** Chéroux, "Ghost Dialectics," 49.

131 **"He had me sit":** Quoted in Monroe, *Laboratories of Faith*, 167.

132 **Buguet increased his studio's turnover:** Chéroux, "Ghost Dialectics," 49.

132 **"That proves nothing":** Monroe, *Laboratories of Faith*, 181.

133 **"It is enough":** Oliver Wendell Holmes, "Doings of the Sunbeam," *Atlantic Monthly*, vol. 12, July 1863, 7–8.

133 **another series of spirit photographs:** On Buguet's "anti-spirit" photographs, see Chéroux, "Ghost Dialectics," 58.

133 **In 1856, David Brewster:** S. Petersen, "Frauds and Fakes," in *Encyclopedia of Nineteenth-Century Photography*.

133 **popular book *Photographic Amusements*:** Fineman notes that the book stayed in print for forty-one years. Fineman, *Faking It*, 124.

133 **"who practice on the gullibility":** Walter E. Woodbury, *Photographic Amusements, Including a Description of a Number of Novel Effects Obtainable with the Camera* (New York: Scovill & Adams Company, 1896), 28.

133 **William H. Mumler:** Crista Clouthier, "Mumler's Ghosts," in Chéroux et al., *The Perfect Medium*, 21–22.

134 **"entertained the whimsical fancy":** "Photographic Spirits," *Humphrey's Journal*, vol. 14, no. 23, Apr. 1, 1863, 312.

134 **"a low swindle":** "The Spirit Photographs," *American Journal of Photography*, vol. 5, no. 14, Jan. 15, 1863, 329.

134 **"The Americans":** "Ghosts Photographed Gratis," *The Spectator*, Feb. 28, 1863, 9.

135 **"never used any trick":** *New York Times*, May 4, 1869, 1.

135 **"morally convinced":** Clouthier, "Mumler's Ghosts," 22–23.

136 **As for Buguet:** Chéroux, "Ghost Dialectics," 55.

136 **"I am just a photographer":** Chéroux, "Ghost Dialectics," 50.

136 **Prime Minister Gladstone:** Jennifer Tucker, "Photography as Witness, Detective, and Impostor: Visual Representation in Victorian Science," in *Victorian Science in Context*, ed. Bernard Lightman (Chicago: University of Chicago Press, 1997), 378–79.

136 **The other photos:** E. A. Jelf, "Photography as Evidence," *The Idler*, 1894, 517.

Chapter 7: In Color

138   **By the 1890s:** W. R. Alschuler, "Color Theory and Practice: 1860–1910 Processes (Positive)," in *Encyclopedia of Nineteenth-Century Photography*.

138   **"Never Despair":** Quoted in Heinz K. Henisch and Bridget A. Henisch, *Positive Pleasures: Early Photography and Humor* (University Park: Penn State University Press, 1998), 34.

138   **he had also designed:** Begley, *The Great Nadar*, 106.

138   **the family formed the Société Antoine Lumière et ses fils:** Bertrand Lavédrine and Jean-Paul Gandolfo, *The Lumière Autochrome: History, Technology, and Preservation*, trans. John McElhone (Los Angeles: Getty Publications, 2013), 8–12.

138   **Antoine saw Edison's Kinetoscope:** Carol Mavor, "Lumière, Auguste (1862–1954) and Louis (1864–1948)," *Encyclopedia of Nineteenth-Century Photography*.

139   **One of their earliest films:** The Lumières filmed at least three versions of this, which you can find on YouTube by searching "lumiere factory workers."

139   **It was over a decade:** Lavédrine and Gandolfo, *The Lumière Autochrome*, 68–70. Coe states that the Lumière brothers tried the Lippmann process in 1893. Coe, *Colour Photography*, 21.

139   **The process by which they:** See Bertrand Lavédrine and Jean-Paul Gandolfo, "The Autochrome Process: From Concept to Prototype," *History of Photography*, vol. 18, no. 2, Summer 1994, 120.

139   **"The spectator was met":** R. Krügener, "Lumiere's New Color-Photography" (from *Photographische Mitteilungen*), *Bulletin of Photography*, vol. 1, no. 6, Sept. 18, 1907, 117.

139   **availability was delayed in England:** Coe, *Colour Photography*, 52.

139   **"Never since I developed":** "My Own Experiences with the Autochrome Plates," R. Child Bayley, *Photography*, vol. 24, no. 975, July 16, 1907, 41–44.

140   **Autochromes made the news:** "Colour-Photography Made Possible at Last: The Lumière Process," *Illustrated London News*, Sept. 7, 1907, 352; "Colour Photography," *The Times*, July 15, 1907, 3; "Nature Reproduced by New Color Plates," *New York Times*, Oct. 5, 1907, 8.

140   **Autochromes had their downsides:** On exposure times and reproduction, see Sylvie Pénichon, "From Potatoes to Pixels, a Short Technical History of Color Photography," in *Color: American Photography Transformed*, ed. John Rohrbach (Austin: University of Texas Press, 2013), 290. Pénichon states that exposures were approximately 60 times slower for autochromes. See also Coe, *Colour Photography*, 52–53, and Gernsheim and Gernsheim, *History of Photography*, 524.

140   **73,000 autochromes:** Lavédrine and Gandolfo, *The Lumière Autochrome*, 2.

140   **autochromes of World War One:** See, for example, "Captured in Color: Rare Photographs from the First World War—The French Autochromists," Australian War Memorial, last accessed May 27, 2024.

140   **first underwater color photograph:** Michel Frizot, *The New History of Photography* (Cologne: Könemann, 1998), 423.

140   **"Soon the world":** Alfred Stieglitz, quoted in Aaron Scharf, *Pioneers of Photography: An Album of Pictures and Words* (New York: Harry N. Abrams, 1976), 176.

140   **William Lang Jr.:** Larry J. Schaaf, "Pleasurable Offerings to Botanical Friends," in Larry J. Schaaf, *Sun Gardens: Cyanotypes by Anna Atkins*, ed. Joshua Chuang (New York: New York Public Library, 2018,) 81–82.

140   **The book contained 377 plates:** William Lang, "A Photographic Blue-Book," *British Journal of Photography*, vol. 36, no. 1538, Oct. 25, 1889, 702.

141   **world's first photographically illustrated book:** Schaaf, *Sun Gardens*, 82.

141   **"the difficulty of making":** Quoted in Lang, "A Photographic Blue-Book," 702.

141   **sodium thiosulfate (hypo):** Newhall, *Latent Image*, 60.

141   **"negative" and "positive":** On this and Herschel's experiments, see Kelley Wilder, "Herschel, Sir John Frederick William, Baronet (1792–1871)," in *Encyclopedia of Nineteenth-Century Photography*.

141   **pet boa constrictor:** Schaaf, *Sun Gardens*, 64.

141   **Herschel also experimented with:** Mike Ware, "Cyanotype," in *Encyclopedia of Nineteenth-Century Photography*, and Newhall, *Latent Image*, 61–62.

141   **Herschel published a paper:** Schaaf, *Sun Gardens*, 63.

141   **Atkins had received an education:** Schaaf, *Sun Gardens*, 39–41.

141   **started making cyanotypes in the autumn:** In a letter from October of that year, Atkins references that she had started "lately." See Schaaf, *Sun Gardens*, 37.

142   **the abolition of the slave trade:** Although the trading in enslaved people was abolished in 1807, slavery was still legal in British colonies until the 1833 Slavery Abolition Act. See Jessica Brain, "The Abolition of Slavery in Britain," Historic UK, June 12, 2019. On John Atkins, see Schaaf, *Sun Gardens*, 53–54, and "Alderman John Atkins," Centre for the Study of the Legacies of British Slavery database, UCL, last accessed May 28, 2024.

142   **Atkins issued her cyanotype prints:** On the more than 6000 cyanotypes, and how she worked on the book more generally, see Schaaf, *Sun Gardens*, 38, 54, and 77.

142   **"are powerful early examples":** Schaaf, *Sun Gardens*, 89.

142   **Atkins also took photographs:** Schaaf, *Sun Gardens*, 60.

142   **"This last summer":** Letter from An Amateur, "The Daguerreotype," *Art-Union*, vol. 1, no. 23, 1840, 187–88, quoted in Margaret Denny, "Royals, Royalties and Remuneration: American and British Women Photographers in the Victorian Era," *Women's History Review*, vol. 18, no. 5, 2009, 802–3.

142   **Queen Victoria set the tone:** B. V. Heathcote and P. F. Heathcote, "The Feminine Influence: Aspects of the Role of Women in the Evolution of Photography in the British Isles," *History of Photography*, vol. 12, no. 3, Jan.–Mar. 1988, 269, and Denny, "Royals," 803.

142   **Constance Talbot:** Naomi Rosenblum, *A History of Women Photographers* (Paris: Abbeville Press, 1994), 40.

142   **Cecilia Glaisher:** She used Talbot's photogenic drawing process to create images of ferns. Her work is held at the Fitzwilliam Museum in Cambridge and can be viewed online at the Cecilia Glaisher website.

142   **Mrs. Ann Cooke:** Heathcote, "The Feminine Influence," 260–61.

143   **"an unerring Likeness":** *Hull Advertiser*, Sept. 4, 1846, 1.

143   **Jane Nina Wigley:** Bill Jay, "Women in Photography, 1840–1900," *British Journal of Photography*, vol. 129, Mar. 20, 1982.

143   **From a legal standpoint:** Steinbach, *Understanding the Victorians*, 166–69. On the gradual legal changes, such as from the Married Women's Property Acts of 1870 and 1882, see 177.

144   **Women in photography often used their husband's name:** Catlin Langford, "Lost and Found," Photo Oxford blog, last accessed May 28, 2024.

145   **"In photography":** "An Occupation for Young Ladies," *The Young Englishwoman: A Volume of Pure Literature, New Fashions, and Pretty Needlework Designs*, vol. 1 (London: Ward, Lock and Tyler, 1867), 441.

145   **"Miss Bond of Southsea":** Henisch and Henisch, *The Photographic Experience*, 99.

146   **Amélie Guillot-Saguez:** John F. McGuigan Jr. and Frank H. Goodyear III, *In Light of Rome: Early Photography in the Capital of the Art World, 1842–1871* (University Park: Penn State University Press, 2022), 23.

146   **"essentially well suited"**: "The Employment of Women in Photography," *Photographic News*, vol. 18, no. 766, May 9, 1873, 223.

146   **"respectable, yet bereaved"**: "Female Employment in Coloring Photographs," *Humphrey's Journal*, vol. 17, no. 2, May 15, 1865, 31, quoted in Coe, *Colour Photography*, 14.

147   **"the art of colouring photographs"**: Una Howard, "Coloured Photographs," Letter to the Editor, *London Evening Standard*, Feb. 25, 1865, 5.

147   **"Self-Help Institute"**: On Howard's system and her institute, see "Answers to Correspondents," *British Journal of Photography*, vol. 13, no. 333, Sept. 21, 1866, 458; "Organic Bromides in Collodion," *British Journal of Photography*, vol. 17, no. 551, Nov. 25, 1870, 552; "Notes and Queries," *British Journal of Photography*, vol. 31, no. 1273, Sept. 26, 1884, 623.

147   **"enormous and lucrative scope"**: C. Jabez Hughes, "Photography as an Industrial Occupation for Women," *Photographic News*, May 9, 1873, 217, partly quoted in Nicole Hudgins, "The Gender of Coloring," in *The Gender of Photography: How Masculine and Feminine Values Shaped the History of Nineteenth-Century Photography* (New York: Routledge, 2020), 173.

147   **"not that they do their work better"**: Jabez Hughes, "Photography," 217.

147   **"the acceptability of coloring"**: Hudgins, *The Gender of Photography*, 172.

147   **James Clerk Maxwell**: On Maxwell, see Laidler, *To Light Such a Candle*, 160–61, 169; for his first paper, see Basil Mahon, *The Man Who Changed Everything: The Life of James Clerk Maxwell* (Chichester: John Wiley & Sons, 2004), 57–58; for his color photography presentation, see J. C. Maxwell, "On the Theory of Three Primary Colours," *Proceedings of the Royal Institution of Great Britain*, May 17, 1861, 370–74.

148   **Sutton used solutions**: Ralph M. Evans, "Maxwell's Color Photograph," *Scientific American*, vol. 205, no. 5, Nov. 1961, 118. Further details of the experiment are from Alschuler, "Color Theory"; Jordi Cat, "A Tale of Two Experiments: From Professional to Cognitive Autonomy," in *Maxwell, Sutton, and the Birth of Color Photography: A Binocular Study* (New York: Palgrave Macmillan, 2013), 132–39; Pénichon, "From Potatoes to Pixels," 287–89.

148   **highly sensitive to the blue**: Kelley Wilder, "Science, Art, and the Business of Color," in *The Colors of Photography*, ed. Bettina Gockel (Berlin: De Gruyter, 2020), 311.

148   **According to Sutton's notes**: Coe, *Colour Photography*, 32.

148   **Evans discovered that Sutton's red filter**: Ralph M. Evans, "Maxwell's Color Photograph," *Scientific American*, vol. 205, no. 5, Nov. 1961, 125; Coe, *Colour Photography*, 32; and Alschuler, "Color Theory."

148   **"happy accident"**: Evans, "Maxwell's Color Photograph," 125.

149   **Later the same decade**: Frizot, *The New History*, 413.

149   **Thanks to experiments by Hermann Wilhelm Vogel**: Coe, *Colour Photography*, 34.

149   **"Panchromatic" plates**: Gernsheim and Gernsheim, *History of Photography*, 523.

149   **Sarah Angelina Acland**: Information about Acland's life is from Giles Hudson, *Sarah Angelina Acland: First Lady of Colour Photography, 1849–1930* (Oxford: The Bodleian Library, 2012), 19–26, 37.

149   **her father's caregiver**: Patricia Jalland, *Death in the Victorian Family* (Oxford: Oxford University Press, 1996), 159.

149   **Acland's "beautiful portraits"**: "Home Portraiture," *British Journal of Photography*, vol. 46, no. 2027, Mar. 10, 1899, 151.

149   **get "rid of all ideas"**: "Home Portraiture."

150   **"Sanger Shepherd Process"**: Hudson, *Sarah Angelina Acland*, 34, 198.

150   **"I am proud to say"**: Catharine Weed Barnes Ward, "The Photographic Convention of the United Kingdom," *International Annual of Anthony's Photographic Bulletin*, vol. 24, 1902, 123.

150   **she traveled to Gibraltar:** Hudson, *Sarah Angelina Acland*, 35.

150   **"a more vivid and correct idea":** Miss Acland, "A Visit to Gibraltar," *Photographic Journal*, vol. 45, no. 7, July 1905, 232.

150   **"most untiring and successful exponent":** "Three-Colour Work from Nature," *Photography*, vol. 19, no. 846, Jan. 24, 1905, 154.

150   **"quite the best":** Catharine Weed Ward, "Woman in Photography," *Photogram*, vol. 12, no. 144, Dec. 1905, 373–74, quoted in Hudson, *Sarah Angelina Acland*, 10.

150   **"is difficult to overstate":** Hudson, *Sarah Angelina Acland*, 41.

151   **Unfortunately, much of Acland's work:** Hudson, *Sarah Angelina Acland*, 233.

151   **Autochromes, as with other early color processes:** "Deterioration and Preservation of Negatives, Autochromes, and Lantern Slides," Library of Congress Genthe Collection Online, last accessed May 28, 2024.

151   **Metropolitan Museum of Art wanted to exhibit:** Luisa Casella, "On View January 25–30: Original Autochromes Produced Using the First Color Photographic Process," The Metropolitan Museum of Art, Jan. 20, 2011. See also Luisa Casella and Katherine Sanderson, "Display of Alfred Stieglitz and Edward Steichen Autochrome Plates: Anoxic Sealed Package and Lighting Conditions," *Topics of Photographic Preservation*, vol. 14, 2011, 162–67.

Chapter 8: Caught!

152   **"may perhaps not be uninteresting":** Richard Leach Maddox, "An Experiment with Gelatino-Bromide," *British Journal of Photography*, vol. 18, no. 592, Sept. 8, 1871, 422, quoted in Gernsheim and Gernsheim, *History of Photography*, 328.

153   **The adoption of dry plates:** Gernsheim and Gernsheim, *History of Photography*, 331.

153   **"as one of the most noteworthy":** "Summary of the Past Year," *British Journal of Photography*, vol. 26, no. 1025, Dec. 26, 1879, 611, quoted in Newhall, *The History of Photography*, 88.

153   **Exposure times were less than a second:** Gernsheim and Gernsheim, *History of Photography*, 332.

153   **Tripods . . . portable darkrooms:** Newhall, *The History of Photography*, 89.

153   **This drove changes in camera design:** Gernsheim and Gernsheim, *History of Photography*, 410.

153   **"The noble army of amateurs":** G. E. Latine, "Characters We Have Met," *British Journal Photographic Almanac*, 1885, 184.

153   **"The amateurs are insatiable":** "Camera Amateurs," *Photographic Times and American Photographer* (reprinted from *San Francisco Chronicle*), Sept. 4, 1885, 513.

153   **"The ease with which":** Sigma Smith, "Amateur Photography," *Photographic News*, vol. 28, no. 1350, July 18, 1884, 461.

153   **"Above all things":** *Amateur Photographer*, vol. 1, Oct. 10, 1884, 3.

154   **the number of photographic clubs:** Gernsheim and Gernsheim, *History of Photography*, 424.

154   **"enormous increase" in amateur clubs:** "Our Views," *Amateur Photographer*, Aug. 4, 1893, 72

155   **George Eastman:** Information on his early years and experiments is from Roger Butterfield, "The Prodigious Life of George Eastman," *Life*, Apr. 26, 1954, 154–68, and Elizabeth Brayer, "Eastman, George (1854–1932)," in *Encyclopedia of Nineteenth-Century Photography*.

156   **No. 1 Kodak:** At first it was known as the Kodak Original, but when Eastman launched the No. 2 Kodak a year later, he renamed this model the No. 1. Gustavson, *Camera*, 133.

156 **the name "Kodak":** Butterfield, "The Prodigious Life," 158.

156 **In 1892, Eastman renamed:** "About George Eastman," Eastman Museum Online, last accessed May 28, 2024.

156 **"Kodak Girl":** "The Kodak Girl: Women in Kodak Advertising," Toronto Metropolitan University Archives and Special Collections blog, October 7, 2013.

156 **1.5 million cameras:** Mia Fineman, "Kodak and the Rise of Amateur Photography," Heilbrunn Timeline of Art History, The Metropolitan Museum of Art, last accessed June 12, 2024. The figure of $100 million in Eastman's donations is from "About George Eastman," Eastman Museum Online, last accessed May 28, 2024.

156 **$3 million in profits:** Butterfield, "The Prodigious Life," 154.

156 **37-room mansion:** Butterfield, "The Prodigious Life," 154, 156.

156 **she owned several cameras:** Ann Novotny, *Alice's World: The Life and Photography of an American Original: Alice Austen, 1866–1952* (Old Greenwich, CT: Chatham Press, 1976), 44–46, 147.

156 **Alice Austen became fascinated:** "About Alice Austen," Alice Austen House Online, last accessed May 29, 2024.

157 **Austen lived in the comfort:** Family home details are from "History of the Home and Museum," "About Alice Austen."

157 **interior of a friend's bedroom:** Novotny, *Alice's World*, 158.

159 **Daisy Elliott:** On the relationship between Austen and Elliott, see Bonnie Yochelson, "Miss Alice Austen and Staten Island's Gilded Age," The Gotham Center for New York City History Online, 2021.

159 **album as a gift to Tate:** Yochelson, "Miss Alice Austen."

159 **"wrong devotion":** Novotny, *Alice's World*, 60.

159 ***Street Types of New York*:** Yochelson, "Miss Alice Austen"; Rosenblum, *A History*, 109; Novotny, *Alice's World*, 159.

159 **"Men can, as a rule":** Catharine Weed Barnes Ward, "Illustration for the Press. Abstract of a Lecture Delivered before the Society of Women Journalists by Mrs. Catharine Weed Barnes Ward," *Photographic News*, vol. 49, no. 483, Mar. 31, 1905, 200.

159 **And Austen was methodical:** Novotny, *Alice's World*, 139.

159 **Of an estimated 8000 images:** Kristine Allegretti, Director of Collections and Operations, Alice Austen House, emails to author, Mar. 12, 2024, and May 30, 2024; Carli DeFillo, Collections Manager, Historic Richmond Town, email to author, Mar. 1, 2024.

160 **In the mid-1940s:** "The Newly Discovered Picture World of Alice Austen: Great Woman Photographer Steps Out of the Past," *Life*, Sept. 24, 1951, 144, and Novotny, *Alice's World*, 182–92.

160 **They were rediscovered in 1951:** On Austen and Jensen, see Novotny, *Alice's World*, 10–12, 195.

160 **experimented with autochromes:** Novotny, *Alice's World*, 139. The only surviving color images in the Austen archive are hand-colored, per email from Allegretti, Mar. 6, 2024.

161 **Austen died the following year:** Hilton Kramer, "E. Alice Austen Photographed Earlier Gracious Days of S.I.," *New York Times*, Apr. 9, 1976, 63, and Julie Besonen, "She Did It Her Way," *New York Times*, June 27, 2014.

161 **"This fiend in human shape":** Latine, "Characters," 184.

161 **One report complained:** *Amateur Photographer*, Aug. 31, 1894, 138.

161 **In an anonymously written:** "Confessions of a Snap-Shot Fiend," *Pearson's Weekly*, Aug. 19, 1893, 72.

161 **"Many estimable young women":** "Hand Camera Snap Shots," *American Journal of Photography*, vol. 12, no. 141, Sept. 1891, 433.

161 **"is highly reprehensible":** "Objectionable Photography," *British Journal of Photography*, vol. 47, no. 2098, July 20, 1900, 451.

162 **"brotherhood of Peeping Toms":** "The Camera Fiends," *Photographic Times and American Photographer*, vol. 19, no. 422, Oct. 18, 1889, 524.

162 **Some fiends used their photos:** Jay, *Cyanide*, 228. See also "Ex Cathedra," *British Journal of Photography*, Dec. 23, 1898, 818.

162 **"haunting" Fifth Avenue:** "Blackmailing Photographers at Work," *St. Louis and Canadian Photographer*, vol. 9, no. 3, Mar. 1891, 116.

163 **"doing a blackmailing business":** "Notes and News," *Photographic Times and American Photographer*, vol. 21, no. 518, Aug. 21, 1891, 422.

163 **"unfortunate incident":** "Another Social Pest," *British Journal of Photography*, vol. 35, no. 1459, Apr. 20, 1888, 256, quoted in Jay, *Cyanide*, 229; "The Queen Photographed While Laughing," *Edinburgh Evening News*, Nov. 22, 1887, 4, and "The Queen Amused," *Sheffield Evening Telegraph*, Dec. 14, 1887, 2.

163 **In the United States:** "An Emphatic Protest," *Photographic News*, vol. 42, no. 187, July 28, 1899, 465.

163 **President Roosevelt:** "Notes," *American Amateur Photographer*, vol. 13, no. 12, Dec. 1901, 559.

163 **At its simplest:** Frizot, *The New History*, 237, and Colin Harding, "Camera Design: 5 Portable Hand Cameras (1880–1900)," in *Encyclopedia of Nineteenth-Century Photography*.

163 **"It requires no skill":** "Scoville Detective Cameras" advertisement, *The American Annual of Photography*, 1887, lxix.

163 **"Deceptive Angle Graphic":** Gustavson, *Camera*, 109.

163 **Henry Slater Detective Agency:** "Lost, Found and Personal," *Surrey Mirror*, Aug. 9, 1895, 4.

163 **"Midnight Hunt for Divorce":** "With a Detective Camera: Photographing a Faithless Wife," *Sheffield Evening Telegraph*, Sept. 24, 1888, 2.

163 **These cameras were typically:** Gernsheim and Gernsheim, *History of Photography*, 417.

163 **"Most Deceptive in Appearance":** Advertisement for "Sharp & Hitchmough's 'Aptus' Pic-Nic Basket Detective Camera," *British Journal Photographic Almanac*, 1889, 763.

163 **Other hidden camera disguises:** Gernsheim and Gernsheim, *History of Photography*, 417; Frizot, *The New History*, 238; Flukinger and Schaaf, *Paul Martin*, 21; and "Bloch's Photographic Scarf" advertisement, *The Beacon*, vol. 3, no. 25, Jan. 1891, 19.

164 **One lawyer even had:** "Camera Amateurs," *Photographic Times and American Photographer* (reprinted from *San Francisco Chronicle*), Sept. 4, 1885, 514.

164 **The camera, which was patented:** Gernsheim and Gernsheim, *History of Photography*.

164 **Størmer had initially bought:** Details on Størmer from "Frederik Carl Mülertz Størmer," *Biographical Memoirs of Fellow of the Royal Society*, vol. 4, no. 4, Nov. 1958, 273, and Jessica Stewart, "Young Student Secretly Photographs People with Hidden Spy Cam in the 1890s," My Modern Met Online, last accessed May 29, 2024.

166 **Størmer's archive of circular:** Email to author from Arne B. Langleite, Photo Archivist, Norwegian Museum of Science and Technology, Oct. 8, 2024.

166 **"one or two books":** *Photography*, May 23, 1889, quoted in Bill Jay, *Victorian Candid Camera* (Newton Abbot: W. J. Holman Limited, 1973), 20.

166 **"It is impossible to describe":** This quotation and details of Martin's modifications are from Martin, *Victorian Snapshots*, 20–23.

166 **"careful composition" . . . "people and things":** Martin, *Victorian Snapshots*, 16–18.

166 **"The first thing to do":** Martin, *Victorian Snapshots*, 3.

168   **Eastman Kodak also offered:** Martin, *Victorian Snapshots*, 30–31.

168   **various forms of punishment:** See, for example, "Notes and News," *Photographic Times and American Photographer*, vol. 21, no. 519, Aug. 28, 1891, 43; and "Blackmailing Photographers," 116. On the use of a brick, see Jay, *Cyanide*, 228.

168   **photograph "fat men":** "Flotsam and Jetsam," *Amateur Photographer*, July 28, 1893, 56.

168   **journal had expressed regret:** Jay, *Cyanide*, 228.

168   **editors readily suggested pseudonyms:** "Our Views," *Amateur Photographer*, Aug. 4, 1893, 72.

168   **guidelines for photographic etiquette:** These were reprinted in *American Amateur Photographer*, vol. 16, no. 9, Sept. 1904, 402, quoted in Jay, *Cyanide*, 233.

168   **"Taken Unawares":** Mary Warner Marien, *Photography: A Cultural History*, 3rd ed. (Englewood Cliffs, NJ: Prentice Hall, 2011), 167.

169   **The August issue:** *Penny Pictorial*, Aug. 5, 1899, 431. Thanks to Julian Carr for accessing this issue from the British Library.

169   **the "snapshottist":** "Unconventional Portraits of Celebrities," *The Graphic*, Nov. 12, 1910, 742.

169   **He later recalled:** Arthur Barrett, "Thirty Years of Press Photography," *Photographic Journal*, vol. 77, June 1937, 375.

169   **The three suffragists had been arrested:** On this case, see Ian Christopher Fletcher, "'A Star Chamber of the Twentieth Century': Suffragettes, Liberals, and the 1908 'Rush the Commons,'" *Journal of British Studies*, vol. 35, no. 4, Oct. 1996, 504–39.

169   **"a stranger passing the police court":** "Three Suffragette Leaders in the Dock at Bow Street," *Daily Mirror*, Oct. 15, 1908, 8.

169   **two-hour closing argument:** "The Charges against Suffragists," *The Times*, Oct. 26, 1908, 3.

169   **When they refused the order:** Fletcher, "'A Star Chamber,'" 523.

169   **"a cheer was raised":** "The Charges against Suffragists," 3.

170   **"Suffragette Attack":** *Lloyd's Weekly Newspaper*, Oct. 18, 1908.

170   **"suffrage meeting attended":** Emmeline Pethick Lawrence, *My Part in a Changing World* (London: Victor Gollancz, 1938), 205, quoted in Fletcher, "'A Star Chamber,'" 506.

170   **By this point, anyone who was arrested:** Gernsheim and Gernsheim, *History of Photography*, 515.

170   **But not all prisoners complied:** Linda Mulcahy, "Docile Suffragettes? Resistance to Police Photography and the Possibility of Object-Subject Transformation," *Feminist Legal Studies*, vol. 23, 2015, 86.

170   **suffragist Evelyn Manesta:** Mulcahy, "Docile Suffragettes?," and "Judge's Remedy for Suffragettes," *Pall Mall Gazette*, Apr. 23, 1913, 4.

171   **"If the law allowed":** "Judge's Remedy," 4. The third defendant was found not guilty.

171   **"The photographer informs":** *Memorandum, from the Prison Commissioner to Holloway Prison*, May 21, 1913, HO 45/12915, National Archives, quoted in Mulcahy, "Docile Suffragettes?," 94.

172   **"Cat and Mouse" Act:** Mulcahy, "Docile Suffragettes?," 88, and Elizabeth Crawford, "Police, Prisons and Prisoners: The View from the Home Office," *Women's History Review*, vol. 14, no. 3–4, 2005, 493.

172   **"in a state of collapse":** "Prison News: Hunger Strikers. More Releases under the Cat and Mouse Bill," *The Suffragette*, May 30, 1913, 548.

172   **"As, however, photographs cannot":** *Holloway*, Apr. 26, HO 45/12915, National Archives.

172   **Barrett attempted some photographs:** Jill Liddington, "Era of Commemoration: Celebrating the Suffrage Centenary," *History Workshop Journal*, no. 59, Spring 2005, 205.

172 **"evidently the Government":** "Photographed in Prison," *The Suffragette*, May 30, 1913, 548.

172 **New Scotland Yard purchased:** Liddington, "Era of Commemoration," 205, and Mulcahy, "Docile Suffragettes?," 89.

173 **offered a "decided advantage":** Walter L. Beasley, "Wild Animal Photography," *American Annual of Photography*, vol. 27, 1913, 286.

173 **"absolutely epoch-making":** Adolphe Abrahams, "Some Recent Advances in High-Speed Photography," *Photographic Journal*, vol. 54, no. 3, Mar. 1914, 143.

173 **Wild Cats:** On code names, see Liddington, "Era of Commemoration," 205, and Mulcahy, "Docile Suffragettes?"

173 **Nettles and Thorns:** *Message codes*, DPP 1/23 f205, National Archives, last accessed June 30, 2024.

173 ***Rokeby Venus*:** In 2023, two climate activists smashed the glass of this very same painting. See Louis Jebb, "How Velázquez's 'Rokeby Venus' Became a Symbol of Public Pride—and Political Protest," *Art Newspaper*, Nov. 8, 2023.

173 **the National Gallery closed:** Mulcahy, "Docile Suffragettes?," 83–84, and Liddington, "Era of Commemoration," 208–9.

173 **press reaction to the exhibition:** Dominic Casciani, "Spy Pictures of Suffragettes Revealed," BBC News, Oct. 3, 2003, and Alan Travis, "Big Brother and the Sisters," *The Guardian*, Oct. 9, 2003.

173 **Mary Richardson was arrested:** See Hilda Kean, "Some Problems of Constructing and Reconstructing a Suffragette's Life: Mary Richardson, Suffragette, Socialist and Fascist," *Women's History Review*, vol. 7, no. 4, 1998, 475–93.

174 **"secretly spied on women":** Liddington, "Era of Commemoration," 209.

174 **British Pathé newsreel:** To find this on YouTube, search "Suffragettes Meet Again."

Chapter 9: Cartomania

175 **300–400 million cartes de visite:** Gernsheim and Gernsheim, *History of Photography*, 301.

175 **"the opportunity of distributing":** "The Carte de Visite," *All the Year Round*, no. 158, Apr. 26, 1862, 165.

175 **Who invented cartes de visite:** Gernsheim and Gernsheim, *History of Photography*, 293–94.

176 **wealthy and the fashionable of Paris:** Elizabeth Anne McCauley, *A.A.E. Disdéri and the Carte de Visite Portrait Photograph* (New Haven: Yale University Press, 1985), 45–47.

176 **cartes de visite arrived in the UK:** William C. Darrah, *Cartes de Visite in Nineteenth-Century Photography* (Gettysburg, PA: W. C. Darrah Publisher, 1981), 5–6, and Audrey Linkman, *The Victorians: Photographic Portraits* (London: Tauris Park Books, 1993), 61–62.

176 **richest photographer in the world:** Gernsheim and Gernsheim, *History of Photography*, 295.

177 **"the cheapest, the most portable":** Holmes, "Doings of the Sunbeam," 7–8.

177 **One journal advocated:** S. R. Devine, "Hints on Card Pictures," *Humphrey's Journal*, vol. 13, no. 16, Dec. 15, 1861, 242.

177 **the woman stood while her husband sat:** Andrea L. Volpe, "Cheap Pictures: Cartes de Visite Portrait Photographs and Visual Culture in the United States, 1860–1877" (PhD dissertation, Rutgers University, 1999), 62.

177 **cartes de visite of horses:** C. Peter, "André-Adolphe-Eugène Disdéri (1819–1899)," in *Ency-*

clopedia of Nineteenth-Century Photography, and Robert Taft, *Photography and the American Scene: A Social History, 1839–1889* (New York: Dover Publications, 1938), 150.

177 **If you were in London:** In 1862, Wynter noted that there were "no less" than 35 photographers in Regent Street alone. A. Wynter, "Cartes de Visite," *American Journal of Photography* (reprinted from *Living Age*), vol. 4, no. 21, Apr. 1, 1862, 482.

177 **props could range:** Darrah, *Cartes de Visite*, 33; Elizabeth Siegel, *Galleries of Friendship and Fame: A History of Nineteenth-Century American Photograph Albums* (New Haven: Yale University Press, 2010), 45–46; Gernsheim and Gernsheim, *History of Photography*, 299–300.

177 **The background scenery:** Darrah, *Cartes de Visite*, 31.

179 **"Nothing can be more unpleasing":** "'Cartes De Visite' Portraits," *Photographic News*, vol. 5, no. 157, Sept. 6, 1861, 423.

179 **"What is the profession":** "The Carte de Visite," *All the Year Round*, no. 158, Apr. 26, 1862, 167.

179 **"It will be observed":** Andrew Wynter, "Cartes de Visite," *British Journal of Photography* (republished from *Good Words*), vol. 16, no. 462, Mar. 12, 1869, 126.

179 **the posing stand:** Volpe, "Cheap Pictures," 90–92.

179 **"Head-rest is written":** "Cartes de Visite of Celebrities," *Photographic Journal* (reprinted from *Saturday Review*), vol. 8, no. 128, Dec. 15, 1862, 189.

179 **Children also sat:** On children and cartes de visite, see Darrah, *Cartes de Visite*, 38.

181 **incorporated a music box:** Linkman, *The Victorians*, 72.

181 **"No drawing-room table":** "Lord Derby's Carte de Visite," *Saturday Review*, May 4, 1861, 446.

181 **one British firm claimed:** Linkman, *The Victorians*, 71.

181 **The first time a photographic portrait:** J. Plunkett, "Celebrity and Royalty," in *Encyclopedia of Nineteenth-Century Photography*.

181 **Queen Victoria approved the sale:** Frances Dimond and Roger Taylor, *Crown and Camera: The Royal Family and Photography, 1842–1910* (New York: Viking Books, 1987), 20.

181 **"exquisite studies":** "Our Weekly Gossip," *Athenaeum*, no. 1712, Aug. 18, 1860, 230.

181 **between three and four million copies:** Plunkett, "Celebrity and Royalty."

181 **"I have been writing":** Eleanor Stanley letter from Nov. 24, 1860, quoted in Darrah, *Cartes de Visite*, 6.

181 **received 70,000 orders:** Wynter, "Cartes de Visite," 482.

182 **"crowds around the photograph shops":** diary entry of A. J. Munby, quoted in Linkman, *The Victorians*, 67.

182 **300,000 copies:** Darrah, *Cartes de Visite*, 43.

182 **"The commercial value":** Wynter, "Cartes de Visite," 481.

183 **"might say with certainty":** "Looking in at Shop Windows," *All the Year Round*, vol. 28, June 12, 1869, 42; see also Rachel Teukolsky, "Cartomania: Sensation, Celebrity, and the Democratized Portrait," *Victorian Studies*, vol. 57, no. 3, Papers and Responses from the Twelfth Annual Conference of the North American Victorian Studies Association (Spring 2015), 465–66.

183 **At least one public figure:** Gernsheim and Gernsheim write that Lord Brougham, when passing his own carte de visite exhibited in a shop window, would ask about sales. Gernsheim and Gernsheim, *History of Photography*, 296.

183 **For celebrities:** Plunkett, "Celebrity and Royalty."

183 **public figures entered into financial arrangements:** Plunkett, "Celebrity and Royalty"; see also, Darrah, *Cartes de Visite*, 43.

183 **Charles Dickens:** Barbara McCandless, "The Portrait Studio and the Celebrity: Promoting

the Art," in *Photography in Nineteenth-Century America*, ed. Martha A. Sandweiss (Fort Worth, TX: Amon Carter Museum, 1991), 67.

183 **photographers and publishers regularly copied:** *Photographic Journal*, no. 113, Sept. 16, 1861, 257–58; Darrah, *Cartes de Visite*, 18; Tim Padfield, "Copyright," in *Encyclopedia of Nineteenth-Century Photography*; and Jason Lee Guthrie, "Entered According to Act of Congress: Copyright and Photography in 19th Century America," *Military Images*, vol. 38, no. 1, Winter 2020, 72–75.

183 **"The workmen in large establishments":** Holmes, "Doings of the Sunbeam," 2.

183 **listed about 2000 different carte de visite portraits:** Siegel, *Galleries of Friendship*, 64.

183 **wholesaler Marion & Company:** Plunkett, "Celebrity and Royalty."

184 **One manufacturer in England:** Gernsheim and Gernsheim, *History of Photography*, 196.

184 **needed to store the glass negative:** Darrah, *Cartes de Visite*, 14–15.

184 **"new style papier mache rock":** Advertisement for Lafayette W. Seavey, *Photographic Mosaics*, 1875, 150.

184 **"a baby cannot fall out":** Advertisement in Sheen & Simpkinson, *Illustrated Catalogue and Price List of Everything Pertaining to Photography* (Cincinnati: Spencer & Craig, 1888), 140.

184 **97 negatives in eight hours:** Gernsheim and Gernsheim, *History of Photography*, 298.

184 **including Nadar:** On cartes de visite, Nadar wrote, "One had to submit, that is, to go with the flow, or resign." Nadar, *When I Was a Photographer*, 151. According to McCauley, Nadar "categorized Disdéri as a figure without talent who produced standardized portraits with no lasting appeal." McCauley, *A.A.E. Disdéri*, 5.

184 **Camille Silvy:** On his clientele, see Juliet Hacking, "Camille Silvy's Repertory: The Carte-de-Visite and the London Theatre," *Art History*, vol. 33, no. 5, Dec. 2010, 856.

184 **"That sense of beauty":** Wynter, "Cartes de Visite," 484.

184 **Silvy's output:** Linkman, *The Victorians*, 73.

184 **Silvey's later ill health:** Mark Haworth-Booth, *Camille Silvy: River Scene, France* (Malibu, CA: J. Paul Getty Museum, 1992), 89–90.

184 **one couple had fallen in love:** "'Love at First Sight' of a Carte," *British Journal of Photography*, vol. 14, no. 386, Sept. 27, 1867, 465.

184 **At a New York brothel:** Volpe, "Cheap Pictures," 234.

185 **introducing a governess or maid:** Siegel, *Galleries of Friendship*, 33.

185 **"The soldier":** "New Publications," *New York Times*, June 2, 1861, 2.

185 **Colonel Elmer E. Ellsworth:** Darrah, *Cartes de Visite*, 75.

185 **around fourteen portrait sessions:** Darcy Grimaldo Grigsby, *Enduring Truths: Sojourner's Shadows and Substance* (Chicago: University of Chicago Press, 2015), 11.

185 **Truth's earliest known carte de visite:** Grigsby, *Enduring Truths*, 28.

185 **It's impossible to know:** On Truth's early years, see Nell Irvin Painter, *Sojourner Truth: A Life, a Symbol* (New York: W. W. Norton, 1996); Nell Irvin Painter, "Sojourner Truth's Knowing and Becoming Known," *Journal of American History*, vol. 81, no. 2, Sept. 1994, 461–62; and Grigsby, *Enduring Truths*, 7–9.

185 **In 1826, as slavery:** On Truth's emancipation, see Painter, *Sojourner Truth*, 22–25.

185 **Sometimes she holds knitting:** Painter, "Sojourner Truth's Knowing," 483, and Grigsby, *Enduring Truths*, 80.

185 **Often, her right hand:** Grigsby, *Enduring Truths*, 58, 107.

185 **her grandson James Caldwell:** Grigsby has noted that there are at least four different cartes des visite with Truth and her grandson's portrait. Grigsby, *Enduring Truths*, 52.

185 **Truth sold her cartes de visite:** Painter, "Sojourner Truth's Knowing," 482–83, and Painter, *Sojourner Truth*, 11.

185   **One report noted:** Kathleen Collins, "Shadows and Substance: Sojourner Truth," *History of Photography*, vol. 7, no. 3, July–Sept. 1983, 200.

186   **Truth's copyright ownership:** Grigsby has not found any other cartes de visite from this period that have the name of the sitter in the copyright field. Grigsby, *Enduring Truths*, 127–39.

186   **"Against law, convention":** Grigsby, *Enduring Truths*, 139.

187   **"I sell the shadow":** On variations of this phrase, see B. A. Henisch and H. K. Henisch, "Secure the Shadow Ere the Substance Fade," *History of Photography*, vol. 28, no. 3, 2004, 211–12.

187   **"were largely missing":** Painter, "Sojourner Truth's Knowing," 488.

188   **"The importance of this criticism":** "The Claims of the Negro, Ethnologically Considered," speech on July 12, 1854, accessed through the Library of Congress, and quoted in Sarah Blackwood, "'Making Good Use of Our Eyes': Nineteenth-Century African Americans Write Visual Culture," *Visual Culture and Race*, vol. 39, no. 2, Summer 2014, 42.

188   **"It is evident":** Frederick Douglass, "Lecture on Pictures," delivered in Boston on Dec. 3, 1861, in John Stauffer, Zoe Trodd, and Celeste-Marie Bernier, *Picturing Frederick Douglass: An Illustrated Biography of the Nineteenth Century's Most Photographed American*, rev. ed. (New York: Liveright, 2019), 120. Stauffer et al. note (113) that Douglass wrote four lectures on photography during the Civil War years.

189   **Douglass's first known portrait:** Stauffer, Trodd, and Bernier, *Picturing Frederick Douglass*, xi–xix. Douglass's last portrait was a post mortem photograph taken in 1895.

190   **"Although they might":** Maurice O. Wallace and Shawn Michelle Smith, eds., *Pictures and Progress: Early Photography and the Making of African American Identity* (Durham, NC: Duke University Press, 2012), 4. See also John Stauffer et al., "Creating an Image in Black: The Power of Abolition Pictures," in *Beyond Blackface: African Americans and the Creation of American Popular Culture, 1890–1930* (Chapel Hill: University of North Carolina Press, 2011), and Deborah Willis and Barbara Krauthamer, *Envisioning Emancipation: Black Americans and the End of Slavery* (Philadelphia: Temple University Press, 2013).

190   **Douglass made use:** Stauffer et al., *Picturing Frederick Douglass*, xv and xix.

190   **Arabella Chapman:** On Chapman's albums, see Keshia Clukey, "Rare African American Family Photo Albums Give Glimpse of 19th Century Albany," *Times Union*, July 27, 2015, and "U-M Students Uncover Story behind 19th-Century African-American 'Selfies,'" May 29, 2015, *Michigan News*, University of Michigan.

191   **What makes them all the more extraordinary:** Clukey, "Rare African American Family Photo Album."

191   **one of four Black photographers:** Stauffer, *Picturing Frederick Douglass*, xx. Ball took two carte de visite portraits of Douglass on January 12, 1867.

191   **"Ball's Great Daguerrean Gallery of the West":** Deborah Willis, *Reflections in Black: A History of Black Photographers 1840 to the Present* (New York: W. W. Norton, 2000), 5–7, and George Sullivan, *Black Artists in Photography* (New York: Cobblehill Books, 1996).

191   **"There is scarcely":** *Gleason's Pictorial Drawing Room Companion*, Apr. 1854, quoted in Sullivan, *Black Artists*, 38. For descriptions of Ball's studio, see James Presley Ball, *Ball's Splendid Mammoth Pictorial Tour of the United States* (Cincinnati: Achilles Pugh, 1855), 8–10.

191   **one that does exist:** Swann Auction Galleries received an inscribed carte de visite of Ball in 2023, which they believe might have been taken by Ball's son, who also worked with his father in the Cincinnati studio. See Swann Auction Gallery, "Printed and Manuscript African Americana," Mar. 30, 2023, Sale 2631—Lot 269.

Chapter 10: Running, Jumping, Galloping

195 **Madison Square:** McShane and Tarr, "The Centrality of the Horse in the Nineteenth-Century American City," in *The Making of Urban America*, ed. Raymond A. Mohl (Lanham, MD: SR Books, 1997), 107.

196 **"I could take the steed apart":** Leland Stanford, from the article "Senator Stanford's Plans," *Chicago Equity*, Dec. 1891, quoted in John Ott, "Iron Horses: Leland Stanford, Eadweard Muybridge, and the Industralised Eye," *Oxford Art Journal*, vol. 28, no. 3, 2005, 414.

196 **At some point afterwards:** Rebecca Solnit, *River of Shadows: Eadweard Muybridge and the Technological Wild West* (New York: Penguin Books, 2003), 78.

196 **Muybridge was an established landscape photographer:** Details on Muybridge are from Solnit, *River of Shadows*, 28–29, 46–47, and and Robert Bartlett Haas, "Eadweard Muybridge, 1830–1904," in Anita Ventura Mozley et al., *Eadweard Muybridge: The Stanford Years, 1872–1882* (Stanford, CA: Stanford University, 1972), 11–13.

196 **Muybridge's early attempts:** Solnit, *River of Shadows*, 82, and Haas, "Eadweard Muybridge, 1830–1904," 14, 18.

197 **Harry Larkyns:** Solnit writes that he was "a rogue whose tales of his life before San Francisco are heroic beyond the reach of credibility." Solnit, *River of Shadows*, 128.

197 **Muybridge also discovered:** Arthur P. Shimamura, "Muybridge in Motion: Travels in Art, Psychology and Neurology," *History of Photography*, vol. 26, no. 4, 2002, 344.

197 **At Muybridge's trial for murder:** Details of Muybridge's trial are from Shimamura, "Muybridge in Motion," 343–46.

197 **"under similar circumstances":** *San Francisco Chronicle*, Feb. 7, 1875, quoted in Shimamura, "Muybridge in Motion," 346.

197 **"moaned and wept convulsively":** Quoted in Shimamura, "Muybridge in Motion," 346.

197 **Muybridge's defense attorney:** Solnit, *River of Shadows*, 141.

197 **In his paper from 2002:** Shimamura, "Muybridge in Motion," 341–50.

197 **Stone tried to divorce:** Solnit, *River of Shadows*, 146–48.

198 **he did paste in the photographed face:** Haas, "Eadweard Muybridge, 1830–1904," 18–19.

198 **Alongside the track:** For a description of the experiment, see Haas, "Eadweard Muybridge, 1830–1904," 22; Gernsheim and Gernsheim, *History of Photography*, 436; "The Stride of a Trotting Horse: Photographing a Horse while Traveling Forty Feet a Second," *Pacific Rural Press*, June 22, 1878, 393; "Photographing a Trotting Horse," *Marin Journal*, June 27, 1878, 1; "Electro-Photography," *Chicago Daily Tribune*, June 30, 1878, 16; and "A Californian Discovery," *Daily Alta California*, July 8, 1878, 1.

199 **forty feet per second:** "Photographing a Trotting Horse," 1.

199 **"The negatives were very small":** "Photographing by Aid of Electricity: How a Trotting Horse Appears When at a 2:20 Gait," *Contra Costa Gazette*, July 20, 1878, 4.

199 **"like a whirlwind":** "The Stride of a Trotting Horse," 393.

200 **"They show the legs":** "Photography by Aid of Electricity," 4.

200 **"the most incongruous attitudes":** "The Photographs of Occident," *New York Times* (reprinted from the *San Francisco Chronicle*), June 30, 1878, 10.

200 **the leg positions as "ludicrous":** "A Californian Discovery," *Daily Alta California*, July 8, 1878, 1.

200 **"so devoid of all naturalness":** "Muybridge's Photographs: Interesting Lecture on Photographing a Horse While Running," *San Francisco Examiner*, July 10, 1878, 3.

200   **"nearly every attitude":** "Echoes of the Week," *Illustrated London News*, vol. 74, no. 2076, Mar. 29, 1879, 294.

200   **"unutterably hideous":** G.A.S., "Echoes of the Week," *Illustrated London News*, vol. 80, no. 2237, Mar. 18, 1882, 251.

200   **For example, when depicting:** Scharf, *Art and Photography*, 211–13.

200   **French artist Ernest Meissonier:** Scharf, *Art and Photography*, 214–15.

200   **As for Stubbs, he showed:** Jonathan Jones, "'He Was Learning about Life by Studying the Dead': The Ghastly Secret behind Artist George Stubbs's Beloved Paintings of Horses," an excerpt from Jonathan Jones, *Sensations: The Story of British Art from Hogarth to Banksy*, in *Art New News*, Apr. 17, 2019.

201   **Once Meissonier recovered:** Françoise Forster-Hahn, "Marey, Muybridge and Meissonier: The Study of Movement in Science and Art," in Mozley et al., *Eadweard Muybridge*, 99.

201   **French painter Rosa Bonheur:** *The Horse Fair*, Metropolitan Museum of Art website, last accessed June 4, 2024.

201   **as fanciful as "unicorns":** "Leland Stanford's Gift to Art and to Science," *San Francisco Examiner*, Feb. 6, 1881, 3, quoted in Ott, "Iron Horses," 416.

201   **"in no single instance":** "Mr. Muybridge at the Royal Institution," *Photographic News*, vol. 26, no. 1228, Mar. 17, 1882, 129.

202   **Bonheur's "glaring faults":** "Muybridge on the Attitudes of Animals in Motion," *Photographic News*, vol. 26, no. 1243, June 30, 1882, 374.

202   **In 1879, he had taken:** Haas, "Eadweard Muybridge, 1830–1904," 24.

202   **The zoopraxiscope was based:** Haas, "Eadweard Muybridge, 1830–1904," 25.

202   **These sequences were projected:** For more on the zoopraxiscope, see Solnit, *River of Shadows*, 202–3; Mozley et al.. *Eadweard Muybridge*, 25–26, 96. According to the Eadweard Muybridge website, run by the Kingston Museum, he also used at least one set of photographs on a zoopraxiscope disk.

202   **"Nothing was wanting":** *San Francisco Call*, May 5, 1880, quoted in Solnit, *River of Shadows*, 203.

202   **"the ugly animals":** G.A.S., "Echoes," 251.

202   **"the essence of life":** "Mr. Muybridge," 129.

203   **"An artist paints":** "Instantaneous Photography v. Eyesight," *British Journal of Photography*, vol. 29, no. 1159, July 21, 1881, 411.

203   **"It is the artist":** Rodin in conversation with Paul Gsell, *On Art and Artists* (London, 1958), quoted in Scharf, *Art and Photography*, 226.

203   **painting by Thomas Eakins:** Ott, "Iron Horses," 424, and Scharf, *Art and Photography*, 224.

203   **"executed and published":** J. D. B. Stillman, *The Horse in Motion as Shown by Instantaneous Photography: With a Study on Animal Mechanics Founded on Anatomy and the Revelations of the Camera* (Boston: James R. Osgood, 1882).

203   **Muybridge sued and lost:** Haas, "Eadweard Muybridge, 1830–1904," 27.

203   **This time he used three banks:** Marta Braun, "The Expanded Present: Photographing Movement," in Thomas, *Beauty of Another Order*, 172–73; Solnit, *River of Shadows*, 220; and Eadweard Muybridge, *Muybridge's Complete Human and Animal Locomotion: All 781 Plates from the 1887 "Animal Locomotion,"* vol. 3 (New York: Dover Publications, 1979), 1585–87.

204   **"the movements appertaining":** Muybridge, *Muybridge's Complete Human and Animal Locomotion*, 1588.

204 **In his notes, Muybridge:** For more about his gender depictions, see Marta Braun, *Picturing Time: The Work of Etienne-Jules Marey (1830–1904)* (Chicago: University of Chicago Press, 1992), 249–51.

205 **The entire volume cost:** Gernsheim and Gernsheim, *History of Photography*, 440.

205 **To advertise it:** Scharf, *Art and Photography*, 218.

205 **"under the guise":** Braun, *Picturing Time*, 251. For the figure of 40 percent and Braun's assessment of how Muybridge sequenced his plates, see 237–51.

206 **Up until he saw Muybridge's work:** Braun, *Picturing Time*, 47.

206 **"From the invisible atom":** E. J. Marey, *Animal Mechanism: A Treatise on Terrestrial and Aerial Locomotion* (London: Henry S. King & Co, 1874), 9, quoted in Braun, *Picturing Time*, 14.

206 **He devised instruments:** Braun, *Picturing Time*, 16–17, 24.

206 **"registering apparatus":** Marey, *Animal Mechanism*, 148–50.

206 **When the horse's hoof:** Braun, *Picturing Time*, 27–28.

206 **His findings validated:** Ott, "Iron Horses," 414.

206 **He also had them translated:** Forster-Hahn, "Marey, Muybridge and Meissonier," 93.

206 **All his devices required:** Braun, *Picturing Time*, 35, 49, 53, 203.

206 **When Muybridge produced:** Braun, *Picturing Time*, 47, 53.

207 **"I have a photographic gun":** Marey quoted in Braun, *Picturing Time*, 57.

207 **The camera looked like a rifle:** Gernsheim and Gernsheim, *History of Photography*, 440, and "Marey's Gun Camera," *Photographic News*, vol. 26, no. 1238, May 26, 1882, 289.

207 **Marey based the design:** John Westfall and William Sheehan, *Celestial Shadows: Eclipses, Transits, and Occultations* (New York: Springer, 2015), 394, 403–4. For the use of photography for the 1874 transit of Venus, see Canales, *A Tenth of a Second*, 121–24.

208 **"The photographer raises the gun":** "Marey's Gun Camera," 289.

208 **But Marey wasn't entirely satisfied:** Braun, "The Expanded Present," 157, and Braun, *Picturing Time*, 62–64.

208 **"The instruments that are not there":** Nadar, *When I Was a Photographer*, 184.

208 **Marey took the idea of the rotating:** Braun, *Picturing Time*, 75–81, and Braun, "The Expanded Present," 159.

208 **He experimented with his subject's attire:** E. J. Marey, "Photography of Moving Objects, and the Study of Animal Movement by Chrono-Photography," *Scientific American*, Supplement No. 579, Feb. 5, 1887, 9245, and Braun, *Picturing Time*, 81.

208 **Throughout the 1880s:** Braun, *Picturing Time*, 104, 124.

210 **Once George Eastman's new roll film:** Braun, "The Expanded Present," 175, and Braun, *Picturing Time*, 151–56.

210 **Marey worked on creating a projector:** Braun, *Picturing Time*, 173–74.

210 **"visible actions and audible words":** Quoted in Forster-Hahn, "Marey, Muybridge and Meisonnier," 98.

210 **Louis Le Prince:** Paul Clee, *Before Hollywood: From Shadow Play to the Silver Screen* (New York: Clarion Books, 2005), 117.

210 **Émile Reynaud projected:** Clee, *Before Hollywood*, 120; and Braun, *Picturing Time*, 156.

210 **"I knew instantly":** Quoted in Paul Fischer, *The Man Who Invented Motion Pictures* (New York: Simon & Schuster, 2022), 263.

211 **"Animated projections":** Étienne-Jules Marey, "The History of Chronophotography," Annual Report of the Board of Regents of the Smithsonian Institution, 1902, 329, quoted in Canales, *A Tenth of a Second*, 149–50.

211 **Marcel Duchamp was inspired:** Scharf, *Art and Photography*, 255–60.

211 **"the movements themselves":** Scharf, *Art and Photography*, 255.

Chapter 11: Cells and Snowflakes

213 **physician Alfred Donné and physicist Léon Foucault:** Ann Thomas, "The Search for Pattern," in *Beauty of Another Order*, 98–99.

213 **A manual on microscopes:** Lionel Smith Beale, *How to Work with the Microscope*, 4th ed. (London: Harrison, 1868), 26–27.

214 **Steadiness was crucial:** Tissandier, *A History and Handbook*, 225.

214 **work at night, or from a basement:** "Mikron," "The Elements of Photo-micrography," *Amateur Photographer*, June 23, 1893, 418; see also J. Connor, "Photomicrography," *Encyclopedia of Nineteenth-Century Photography*.

214 **It was sometimes necessary to rearrange:** Tucker, "Photography of the Invisible," 167.

214 **"It must be evident":** Olly's Patent Micro-Photographic Reflecting Process, *The Wonders of the Microscope, Photographically Revealed* (London: W. Kent and Co, 1857), i.

214 **"We appear to be":** "Microscopic Preparations," *Household Words*, vol. 16, Aug. 8, 1857, 132, quoted in Jude V. Nixon, " 'Lost in the Vast Worlds of Wonder': Dickens and Science," *Dickens Studies Annual*, vol. 35, 2005, 299.

214 **"additional reverence":** Jabez Hogg, *The Microscope: Its History, Construction, and Application* (London: The Illustrated London Library, 1854), viii.

214 **"a most extensive colony":** "The Queckett Microscopical Club," *London Evening Standard*, Jan. 5, 1867, 6.

215 **One Quekett soirée also featured:** *Journal of the Quekett Microscopical Club*, vol. 1, Sept. 1869, 54.

215 **"The milk they yield":** W. Bernard, "Chemistry for the Kitchen," *Eliza Cook's Journal*, Dec. 29, 1849, 132.

216 **"Let it only be":** "Science for Women," *English Woman's Journal*, vol. 9, no. 51, May 1, 1862, 153.

216 **the book *Food Materials*:** Ellen H. Richards, *Food Materials and Their Adulterations* (Boston: Estes and Lauriat, 1886).

216 **The English winter of early 1855:** See "The Weather and the Parks," *London Evening Standard*, Feb. 1, 1855, 1; "The Weather," *Bell's Weekly Messenger*, Feb. 26, 1855, 5; "Bread Riots and the Weather," *The Nonconformist*, Feb. 28, 1855, 171; "Health of London during the Frost," *Daily News*, Feb. 28, 1855, 6; "Health of London," *Illustrated London News*, Jan. 27, 1855, 79; "Health of London," *Daily News*, Feb. 1, 1855, 2.

216 **ice-skate on frozen ponds:** "The Weather, the Parks, etc," *St. James's Chronicle*, Feb. 17, 1855, and "The Weather," *Bell's*, 5.

216 **"almost microscopic power":** James Glaisher, "On the Severe Weather at the Beginning of the Year 1855; and on Snow and Snow-crystals," *British Meteorological Society [Journal]*, 1855, 18. On Glaisher's setup, see James Glaisher, "Snow Crystals," *Illustrated London News*, Feb. 17, 1855, 154.

217 **"It was difficult to conceive" . . . "faithful copies":** Glaisher, "On the Severe Weather," 22.

217 **He gave his rough drawings:** Hunt, "James Glaisher," 325.

217 **He was also aided:** J. Glaisher, "Snow Crystals in 1855," *Quarterly Journal of Microscopical Science*, vol. 3, 1855, 179.

217 **its emblem, a snow crystal:** Hunt, "James Glaisher," 325.

217 **Recognizing this, Glaisher:** "On the Crystals of Snow, as Applied to the Purposes of Design," *Art-Journal*, July 4, 1857, 73.

218 **Professor Snow:** Wilson A. Bentley, "Photomicrographs of Snow Crystals, and Methods of Reproduction," *Monthly Weather Review*, Aug. 1918, 360.

219 **The process was intricate:** See W. A. Bentley, "Photographing Snowflakes," *Popular Mechanics Magazine*, vol. 37, no. 2, Feb. 1922, 309–12; W. A. Bentley and W. J. Humphreys,

Snow Crystals (New York: Dover Publications, 1931); Duncan C. Blanchard, *The Snowflake Man: A Biography of Wilson A. Bentley* (Blacksburg, VA: McDonald & Woodward Publishing Company, 1998), 28–32.

219 **"in all their unrivalled beauty":** Bentley, "Photomicrographs of Snow Crystals," 359.

219 **Bentley adopted a method:** Bentley, "Photomicrographs of Snow Crystals." See also Blanchard, *Snowflake Man*, 48.

219 **"almost completely worthless":** "Aufnahmen von Schneekrystallen," *Photographische Rundschau*, no. 3, Mar. 1900, 60. On Neuhauss and the debates over snow crystals, see also Lorraine Daston and Peter Galison, *Objectivity* (Cambridge, MA: MIT Press, 2007), 150–51.

220 **"few and rather poor":** Gustaf Nordenskiöld, "Photographs of Snow Crystals," *British Journal of Photography*, vol. 41, no. 1775, May 11, 1894, 303. On Nordenskiöld's work, with a sprinkling of more criticism of Neuhauss, see Nordenskiöld, "The Photography of Snow Flakes," *Photographic Times and American Photographer*, vol. 26, no. 2, Feb. 1895, 95–96.

220 **"old habit of seeking the beautiful":** Wilson A. Bentley, "Studies among the Snow Crystals during the Winter of 1901–1902," reprinted from the Annual Summary of the *Monthly Weather Review* for 1902, June 10, 1902, 1.

221 **"Photography certainly cannot lie":** Andrew Pringle, *Practical Photo-Micrography: By the Latest Methods* (New York: Scoville & Adams Company, 1890), 15.

221 **photomicrography's "great value":** Edward C. Bousfield, *Guide to the Science of Photo-Micrography*, 2nd ed. (London: J. & A. Churchill, 1892), 128. Bousfield states that a good print tells the truth but perhaps not the whole truth "in more than one plane."

221 **"ultrascientific viewpoint that insists":** Bentley, "Photomicrographs of Snow Crystals," 359–60.

222 **world's first artificial snow crystals . . . "regrettable":** Ukichiro Nakaya, *Snow Crystals: Natural and Artificial* (Cambridge, MA: Harvard University Press, 1954), vi, 4.

222 **Joseph Janvier Woodward:** On Woodward, see J. S. Billings, "Biographical Memoir of Joseph Janvier Woodward, 1833–1884," read before the National Academy, April 22, 1885, accessed online at the US National Library of Medicine; and Amy V. Rapkiewicz, Alan Hawk, Adrienne Noe, and David M. Berman, "Surgical Pathology in the Era of the Civil War: The Remarkable Life and Accomplishments of Joseph Janvier Woodward, MD," *Archives of Pathology and Laboratory Medicine*, vol. 129, no. 10, Oct. 2005, 1313–16.

222 **"indispensable to the proper presentation":** Dr. J. J. Woodward, "On Photomicrography with the Highest Powers, as Practiced in the Army Medical Museum," *British Journal of Photography*, vol. 13, Oct. 12, 1866, 488.

223 **Civil War's estimated 750,000 deaths:** Research from 2012 found that the Civil War death toll was far higher than previous estimates. See Guy Gugliotta, "New Estimate Raises Civil War Death Toll," *New York Times*, Apr. 2, 2012. Those estimates had calculated that two-thirds of fatalities were from disease. See, for example, Michael Mahr, "Typhoid Fever—One of the Civil War's Deadliest Diseases," National Museum of Civil War Medicine website, July 1, 2021.

223 **Stepping away from the microscope:** Woodward was also one of a team of physicians during Garfield's slow demise from a gunshot wound. The lead physician was Dr. Doctor Willard Bliss. That is not a typo: his first name was Doctor.

223 **Woodward wanted to include:** Details on his setup from Rapkiewicz et al., "Surgical Pathology," 1314–15.

223 **what he called a "practical photographer":** Brevet Lieutenant Colonel J. J. Woodward, "Report to the Surgeon General of the United States Army on the Magnesium and Electric Lights as Applied to Photo-Micrography," War Department, Surgeon General's Office, Jan. 5, 1870.

223 **aniline dyes to stain:** R. S. Henry, *The Armed Forces Institute of Pathology: Its First Century, 1862–1962*, Office of the Surgeon General, Department of the Army (Washington, DC: US Government Printing Office, 1964), 34.

223 **He tried artificial light sources:** Woodward, "Report."

223 **"extremely perfect":** Minutes of the meeting of February 26, 1869, *Journal of the Quekett Microscopical Club*, vol. 1, Sept. 1869, 180.

223 **"No photomicrographs":** Simon Henry Gage, *The Microscope and Microscopical Methods*, 6th ed. (New York: Comstock Publishing Co, 1896), 184.

224 **Photomicrography was the earliest form:** Alison Gernsheim, "Medical Photography in the Nineteenth Century," *Medical and Biological Illustration*, vol. 11, Apr. 1961, 85.

224 **Initially, doctors who wanted:** Daniel M. Fox and Christopher Lawrence, *Photographing Medicine: Images and Power in Britain and America since 1840* (New York: Greenwood Press, 1988), 25; Jeffrey Mifflin, "Visual Archives in Perspective: Enlarging on Historical Medical Photographs," *American Archivist*, vol. 70, no. 1, Spring–Summer 2007, 32–69; Øystein H. Horgmo, "Mirrors in Early Clinical Photography (1862–1882): A Descriptive Study," *Journal of Visual Communication in Medicine*, vol. 38, 2015, 184.

225 **photographs of organs and tissues:** Fox and Lawrence, *Photographing Medicine*, 24.

225 **whatever was in the camera's field of vision:** Fox and Lawrence, *Photographing Medicine*, 26. Martin Kemp writes that the "visual confusion and lack of selective emphasis" in photography was a problem for some fields of medicine. Kemp, "'A Perfect and Faithful Record': Mind and Body in Medical Photography before 1900," in Thomas, *Beauty of Another Order*, 148.

226 **"What is it?":** Quoted in Erin O'Connor, "Camera Medica: Towards a Morbid History of Photography," *History of Photography*, vol. 23, no. 3, 1999, 234.

226 **the lure of objectivity and speed:** Lauren Barnett, "Alienation and Beauty in Medical Photography," *Journal of Visual Communication in Medicine*, vol. 41, no. 3, 2018, 134–35.

227 **"The photographs are singularly perfect representations":** "Notes on Books," *British Medical Journal*, Aug. 31, 1867, 185; also quoted in Fox and Lawrence, *Photographing Medicine*, 26.

227 **a practice known as physiognomy:** Kemp, "'A Perfect and Faithful Record,'" 128.

227 **"central physiognomical type":** Francis Galton, "Photographic Chronicles from Childhood to Age," *Fortnightly Review*, 1882, quoted in Kemp, "'A Perfect and Faithful Record,'" 132. For more on Galton, see Kemp, "'A Perfect and Faithful Record,'" 128–36.

227 **Hugh Welch Diamond:** On Diamond and his patients, see Richard Lansdown, "Photographing Madness," *History Today*, Sept. 2011, 49.

227 **"unerring accuracy":** Diamond, "On the Application of Photography to the Physiognomy and Mental Phenomena of Insanity," quoted in Lansdown, "Photographing Madness," 52.

227 **dressed one patient as Ophelia:** Sharrona Pearl, "Through a Mediated Mirror: The Photographic Physiognomy of Dr Hugh Welch Diamond," *History of Photography*, vol. 33, no. 3, 2009, 294.

227 **"photography unquestionably led":** Diamond, "On the Application of Photography to the Physiognomy and Mental Phenomena of Insanity," 23, quoted in Pearl, "Through a Mediated Mirror," 289.

227 **One example he gave:** Lansdown, "Photographing Madness," 52.

229 **physically restrain patients:** Lansdown writes that physical restraint had largely ended at the Surrey Asylum by the time Diamond arrived. But "while there was an emphasis on moral therapy, socialisation and labour, inmates were still subjected to cold baths and showers, isolation and the rotating chair," in which the patient was spun around until she vomited. Lansdown, "Photographing Madness," 50.

229 **"painful caricaturing":** Diamond, "On the Application of Photography," 24, quoted in Pearl, "Through a Mediated Mirror," 291.

229   **"Diamond's camera":** Pearl, "Through a Mediated Mirror," 301. See also Barnett, "Alienation and Beauty," 133–39; and John Tagg, *The Burden of Representation: Essays on Photographies and Histories* (Minneapolis: University of Minnesota Press, 1988), 77–81.

229   **in 1881, a doctor:** G. Thomson, "Photography in Lunatic Asylums," *British Journal of Photography*, vol. 18, no. 1128, Dec. 16, 1881, 648.

229   **Photographs such as these:** Erin O'Connor writes that some medical photos "told stories of spiritual degeneracy and failure of will." O'Connor, "Camera Medica," 237.

229   **Processes to protect patient anonymity:** Fox and Lawrence, *Photographing Medicine*, 26; Mifflin, "Visual Archives," 56.

229   **three London hospitals:** Jason Bate, "Reconfiguring Amateurism: Hospital Photographic Clubs and the Intersecting Networks of Late Victorian Medical Photography," *The Photo-Historian*, no. 195, Apr. 2023.

230   **many photographers were also dentists:** See Henisch and Henisch, *The Photographic Experience*, 216–17.

230   **"dentist and photo artist":** Henisch and Henisch, *The Photographic Experience*, 217.

230   **"first-class picture":** "The Editorial Dropshutter," *American Journal of Photography*, vol. 14, no. 157, Jan. 1893, 40.

230   **"for thirty years practiced":** "Views Caught with the Drop Shutter," *Anthony's Photographic Bulletin*, vol. 23, no. 3, Feb. 13, 1892, 96.

230   **Before he opened:** Gernsheim and Gernsheim, *History of Photography*, 121, 124.

230   **One report from 1904:** "Photography and Dentistry," *The Photographer*, vol. 2, no. 34, Dec. 17, 1904, 121.

231   **"Photographic Refuse":** "To Artists in Photography, Dentists, Jewellers, etc" advertisement, *British Journal of Photography*, vol. 6, no. 105, Nov. 1, 1859, v.

231   **To bulk out:** J. Traill Taylor, "Ethics of Photography and Photographers," *Photography: The Journal of the Amateur, the Professional, and the Trade*, vol. 2, no. 81, May 29, 1890, 342.

231   **before-and-after portrait:** See Gernsheim, "Medical Photography," 88; Pearl, "Through a Mediated Mirror," 297; Kemp, " 'A Perfect and Faithful Record,' " 146; Erin O'Connor, "Pictures of Health: Medical Photography and the Emergence of Anorexia Nervosa," *Journal of the History of Sexuality*, vol. 5, no. 4, Apr. 1995, 535–72.

Chapter 12: Bones, Skin, Eyes, and Brains

232   **On November 8, 1895:** Accounts of Röntgen's experiment drawn from H. J. W. Dam, "The New Marvel in Photography," *McClure's Magazine*, vol. 6, no. 5, Apr. 1896, 403–15; Graham Farmelo, "The Discovery of X-rays," *Scientific American*, vol. 273, no. 5, Nov. 1995, 86–88; Richard F. Mould, *A Century of X-rays and Radioactivity in Medicine* (Bristol: Institute of Physics Publishing, 1993), 1; Catherine Caufield, *Multiple Exposures: Chronicles of the Radiation Age* (New York: Harper & Row, 1989), 3–4; and George I. Manes, "The Discovery of X-Ray," *Isis*, vol. 47, no. 3, Sept. 1956, 236–38.

234   **Averse to the publicity:** Farmelo, "The Discovery of X-rays," 89–90.

234   **In the rush to understand:** Lawrence Badash, "Becquerel's Blunder," *Social Research*, vol. 72, no. 1, Spring 2005, 31.

234   **"Never has a scientific discovery":** "The Photography of the Invisible," *Quarterly Review*, vol. 183, Jan.–Apr. 1896, 496.

234   **From the very beginning:** Marien, *Photography*, 214.

234   **This irked Röntgen:** Farmelo, "The Discovery of X-rays," 89.

234   **"Unlike most epoch-making results":** Dam, "The New Marvel," 407.

234   **other articles and books about X-rays:** E. J. Ward, ed., *The New Light and the New*

Photography: Full Particulars, Popularly Written, of Prof. Röntgen's Discovery (London: Dawbarn & Ward, 1896); Hector MacLean, "Progress in Shadowgraphy with 'X' Rays," *Practical Photographer*, vol. 7, no. 75, Mar. 1896, 72; Swinton, "The New Shadow Photography," 63.

234   **published their book:** Josef Maria Eder and Eduard Valenta, *Versuche über Photographie mittelst der Röntgen'schen Strahlen*, or *Research on Photography with Röntgen Rays*, trans. Ulrich Riedel, ed. Lita Tirak (n.p., 2012).

234   **A report in 1896:** V. E. Johnson, "The Röntgen Rays at the Royal Veterinary College," *Photography: The Journal of the Amateur, the Professional, and the Trade*, vol. 8, no. 410, Sept. 17, 1896, 610–11.

236   **X-ray of crayfish:** According to curator Corey Keller, Victor Chabaud created the crayfish X-ray to demonstrate his new focus tube. Keller, "X-Rays," in Keller, *Brought to Light*.

236   **X-ray imaging authenticated:** "The X-Rays Test a Painting," *Photographic Life*, Mar. 31, 1897, 194.

237   **"a full line of apparatus":** *Anthony's Photographic Bulletin*, vol. 27, no. 5, May 1896.

237   **"Röntgen Ray Apparatus":** *Illustrated London News*, Dec. 12, 1896, 33.

237   **"the New Rontgen X Ray Photographs":** *Hastings and St Leonards Observer*, Aug. 20, 1898, 1.

237   **Elizabeth Fleischmann:** Peter E. Palmquist, *Elizabeth Fleischmann: X-Ray Pioneer* (Berkeley: Judah L. Magnes Museum, 1990). All quotes are from "The Woman Who Takes the Best Radiographs," *San Francisco Chronicle*, June 3, 1900, reprinted in Palmquist, *Fleischmann*, 11–17.

237   **Staff in Queen Victoria's kitchen:** "Notes," *Photographic Times and American Photographer*, vol. 29, no. 3, Mar. 1897, 155.

237   **"extraordinary deformities":** "Queen v Corset," *Reynold's Newspaper*, Jan. 3, 1897, 5. See also Countess A. Von Bothmer, ed., *The Sovereign Ladies of Europe* (London: Hutchinson & Co.,1899), 128.

237   **In both Germany and France:** Frau Kober, "What Women Are Doing in Germany," *Womanhood*, vol. 11, no. 61, Dec. 1903, 36, and L. Gastine, "Hygiène et Radiographie," *La Photographie Française*, no. 2, Feb. 1897, 17.

237   **In Britain, bootmakers:** Mould, *A Century of X-rays*, 103. For Dorothy Dodd, see, for example, *Woman's Journal*, vol. 34, no. 23, June 6, 1903, 184.

238   **"ped-o-skop":** Mould, *A Century of X-rays*, 107.

238   **"strictly regulate" the use of such devices:** "Shoe X-Rays Scored," *New York Times*, Aug.19, 1960, 10.

238   **an X-ray of a bottle of beer:** "X-Rays and Beer," *Photographic Life*, Apr. 7, 1897.

238   **An English farmer:** "Chit Chat," *Photographic News*, vol. 41, no. 73, May 21, 1897, 880; and "From the Editor's Chair," *Photographic News*, vol. 41, no. 86, Aug. 20, 1897, 86.

238   **A woman in Sunderland:** "The Ladies Page," *Photographic Life*, Mar. 31, 1897, 215; "Chit Chat," *Photographic News*, vol. 41, no. 56, Jan. 22, 1897, 59.

239   **full-blown "X-ray mania":** "Phoebus Junius," "Note from the North," *Photographic News*, vol. 40, no. 7, Feb. 14, 1896.

239   **according to one report:** Allen W. Grove, "Röntgen's Ghosts: Photography, X-Rays and the Victorian Imagination," *Literature and Medicine*, vol. 16, no. 2, 1997, 160.

239   **"YOUR OWN BONES":** *The Graphic*, Oct. 31, 1896, 547.

239   **An exhibition at Earl's Court:** "Marmaduke," "Court and Club," *The Graphic*, May 23, 1896, 626.

240   **daily X-ray demonstrations:** *Photography: The Journal of the Amateur, the Professional, and the Trade*, vol. 8, no. 417, Nov. 5, 1896, 72.

240   **your bones x-rayed on a major London street:** "Our Note Book: Facile Descensus and Co," *Photographic News*, vol. 40, no. 48, Nov. 27, 1896, 769.

240   **By the end of 1896:** "Humours of Photography," *Photography: The Journal of the Amateur, the Professional, and the Trade*, vol. 8, no. 413, Oct. 8, 1896, 666.

240   **Across the Atlantic:** "X Rays as an Advertiser," *Photography: The Journal of the Amateur, the Professional, and the Trade*, vol. 7, no. 415, Oct. 22, 1896, 697. See also Laidler, *To Light Such a Candle*, 243.

240   **X-ray–powered light bulb:** Caufield, *Multiple Exposures*, 8–9.

240   **The fluoroscope was a screen:** "Editorial Notes," *Photographic Times and American Photographer*, vol. 28, no. 6, June 1896, 287, and Sylvia Pamboukian, "'Looking Radiant': Science, Photography and the X-ray Craze of 1896," *Victorian Review*, vol. 27, no. 2, 2001, 60–61.

240   **"The fluoroscope exhibition":** "Fluoroscope a Success," *New York Times*, May 12, 1896, 3.

240   **"revolting indecency":** "Occasional Notes," *Pall Mall Gazette*, Mar. 20, 1896, 2.

240   **"They may show":** "Snap Shots," *Photography: The Journal of the Amateur, the Professional, and the Trade*, vol. 8, no. 424, Dec. 24, 1896, 837.

240   **"an absolute bore" . . . "X-actly So":** "Snap Shots," *Photography: The Journal of the Amateur, the Professional, and the Trade*, vol. 8, no. 384, Mar. 19, 1896, 195.

242   **One article advised women:** "Spirit of the Times," *Photography: The Journal of the Amateur, the Professional, and the Trade*, vol. 8, no. 379, Feb. 13, 1896, 100.

242   **"Röntgen's Curse":** C. H. T. Crosthwaite, "Röntgen's Curse," *Longman's Magazine*, vol. 28, 1896, 469–84.

242   **H. G. Wells's famous novel:** Of *The Invisible Man*, Allen Grove writes, "Perhaps no work of British fiction makes such an astute presentation of the wonder, fear, and confusion surrounding Röntgen's discovery." Grove, "Rontgen's Ghosts," 169.

242   **"Röntgen Pill":** "Our Note Book," *Photographic News*, vol. 40, no. 30, July 24, 1896, 466.

242   **One scientist suggested:** "Dr. Giering's Discovery," *Professional and Amateur Photographer*, vol. 2, no. 12, Dec. 1897, 503.

242   **Others experimented with the notion:** For example, in "Items of Interest," *Anthony's Photographic Bulletin*, vol. 28, no. 1, Jan. 1897, 15, and John Hall-Edwards, "Photography by the Rontgen Rays, Up to Date," *Photographic Times and American Photographer*, vol. 29, no. 9, Sept. 1897, 422.

242   **Because it didn't hurt:** Marien, *Photography*, 214.

242   **Yet accounts of X-ray "burns":** Ernst Amory Codman, *A Study of the Cases of Accidental X-Ray Burns Hitherto Recorded*, reprinted from the *Philadelphia Medical Journal*, Mar. 8, 1902, 5.

242   **"Burned by the X Rays":** "Burned by the X Rays," *New York Times*, July 30, 1897, 1.

243   **"place the patient":** Quoted in Caufield, *Multiple Exposures*, 10.

243   **a woman in Hastings:** "Alleged Death from X-Rays," *Photography: The Journal of the Amateur, the Professional, and the Trade*, vol. 12, no. 627, Nov. 15, 1900, 757; A. Woolsey Blackblock, "The X-ray Case at Hastings," *Photography: The Journal of the Amateur, the Professional, and the Trade*, vol. 12, no. 628, Nov. 22, 1900; "News and Notes," *British Journal of Photography*, vol. 47, no. 2166, Nov. 23, 1900, 749.

243   **Many injuries were to the hands:** Mould, *A Century of X-Rays*, 35, and Daniel S. Goldberg, "Suffering and Death among Early American Roentgenologists: The Power of Remotely Anatomizing the Living Body in Fin de Siècle America," *Bulletin of the History of Medicine*, vol. 85, no. 1, Spring 2011, 11.

243   **The X-ray demonstrators:** "From the Editor's Chair," *Photographic News*, vol. 40, no. 47, Nov. 20, 1896, 767, and Laidler, *To Light Such a Candle*, 241.

243 **Edison's assistant Clarence Dally:** Pamboukian, ""Looking Radiant," 67; Goldberg, "Suffering and Death," 10–11; Percy Brown, *American Martyrs to Science through the Roentgen Rays* (Springfield, IL: Charles C. Thomas, 1936), 40–41.

243 **Elizabeth Fleischmann's symptoms:** Brown, *American Martyrs*, 48–49, and Palmquist, *Elizabeth Fleischmann*, 6, 18.

244 **"Thought, the Invisible":** "Thought, the Invisible, and the Unknown," *Amateur Photographer*, Feb. 21, 1896, 151.

244 **a vital force, sometimes called a "universal fluid":** Peter Geimer, *Inadvertent Images: A History of Photographic Apparitions* (Chicago: University of Chicago Press, 2018), 77, and "Psychography," *Photogram*, vol. 5, no. 57, Sept. 1898, 279.

244 **Karl von Reichenbach:** Clément Chéroux, "Photographs of Fluids: An Alphabet of Invisible Rays," in Chéroux et al., *The Perfect Medium*, 114, and Nicolas Pethes, "Psychicones: Visual Traces of the Soul in Late 19th Century Fluidic Photography," *Medical History Journal*, vol. 60, no. 3, 2016, 328–29.

244 **Hippolyte Baraduc:** Pethes, "Psychicones," 329–31.

244 **"The invisible fluid":** *L'âme humaine, ses mouvements, ses lumières, et l'iconographie de l'invisible fluidique* (Paris: G. Carré, 1896), quoted in Frizot, *The New History*, 281–82.

244 **photographs of what he called "psychicones":** Pethes, "Psychicones," 329.

245 **"fluidico-vital images":** Hippolyte Baraduc, *The Human Soul: Its Movements, Its Lights, and the Iconography of the Fluidic Invisible* (Paris: Librairie Internationale de la Pensée Nouvelle, 1913), 13.

245 **Baraduc's images typically:** Geimer, *Inadvertent Images*, 93, and Chéroux, "Photographs of Fluids," 117.

245 **"vital force" . . . "state of soul":** Baraduc, *The Human Soul*, Explanation XLIII.

245 **Baraduc's beliefs were so strong:** Pethes, "Psychicones," 334.

246 **impressions of its "wild terror" . . . "A wicked and unnecessary":** "Psychicons," *Photogram*, vol. 4, no. 37, Jan. 1897, 10.

246 **Louis Darget:** Andreas Fischer, " 'La lune au front': Remarks on the History of the Photography of Thought," in Chéroux et al., *A Perfect Medium*, 140.

246 **Darget's techniques varied:** "New Photographs of Psychic Radiations," *British Journal of Photography*, vol. 44, no. 1937, June 18, 1897, 394.

246 **after an argument, Darget once pressed:** Stenger, *The March of Photography*, 222.

246 **worked together to create . . . "portable radiographer":** Chéroux, "Photographs of Fluids," 118.

246 **"The brain throws out rays":** "New Photographs of Psychic Radiations," 392.

246 **The word "radiation":** See, for example, Pethes, "Psychicones," 331; "Thought Radiography," *Photography: The Journal of the Amateur, the Professional, and the Trade*, vol. 8, no. 398, June 25, 1896, 424; "New Photographs of Psychic Radiations," 394.

246 **the term "v-rays":** Chéroux writes that after 1896, the terminology focused on "rays" and that, in doing so, "the effluvists sought to gain recognition for their experiments, theories, and disciplines." Chéroux, "Photographs of Fluids," 119.

246 **the experiments of W. Ingles Rogers:** W. Ingles Rogers, "Can Thought Be Photographed? The Problem Solved," *Amateur Photographer*, Feb. 21, 1896, 160; Feb. 28, 1896, 186; Mar. 6, 1896, 207.

247 **"hardly fair to criticize":** A. B. Chatwood, "Can Thought Be Photographed?," *Amateur Photographer*, Mar. 20, 1896, 244.

247 **"the polarity inherent":** "New Photographs of Psychic Radiations," 394.

247 **"familiar accidents":** Adrien Guébhard, "De l'emploi de la plaque voilée," quoted in Geimer, *Inadvertent Images*, 86.

247 **Guébhard had investigated:** Geimer, *Inadvertent Images*, 86–89.

247 **"And now comes the brutal scientist":** "From the Editor's Chair," *Photographic News*, vol. 42, no. 125, May 20, 1898, 315.

248 **In another inquiry:** Geimer, *Inadvertent Images*, 85–86. Geimer also details additional investigations involving iron weights, the red safe light, and the developer bath, and notes that these findings did not always align. "The debate over effluviography is by no means a pitched battle of a group of established scientists against the dubious schemes of two photographic amateurs." Geimer, *Inadvertent Images*, 89.

248 **"The Sympsychograph":** See David Starr Jordan, "The Sympsychograph: A Study in Impressionist Physics," *Appleton's Popular Science Monthly*, vol. 49, Sept. 1896, 597, and Linda Simon, *Dark Light: Electricity and Anxiety from the Telegraph to the X-Ray* (Orlando, FL: Harcourt Books, 2004), 280–81.

249 **"surprised and humiliated":** David Starr Jordan, "The Moral of the 'Sympsychograph,'" *Appleton's Popular Science Monthly*, vol. 50, 1897, 265.

249 **The editors of that magazine:** "An Almost Too Successful Joke," *Appleton's Popular Science Monthly*, vol. 50, 1897, 123.

Conclusion: A New Century

251 ***The British Journal:*** Thomas Bedding, ed., *The British Journal Photographic Almanac and Photographers Daily Companion* (London: Henry Greenwood & Co., 1898). The summary of the year starts on page 501.

251 **"In a retrospective glance":** J. K. Tulloch, "Progress in Photography," *British Journal of Photography*, vol. 33, no. 1385, Nov. 19, 1886, 733.

252 **One summer night:** Em. Touchet, "Coups de foudre sur la Tour Eiffel," *Bulletin de la Société Astronomique de France* (Paris: Au Siège de la Société, 1905), 238–40.

SELECTED BIBLIOGRAPHY

The following books are highlights from my research. I also studied photographic journals from the era, including the *British Journal of Photography*; *Photographic Notes*; *Photography: The Journal of the Amateur, the Professional, and the Trade*; *Amateur Photographer*; *American Journal of Photography*; *Anthony's Photographic Bulletin*; and *Photographic News*. Many of these are accessible online.

Boutan, Louis. *La photographie sous-marine et les progrès de la photographie.* Paris: Schleicher Frères, 1900.

Braun, Marta. *Picturing Time: The Work of Etienne-Jules Marey (1830–1904).* Chicago: University of Chicago Press, 1992.

Canales, Jimena. *A Tenth of a Second: A History.* Chicago: University of Chicago Press, 2009.

Caufield, Catherine. *Multiple Exposures: Chronicles of the Radiation Age.* New York: Harper & Row, 1989.

Chéroux, Clément, Andreas Fischer, Pierre Apraxine, Denis Canguilhem, and Sophie Schmit. *The Perfect Medium: Photography and the Occult.* New Haven: Yale University Press, 2005.

Coe, Brian. *Colour Photography: The First Hundred Years 1840–1940.* London: Ash & Grant, 1978.

Darrah, William C. *Cartes de Visite in Nineteenth-Century Photography.* Gettysburg, PA: W. C. Darrah Publisher, 1981.

Fineman, Mia. *Faking It: Manipulated Photography before Photoshop.* New Haven: Yale University Press, 2012.

Fineman, Mia, and Beth Saunders. *Apollo's Muse: The Moon in the Age of Photography.* New York: Metropolitan Museum of Art, 2019.

Flukinger, Roy, Larry Schaaf, and Standish Meacham. *Paul Martin: Victorian Photographer.* Austin: University of Texas Press, 1977.

Frizot, Michael. *The New History of Photography.* Cologne: Könemann, 1998.

Geimer, Peter. *Inadvertent Images: A History of Photographic Apparitions.* Chicago: University of Chicago Press, 2018.

Gernsheim, Helmut, in collaboration with Alison Gernsheim. *The History of Photography from the Camera Obscura to the Beginning of the Modern Era*. London: Thames and Hudson, 1955.

Goodman, Ruth. *How to Be a Victorian: A Dawn-to-Dusk Guide to Victorian Life*. New York: Liveright, 2014.

Green, Tyler. *Carleton Watkins: Making the West American*. Oakland: University of California Press, 2018.

Grimaldo Grigsby, Darcy. *Enduring Truths: Sojourner's Shadows and Substance*. Chicago: University of Chicago Press, 2015.

Gustavson, Todd. *Camera: A History of Photography from Daguerreotype to Digital*. New York: Sterling Publishing Co., 2009.

Hambourg, Maria Morris, Françoise Heilbrun, and Philippe Néagu. *Nadar*. New York: Metropolitan Museum of Art, distributed by H. N. Adams, 1995.

Henisch, Heinz K., and Bridget A. Henisch. *The Photographic Experience, 1839–1914: Images and Attitudes*. University Park: Penn State University Press, 1994.

Howes, Chris. *To Photograph Darkness: The History of Underground and Flash Photography*. Carbondale: Southern Illinois University Press, 1989.

Hudson, Giles. *Sarah Angelina Acland: First Lady of Colour Photography, 1849–1930*. Oxford: The Bodleian Library, 2012.

Jay, Bill. *Cyanide and Spirits*. Munich: Nazraeli Press, 1991.

Keller, Corey, ed. *Brought to Light: Photography and the Invisible, 1840—1900*. New Haven and London: San Francisco Museum of Modern Art in association with Yale University Press, 2008.

Lavédrine, Bertrand, and Jean-Paul Gandolfo. *The Lumière Autochrome: History, Technology and Preservation*. Translated by John McElhone. Los Angeles: Getty Publications, 2013.

Lightman, Bernard, ed. *Victorian Science in Context*. Chicago: University of Chicago Press, 1997.

Linkman, Audrey. *The Victorians: Photographic Portraits*. London: Tauris Park Books, 1993.

Luther, Frederic. *Microfilm: A History, 1839–1900*. Annapolis, MD: The National Microfilm Association, 1959.

Marey, E. J. *Animal Mechanism: A Treatise on Terrestrial and Aerial Locomotion*. London: Henry S. King & Co, 1874.

Martin, Paul. *Victorian Snapshots*. New York: Arno Press, 1939, reprint 1973.

McCauley, Elizabeth Anne. *A.A.E. Disdéri and the Carte de Visite Portrait Photograph*. New Haven: Yale University Press, 1985.

Mould, Richard F. *A Century of X-rays and Radioactivity in Medicine*. Bristol: Institute of Physics Publishing, 1993.

Mozley, Anita Ventura, Robert Bartlett Haas, and Françoise Forster-Hahn. *Eadweard Muybridge: The Stanford Years, 1872–1882*. Stanford, CA: Stanford University, 1972.

Nadar. *When I Was a Photographer*. Translated by Eduardo Cadava and Liana Theodoratou. Cambridge, MA: MIT Press, 2015; originally published in 1900.

Naef, Weston, and Christine Hult-Lewis. *Carleton Watkins: The Complete Mammoth Photographs*. Los Angeles: J. Paul Getty Museum, 2011.

Nead, Lynda. *Victorian Babylon: People, Streets and Images in Nineteenth-Century London*. New Haven: Yale University Press, 2000.

Newhall, Beaumont. *Airborne Camera: The World from the Air and Outer Space*. New York: Hastings House, 1969.

Newhall, Beaumont. *The History of Photography: From 1839 to the Present Day*. Revised and enlarged ed. London: Secker & Warburg, 1964.

Newhall, Beaumont. *Latent Image: The Discovery of Photography*. Albuquerque: University of New Mexico Press, 1983.

Nickel, Douglas R. *Carleton Watkins: The Art of Perception*. San Francisco and New York: San Francisco Museum of Modern Art and Harry N. Abrams, 1999.

Novotny, Ann. *Alice's World: The Life and Photography of an American Original: Alice Austen, 1866–1952*. Old Greenwich, CT: Chatham Press, 1976.

Painter, Nell Irvin. *Sojourner Truth: A Life, a Symbol*. New York: W. W. Norton, 1996.

Palmquist, Peter E. *Carleton E. Watkins: Photographer of the American West*. Albuquerque: Published for Amon Carter Museum by University of New Mexico Press, 1983.

Pauli, Lori. *Oscar G. Rejlander: Artist Photographer*. New Haven: Yale University Press, 2018.

Pritchard, Michael, ed. *Technology and Art: The Birth and Early Years of Photography*. Bath: Royal Photographic Society Historical Group, 1990.

Prodger, Phillip. *Darwin's Camera: Art and Photography in the Theory of Evolution* (New York: Oxford University Press, 2009) .

Prodger, Phillip. *Time Stands Still: Muybridge and the Instantaneous Photography Movement* (New York: Oxford University Press, 2003).

Rosenblum, Naomi. *A History of Women Photographers*. Paris: Abbeville Press, 1994.

Rosenblum, Naomi. *A World History of Photography*. 3rd ed. New York: Abbeville Press, 1997.

Schaaf, Larry J., and Joshua Chuang, eds. *Sun Gardens: Cyanotypes by Anna Atkins*. New York: New York Public Library, 2018.

Scharf, Aaron. *Art and Photography*. Middlesex: Penguin Books, 1974.

Schivelbusch, Wolfgang. *Disenchanted Night: The Industrialization of Light in the Nineteenth Century*. Berkeley: University of California Press, 1988.

Scott, Jean. *Stanhopes: A Closer View*. Essex: Greenlight Publishing, 2002.

Seiberling, Grace. *Amateurs, Photography, and the Mid-Victorian Imagination*. Chicago: University of Chicago Press, 1986.

Siegel, Elizabeth. *Galleries of Friendship and Fame: A History of Nineteenth-Century American Photograph Albums*. New Haven: Yale University Press, 2010.

Solnit, Rebecca. *River of Shadows: Eadweard Muybridge and the Technological Wild West*. New York: Penguin Books, 2003.

Stauffer, John, Zoe Trodd, and Celeste-Marie Bernier. *Picturing Frederick Douglass: An Illustrated Biography of the Nineteenth Century's Most Photographed American*. Revised edition. New York: Liveright, 2019.

Steinbach, Susie L. *Understanding the Victorians: Politics, Culture and Society in Nineteenth-Century Britain*. 2nd ed. London: Routledge, 2017.

Sullivan, George, *Black Artists in Photography*. New York: Cobblehill Books, 1996.

Tagg, John. *The Burden of Representation: Essays on Photographies and Histories*. Minneapolis: University of Minnesota Press, 1988.

Taft, Robert. *Photography and the American Scene: A Social History, 1839–1889*. New York: Dover Publications, 1938.

Thomas, Ann, ed. *Beauty of Another Order: Photography in Science*. New Haven: Yale University Press in association with the National Gallery of Canada, 1997.

Tissandier, Gaston. *A History and Handbook of Photography*. Translated from the French. New York: Scovill Manufacturing Company, 1977.

Trachtenburg, Alan, ed. *Classic Essays on Photography*. New Haven: Leete's Island Books, 1980.

Tucker, Jennifer. *Nature Exposed: Photography as Eyewitness in Victorian Science*. Baltimore: Johns Hopkins University Press, 2005.

Wallace, Maurice O., and Shawn Michelle Smith, eds. *Pictures and Progress: Early Photography and the Making of African American Identity*. Durham, NC: Duke University Press, 2012.

Warner Marien, Mary. *Photography: A Cultural History*. 3rd ed. Upper Saddle River, NJ: Prentice Hall, 2011.

Watson, Roger, and Helen Rappaport. *Capturing the Light: The Birth of Photography, a True Story of Genius and Rivalry.* New York: St. Martin's Press, 2013.

Willis, Deborah. *Reflections in Black: A History of Black Photographers 1840 to the Present.* New York: W. W. Norton, 2000.

Yochelson, Bonnie, and Daniel Czitrom. *Rediscovering Jacob Riis: Exposure Journalism and Photography in Turn-of-the-Century New York.* Chicago: University of Chicago Press, 2007.

PHOTOGRAPHY CREDITS

Frontispiece

Museum of Fine Arts, Houston, Texas / Bridgeman Images.

Introduction

Digital images courtesy of Getty's Open Content Program (pp. 4, 6); The Metropolitan Museum of Art, New York (p. 7); The Library Company of Philadelphia (p. 9); LSE Library (p. 10); Science Museum Group (p. 12); Gale Nineteenth Century Collections (p. 13).

Chapter 1

Bibliothèque nationale de France (pp. 22, 26); Digital image courtesy of Getty's Open Content Program (p. 24); Gale Nineteenth Century Collections (pp. 28, 29, 31); Jacob Riis/90.13.1.158 / Museum of the City of New York (p. 34).

Chapter 2

Division of Work & Industry, National Museum of American History, Smithsonian Institution (p. 43); San Francisco Museum of Modern Art, Accessions Committee Fund purchase / photo: Don Ross (p. 44); © RMN-Grand Palais / Art Resource, NY (p. 45); courtesy National Gallery of Art, Washington (p. 47); Science Museum Group (pp. 50, 51, 54); images provided by The John Rylands Research Institute and Library, The University of Manchester (p. 52); digital image courtesy of Getty's Open Content Program (p. 55, top); Library of Congress/ LC-USZC4-9601 (p. 55, bottom).

Chapter 3

Paris Musées / Musé Carnavalet—Histoire de Paris (p. 61); The Metropolitan Museum of Art, New York (p. 62); Bibliothèque nationale de France (p. 63); Historic England Archive (p. 65); Smithsonian National Air and Space Museum (NASM 00181712) (p. 67); © IWM (pp.

69, 70); Deutsches Museum, Munich, Archive, BN14510 (p. 71); German Museum of Technology Berlin, Historical Archive I.4.052 199 (p. 72), I.4.052 195 (p. 73, top), I.4.052 190 (p. 73, bottom).

Chapter 4

Charles Atwood Kofoid Papers, Special Collections & Archives, UC San Diego (p. 77); © Bibliothèque du Laboratoire Arago / Sorbonne Université (pp. 79, 81, 83, 86, 87, 88).

Chapter 5

Wh.1875 / Whipple Museum of the History of Science, University of Cambridge (p. 98); courtesy of the George Eastman Museum (p. 100); Special Collections, Princeton University Library (p. 101); Division of Work & Industry, National Museum of American History, Smithsonian Institution (p. 102); Ville de Chalon-sur-Saône, musée Nicéphore Niépce / MNN 1986.116.135.1 (p. 103); from the Collections of the Kinsey Institute, Indiana University, all rights reserved (p. 105); digital images courtesy of Getty's Open Content Program (pp. 108, 113, 114); National Portrait Gallery, Smithsonian Institution (p. 116); courtesy of UC Berkeley, Bancroft Library (p. 117).

Chapter 6

Ville de Chalon-sur-Saône, musée Nicéphore Niépce/1986.116.232 (p. 120); Harry Ransom Center, The University of Texas at Austin (p. 121); digital image courtesy of Getty's Open Content Program (p. 125, top); Bibliothèque nationale de France (p. 125, bottom); Sterling Memorial Library, Yale, via Gale Nineteenth Century Collections Online (p. 126); Gale Nineteenth Century Collections (p. 129); American Philosophical Society (p. 131); Bibliothèque nationale de France (p. 132); Paul Frecker at The Library of Nineteenth-Century Photography (p. 134); Jeffrey Kraus Collection (p. 135); The Metropolitan Museum of Art, New York (p. 136); from the Lincoln Financial Foundation Collection (p. 137).

Chapter 7

Courtesy of the Nelson-Atkins Museum of Art Media Services (p. 143); © RMN-Grand Palais / Art Resource, NY (p. 144); Nordiska museet (p. 145); courtesy of Lynns Waffles (p. 146).

Chapter 8

Courtesy of the George Eastman Museum (pp. 154, 155); Collection of Historic Richmond Town (pp. 157, 158, 160); Library of Congress/LC-USZ62-3052 (p. 162); Gale Nineteenth Century Collections Online (p. 164); Carl Størmer / Norwegian Museum of Science and Technology (p. 165); HIP / Art Resource, NY (p. 167, top); digital image © The Museum of Modern Art / Licensed by SCALA / Art Resource, NY (p. 167, bottom); LSE Library/7JCC/O/02/061 (p. 170); The National Archives (pp.171, 174).

Chapter 9

Digital image courtesy of Getty's Open Content Program (p. 176); The Bernard Howarth-Loomes Collection / National Museums Scotland (p. 178, top; p. 180, top left); Carl Mautz Collection, Beinecke Rare Book and Manuscript Library (p. 178, bottom left); William L. Clements Library, The University of Michigan (p. 178, bottom right); CDV #6330, CDV #39762, CDV #30810: William C. Darrah collection of cartes-de-visite (01515), from the Eberly Family Special Collections Library, Penn State University Libraries (p. 180); The Metropolitan Museum of Art, New York (p. 182); Library of Congress/LC-DIG-ppmsca-52069 (p. 186); National Portrait Gallery, Smithsonian Institution (p. 187); Library of Congress (p. 188); James Presley Ball / Cincinnati Museum Center / Getty Images (p. 189); William L. Clements Library, The University of Michigan (p. 190); courtesy of Swann Auction Galleries (p. 192).

Chapter 10

Library of Congress/LC-USZC4-13701 (p. 198); Library of Congress/LC-DIG-ppmsca-05952 (p. 199); Cantor Arts Center at Stanford University; Museum Purchase Fund, 1974.190.1.a-p (p. 201); Philadelphia Museum of Art/1962-135-229 (p. 204); Philadelphia Museum of Art/1962-135-375 (p. 205); Biodiversity Heritage Library, contributed by Smithsonian Libraries and Archives (p. 207); Collège de France (p. 209, top); La Cinémathèque française (p. 209, bottom; p. 211).

Chapter 11

The Metropolitan Museum of Art, New York (p. 215); Smithsonian Institution Archives, Record Unit 31, Image No. SIA2013-09136 and Image No. SIA2013-09132 (p. 218); The Albertina Museum, Vienna. "Höhere Graphische Bundes-Lehr- und Versuchsanstalt" (p. 220); American Meteorological Society online journals (p. 221); OHA 79 Photomicrography Collection, Otis Historical Archives, National Museum of Health and Medicine (p. 224); Biodiversity Heritage Library, contributed by the Getty Research Institute (p. 225); SP 177, Otis Historical Archives, National Museum of Health and Medicine (p. 226); by kind permission of the Royal Society of Medicine (p. 228); Steve Heselton Collection (p. 230).

Chapter 12

Wellcome Collection (p. 233); The Albertina Museum, Vienna. "Höhere Graphische Bundes-Lehr- und Versuchsanstalt" (p. 235); © Musée des Arts et Métiers-Cnam, Paris / photo: Pascal Faligot (p. 236); from the British Library Collection: *The Londoner*, p. 7 / Gale KHJZWK343009718 (p. 239); National Park Service (p. 241); digital image courtesy of Getty's Open Content Program (p. 248).

Insert

Paris Observatory Library (p. 1); Christie's Images / Bridgeman Images (p. 2); John G. Wolbach Library, Harvard College Observatory (p. 3, top left); © National Museums Scotland (p. 3, top right); © Bibliothèque du Laboratoire Arago / Sorbonne Université (p. 3, bottom; p. 4); from the New York Public Library (p. 5, top left); courtesy of Lynns Waffles (p. 5, top right); digital image courtesy of Getty's Open Content Program (p. 5, bottom left); © National Museums Scotland (p. 5, bottom right); Inv. 14780 © History of Science Museum, University of Oxford (p. 6, bottom); courtesy of the National Gallery of Art, Washington (p. 7, top left); © Fitzwilliam Museum / Bridgeman Images (p. 7, top right); Collège de France (p. 7, bottom left); National Library of Medicine (p. 7, bottom right); *X-Ray of a Snake Head*, Arthur Radiguet, gelatin silver transparency on glass, 3 5/16 × 3 7/8 in. (8.4 × 9.9 cm), gift of Wm. B. Becker, 2020 (2020.275.1), © The Metropolitan Museum of Art, New York, image source: Art Resource, NY (p. 8, top); © Musée des Arts et Métiers-Cnam, Paris / photo: L. Karleskind (p. 8, bottom).

INDEX

Page numbers in *italics* refer to illustrations and accompanying captions.
Page numbers after 254 refer to endnotes.